Sustainable Agriculture for Climate Change Adaptation

Sustainable Agriculture for Climate Change Adaptation

Special Issue Editors

Kathy Lewis
Douglas Warner

MDPI • Basel • Beijing • Wuhan • Barcelona • Belgrade • Manchester • Tokyo • Cluj • Tianjin

Special Issue Editors

Kathy Lewis
University of Hertfordshire
UK

Douglas Warner
University of Hertfordshire
UK

Editorial Office
MDPI
St. Alban-Anlage 66
4052 Basel, Switzerland

This is a reprint of articles from the Special Issue published online in the open access journal *Climate* (ISSN 2225-1154) (available at: https://www.mdpi.com/journal/climate/special_issues/sustainable_agriculture).

For citation purposes, cite each article independently as indicated on the article page online and as indicated below:

LastName, A.A.; LastName, B.B.; LastName, C.C. Article Title. *Journal Name* **Year**, *Article Number*, Page Range.

ISBN 978-3-03936-382-7 (Pbk)
ISBN 978-3-03936-383-4 (PDF)

Contents

About the Special Issue Editors

Kathy Lewis (Prof.)   Interests: environmental impacts of agriculture and land use; agri-environmental management; agriculture and climate change; fate and toxicity of agricultural chemicals; agricultural risk assessment and regulation. agri-environmental management; agriculture and climate change; fate and toxicity of agricultural chemicals; agricultural risk assessment and regulation.

Doug Warner (Dr.) Interests: agricultural greenhouse gas emissions and their mitigation; carbon sequestration; crop nutrition; integrated farm management and ecologically based methods of pest control; precision agriculture; farmland biodiversity and conservation

Editorial

Editorial for the Special Issue "Sustainable Agriculture for Climate Change Adaptation"

Kathy Lewis and Douglas Warner *

Agriculture & Environment Research Unit, School of Life & Medical Sciences, University of Hertfordshire, Hatfield AL10 9AB, UK; k.a.lewis@herts.ac.uk
* Correspondence: d.j.warner@herts.ac.uk

Received: 9 April 2020; Accepted: 23 April 2020; Published: 29 April 2020

As we lie firmly entrenched within what many have termed the Anthropocene, the time of humans, human influence on the functioning of the planet has never been greater or in greater need of mitigation. Climate change, the accelerated warming of the planet's surface attributed to human activities, is now at the forefront of global politics. The 21st United Nations Climate Change Conference of the Parties (COP21) Paris Agreement saw a landmark agreement reached between countries belonging to the United Nations Framework Convention on Climate Change (UNFCCC). The agreement seeks to arrest climate change and maintain the global temperature rise below a 2 °C increase compared to pre-industrial levels, and to devise means and ways to adapt to its effects.

The agriculture sector not only contributes to climate change but, as a land-based industry, is also greatly affected by climate change. Agriculture has a key function in the role of the carbon and nitrogen cycles, contributing a significant proportion of methane and nitrous oxide toward global greenhouse gas (GHG) emissions, more than any other sector. The Organisation for Economic Co-operation and Development (OECD) states that 17% of GHGs arise from agricultural activities directly, with a further 7% to 14% due to changes in land use. Agriculture will be affected by climate change, particularly in some parts of the world, where the extremes of its impact will be felt severely. Flooding and droughts are predicted to increase in frequency with an associated detrimental impact on crop productivity either due to prolonged water shortages or the creation of anoxic soil conditions and crop hypoxia. Flooded soils also promote the denitrification process and an increase in the release of nitrous oxide.

The type of risk and the severity of its impact is spatially explicit, with different parts of the planet and their associated crop production systems subject to more intense effects and levels of threat, as illustrated for Iran by Alamgir et al. [1] and Bangladesh by Mirgol et al. [2]. The sub-Saharan region of Africa is becoming increasingly vulnerable to drought and temperature rises and farmers will need to adapt the types of crops they grow and their associated management practices [3–6]. Other parts of the world, including North America, may experience warmer winters, resulting in diminished vernalisation [7,8], a process required to promote flowering in certain types of crops. It is not all bad news, however. Significant potential exists to both adapt to and mitigate climate change within the agricultural sector. Any changes will need to be implemented in a sustainable manner to ensure that the solution does not cause other socio-economic or environmental problems. Each potential solution must also be tailored to individual regions and farming systems, as highlighted by Zheng et al. [9] in Australia. The introduction of Climate-Smart Agriculture and technology for use by smallholder farmers in South America, Africa and Asia [10–12] and the provision of farming subsidies to promote further engagement with these techniques is demonstrated by Arunrat et al. [13]. The growing of novel crops such as *Cannabis sativa* for energy production in Europe [14] or the utilisation of plant breeding to develop novel wheat varieties capable of reducing nitrous oxide emissions [15] are other examples.

All these factors are explored in this Special Issue. We are pleased to include a range of quality academic contributions from across the five continents, providing a truly global perspective. Multiple crops and production systems are represented, including studies that utilise valuable research completed with limited resources available.

Author Contributions: The guest editors contributed equally to all aspects of this editorial. All authors have read and agreed to the published version of the manuscript.

Acknowledgments: The guest editors would like to extend their thanks to the authors who contributed to this Special Issue and to the reviewers who dedicated their time providing the authors with valuable and constructive recommendations.

Conflicts of Interest: The guest editors declare no conflict of interest.

References

1. Alamgir, M.; Furuya, J.; Kobayashi, S.; Binte, M.; Salam, M. Farmers' Net Income Distribution and Regional Vulnerability to Climate Change: An Empirical Study of Bangladesh. *Climate* **2018**, *6*, 65. [CrossRef]
2. Mirgol, B.; Nazari, M. Possible Scenarios of Winter Wheat Yield Reduction of Dryland Qazvin Province, Iran, Based on Prediction of Temperature and Precipitation Till the End of the Century. *Climate* **2018**, *6*, 78. [CrossRef]
3. Bossa, A.; Hounkpè, J.; Yira, Y.; Serpantié, G.; Lidon, B.; Fusillier, J.; Sintondji, L.; Tondoh, J.; Diekkrüger, B. Managing New Risks of and Opportunities for the Agricultural Development of West-African Floodplains: Hydroclimatic Conditions and Implications for Rice Production. *Climate* **2020**, *8*, 11. [CrossRef]
4. Egbebiyi, T.; Crespo, O.; Lennard, C. Defining Crop–Climate Departure in West Africa: Improved Understanding of the Timing of Future Changes in Crop Suitability. *Climate* **2019**, *7*, 101. [CrossRef]
5. Egbebiyi, T.; Lennard, C.; Crespo, O.; Mukwenha, P.; Lawal, S.; Quagraine, K. Assessing Future Spatio-Temporal Changes in Crop Suitability and Planting Season over West Africa: Using the Concept of Crop-Climate Departure. *Climate* **2019**, *7*, 102. [CrossRef]
6. Ibn Musah, A.; Du, J.; Bilaliib Udimal, T.; Abubakari Sadick, M. The Nexus of Weather Extremes to Agriculture Production Indexes and the Future Risk in Ghana. *Climate* **2018**, *6*, 86. [CrossRef]
7. Parker, L.; Abatzoglou, J. Warming Winters Reduce Chill Accumulation for Peach Production in the Southeastern United States. *Climate* **2019**, *7*, 94. [CrossRef]
8. Petersen, L. Impact of Climate Change on Twenty-First Century Crop Yields in the U.S. *Climate* **2019**, *7*, 40. [CrossRef]
9. Zheng, B.; Chapman, S.; Chenu, K. The Value of Tactical Adaptation to El Niño–Southern Oscillation for East Australian Wheat. *Climate* **2018**, *6*, 77. [CrossRef]
10. Haworth, B.; Biggs, E.; Duncan, J.; Wales, N.; Boruff, B.; Bruce, E. Geographic Information and Communication Technologies for Supporting Smallholder Agriculture and Climate Resilience. *Climate* **2018**, *6*, 97. [CrossRef]
11. Hellin, J.; Fisher, E. Climate-Smart Agriculture and Non-Agricultural Livelihood Transformation. *Climate* **2019**, *7*, 48. [CrossRef]
12. Matewos, T. Climate Change-Induced Impacts on Smallholder Farmers in Selected Districts of Sidama, Southern Ethiopia. *Climate* **2019**, *7*, 70. [CrossRef]
13. Arunrat, N.; Sereenonchai, S.; Pumijumnong, N. On-Farm Evaluation of the Potential Use of Greenhouse Gas Mitigation Techniques for Rice Cultivation: A Case Study in Thailand. *Climate* **2018**, *6*, 36. [CrossRef]
14. Asquer, C.; Melis, E.; Scano, E.; Carboni, G. Opportunities for Green Energy through Emerging Crops: Biogas Valorization of Cannabis sativa L. Residues. *Climate* **2019**, *7*, 142. [CrossRef]
15. Demone, J.; Wan, S.; Nourimand, M.; Hansen, A.; Shu, Q.; Altosaar, I. New Breeding Techniques for Greenhouse Gas (GHG) Mitigation: Plants May Express Nitrous Oxide Reductase. *Climate* **2018**, *6*, 80. [CrossRef]

Article

On-Farm Evaluation of the Potential Use of Greenhouse Gas Mitigation Techniques for Rice Cultivation: A Case Study in Thailand

Noppol Arunrat *, Sukanya Sereenonchai and Nathsuda Pumijumnong

Faculty of Environment and Resource Studies, Mahidol University, Nakhon Pathom 73170, Thailand; sukanya.ser@mahidol.ac.th (S.S.); nathsuda.pum@mahidol.ac.th (N.P.)
* Correspondence: noppol.aru@mahidol.ac.th; Tel.: +66-2-441-5000

Received: 27 March 2018; Accepted: 25 April 2018; Published: 2 May 2018

Abstract: Environmental and socio-economic evaluations that imply techniques for mitigating greenhouse gas (GHG) emissions from rice cultivation are a challenging and controversial issue. This study was designed to investigate the potential use of mitigation techniques for rice cultivation. Mid-season drainage (MD), using ammonium sulfate instead of urea (AS), and site-specific nutrient management (SSNM) were chosen as mitigation techniques. Data were collected using field surveys and structured questionnaires at the same 156 farms, covering four crop years. The GHG emissions were evaluated based on the concept of the life cycle assessment of the GHG emissions of products. The farmers' assessments of mitigation techniques, with multiple criteria evaluation, were obtained by face-to-face interviews. Opinions on all mitigation techniques were requested two times covering four years with the same 156 farm owners. The multinomial logistic regression model was used to examine the factors influencing the farmers' decisions. The results show that SSNM was evaluated as the highest abatement potential (363.52 kgCO$_2$eq ha^{-1}), the negative value of abatement cost (−2565 THB ha^{-1}), and the negative value of the average abatement cost (−14 THB kgCO$_2$eq^{-1}). Among the different techniques, SSNM was perceived as the most suitable one, followed by MD and AS. Highly significant factors influencing decision making consisted of planted area, land size, farmer liability, farmer perception of yield, and GHG emissions. Subsidies or cost-sharing measures to convince farmers to adopt new techniques can enhance their practices, and more support for the development of water systems can increase their availability.

Keywords: rice field; mitigation techniques; greenhouse gas emissions; life cycle assessment; farmer acceptance; incentive measures

---

1. Introduction

Rice paddies are considered to be one of the most important sources of anthropogenic emissions of greenhouse gases (GHGs), particularly nitrous oxide (N$_2$O), methane (CH$_4$), and carbon dioxide (CO$_2$) [1] and therefore play an important role in climate change [2,3]. Notably, many studies state that N$_2$O emissions are associated with nitrogen (N) fertilizer application and dry land conditions [4,5], while flooded fields are a significant source of CH$_4$ and contribute little to N$_2$O emissions [6–8]. The use of agricultural machines requires the use of fossil fuels, resulting in CO$_2$ emissions. Projected increases in the demand for rice have raised considerable concerns about increasing greenhouse gas (GHG) emissions [9]. Thus, knowledge about trade-offs between rice yield increases and GHG emission reductions is urgently needed for the development of effective mitigation and adaptation strategies.

Considering possible strategies for mitigating GHG emissions from rice cultivation, those having no effect on rice yield would be the best techniques. Methane emissions vary markedly with water management. In particular, mid-season drainage, with the short-term removal of irrigation water, is one of the most promising strategies for reducing CH_4 emissions [10–12]. Several field measurements indicate that mid-season drainage (MD) significantly reduces CH_4 emissions and exerts a positive impact on rice yields by increasing N mineralization in the soil and increasing rice plant root development [13–17]. However, it also increases N_2O emissions by creating nearly saturated soil conditions, which promote N_2O production [18–20]. Fertilizer management has frequently been suggested as a mitigation option by substituting urea as N fertilizer with ammonium sulfate $(NH_4)_2SO_4$ (inhibits methanogens) and ammonium phosphate (promotes rice plant growth) [21]. Ammonium sulfate has a significant effect on N_2O reduction and slightly depresses CH_4 production by 10–67% [22], because sulfate-reducing bacteria can outcompete CH_4-producing bacteria under these conditions [23]. Moreover, site-specific nutrient management (SSNM) has been suggested as a method to reduce N_2O emissions by controlling the use of fertilizers with synchronization and precise farming techniques, using slow-release nutrients (including nitrification inhibitors) [24,25] and avoiding their overuse [26]. Dobermann and Cassman [27] state that an N recovery of over 70% can be achieved for many cereal crops by using intensive site-specific nutrient management, based on the principles of the 4R nutrient stewardship—the right source at the right rate, time, and place [28]. However, the sources of CH_4 and N_2O from rice fields cannot be reliably identified and discriminated in various areas.

There is an urgent need to quantify the effects and costs of mitigation strategies in rice fields, which, at present, remain difficult to enumerate, and could result as being speculative. A significant problem is that most farmers do not apply these mitigation strategies, for various reasons such as no ownership on farmland [29,30], less education or training on mitigation strategies [30,31], low income and access to credit [30–32], or less farming experience [33]. An evaluation method is therefore required that highlights decision factors and provides insight into the balance between environmental impacts, economic productivity, and social acceptance regarding mitigation strategies. Another significant problem is that the decision-making processes in terms of employing mitigation strategies are complicated by financial incentives and because agricultural activities depend on, and have a large impact on, natural resources [34]. These factors indicate the need to better understand decision making by farmers and the barriers inhibiting the adoption of mitigation and adaptation strategies.

Mitigation and adaptation are two basic, but distinctly different responses. Farmers' attitudes towards these two general responses to tackle changing climate conditions must be understood if scientists, policy makers, and others are to effectively support adaptive and mitigative actions [35,36]. Moreover, integrating mitigation and adaptation are win-win actions because they can mitigate the causes of climate change (mitigation) and adapt to changing climatic conditions (adaptation) [37]. Many studies have investigated farmer behavior and the associated socio-economic characteristics (e.g., [38–40]). Until now, mitigation costs caused by improvements in farming practices have rarely been reported, and information on the socio-economic feasibility of these mitigation techniques are still lacking, while their social acceptance and the minimization of their costs have not been discussed at any length. Therefore, the objectives of this study are: (1) to evaluate the GHG emissions of each mitigation technique for rice cultivation; (2) to clarify the farmers' assessment with multiple criteria evaluation of each mitigation technique; and (3) to examine the factors influencing the farmers' decisions to use a mitigation technique. The knowledge provided by this study can aid policy makers and other related agencies in their efforts to design and compare mitigation policies and reach mitigation goals.

2. Materials and Methods

2.1. Mitigation Technique Selection

Mitigation techniques were selected based on a literature review and on the recommendations of experts, provided in a report by the Office of Agricultural Economics [41], Ministry of Agriculture and Cooperatives, Thailand. Moreover, we expected that any mitigation techniques suggested to government agencies would be likely to be promoted and supported by the government in the near future. Based on these criteria, mid-season drainage (MD), replacement of urea with ammonium sulfate $((NH_4)_2SO_4)$ (AS), and site-specific nutrient management (SSNM) were chosen as mitigation techniques for this study.

2.2. Site Selection

Multi-stage sampling was employed for this study as follows. Firstly, at the provincial level, purposive sampling was used, focusing on farmers who have grown rice. They voluntarily participated and provided their information and opinions. Secondly, at the district and sub-district levels, cluster sampling was used to determine two clusters: irrigated areas and rain-fed areas. Moreover, farmers' average net household incomes (calculated by subtracting expenses from total revenue) for each district and sub-district were set as the criterion, based on the assumption that money is the major factor that can improve their livelihood and is the major factor likely to convince them to change their behavior. The four districts with the highest net incomes (Bang Mun Nak, Taphan Hin, Bueng Na Rang, and Pho Prathap Chang districts) and the four districts with the lowest net incomes (Sam Ngam, Wachira Barami, Wang Sai Phun, and Thap Khlo districts) in Phichit province were selected as samples.

2.3. Data Collection

Data were obtained from participatory observation, in-depth interviews, and a questionnaire survey at the same 156 farms (in irrigated and rain-fed areas of 78 farms, respectively) in four crop years (2012/2013, 2013/2014, 2014/2015 and 2015/2016) to avoid data variation. Data throughout the crop years from each crop, consisting of cultivation practices, agricultural inputs (e.g., fossil fuels, fertilizers, insecticides, herbicides, and water sources), yields, transportation costs, and benefits were collected from the farm owners. Data were also obtained from the record books for the standards for good agricultural practices (GAP) for farm owners, which was disseminated to the farmers by the Department of Agricultural Extension, Ministry of Agriculture and Cooperatives, Thailand.

2.4. Estimation of GHG Emissions

2.4.1. System Boundary and Functional Unit

The concept of the life cycle assessment of the greenhouse gas emissions of products, based on cradle-to-gate, was employed. It is because this approach is widely used for evaluating and comparing the environmental impacts of various products, and also to identify, quantify, and track the sources of GHG emissions throughout production process [42]. System boundary covers raw material production, transport of agricultural inputs (diesel fuel, gasoline fuel, chemical fertilizers, insecticides and herbicides) to the farm, land preparation, planting, harvesting, storing and post-harvest burning of crop residues (Figure 1). The transportation data were considered for two distances: the average distance from the farms to the retailer in the municipality of each sub-district and the average distance from the farms to the retailer in the community of each farm. Burning crop residues in the paddy field were included in this study because it is a common way to eliminate rice residues in Asia, including Thailand [43,44], and GHG emissions from open burning concentrated in the harvest season [45]. It is indicated that emissions from burning crop residues play an important role in the air pollution and climate change [46]. To assess the combined global warming potential (GWP), CH_4, and N_2O were calculated as CO_2 equivalents over a 100-year time scale, using a radiative forcing potential relative to CO_2 of 28 for CH_4 and 265 for N_2O [47]. The functional unit used in assessments was kg CO_2eq ha^{-1} for each technique.

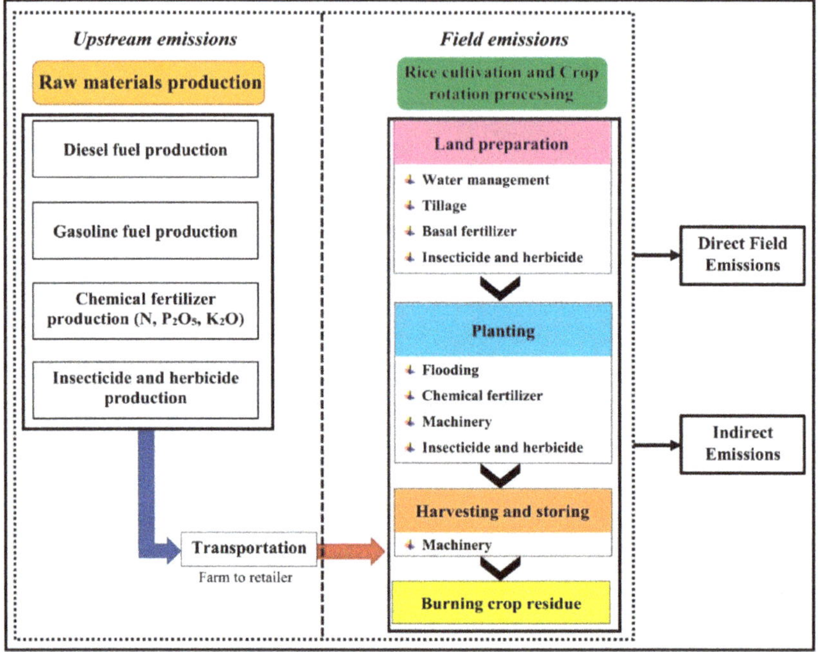

Figure 1. System boundary from cradle to farm gate of the study (adapted from Arunrat et al. [48]).

2.4.2. Calculation of GHG Emissions

The GHG emissions were calculated for each farm using four scenarios, including the business as usual (BAU) case, and the use of MD, AS, and SSNM techniques. Upstream emissions were accounted for in terms of raw material production and the transportation of agricultural inputs to the farm. Fossil fuels, chemical fertilizers, as well as insecticide and herbicide production were estimated using specific emission factors, as characterized in Ecoinvent 3.2 [49]. Emissions from the transportation of agricultural inputs to the farm were estimated based on diesel fuel consumption, using the emission factors from the National Technical Committee on Product Carbon Footprinting (Thailand) [50]. In some cases, specific emission factors for gasoline or insecticides and herbicides were not available in Ecoinvent 3.2, so country-specific emission factors for Thailand from the National Technical Committee on Product Carbon Footprinting (Thailand) [50] were used instead.

Field CH_4 emissions from rice cultivation were used as the model for the calculations, according to the 2006 Intergovernmental Panel on Climate Change (IPCC) Guidelines for National Greenhouse Gas Inventories [51]. The baseline emission factor was taken from Yan et al. [16], who adjusted region-specific emission factors for rice fields in east, southeast, and south Asian countries, and all scaling factors used were derived from the IPCC [51]. Direct and indirect N_2O emissions and CO_2 emissions from urea applications were also estimated using the methodology proposed by the IPCC [50]. The GHG emission calculations and parameters and emission factors for diesel and gasoline usage in stationary combustion were taken from the IPCC [51]. The GHG emissions from the mobile combustion of diesel fuel by farm tractors and harvesters were estimated from the emission factors of Maciel et al. [52], and GHG emissions from gasoline fuel were estimated following the EPA [53]. Figures for insecticides and herbicides were provided by the emission factors from Lal [54]. Equations, parameters, and emission factors for the calculation of GHG emissions are presented in the Supplementary Material by Arunrat et al. [48].

2.5. Economic Analysis

2.5.1. Estimation of the Costs of Each Technique

The production input of each technique consists of water (W), tillage (T), seed (S), labor (L), fertilizer (F), insecticide (P), herbicide (H), harvest (V), and land rental (R). The total production cost $[C(Q_i)]$ for each technique is the sum of production input costs Equation (1).

$$C(Q_i) = (C_W \times W_i) + (C_T \times T_i) + (C_s \times S_i) + (C_L \times L_i) + (C_F \times F_i) + (C_P \times P_i) + (C_H \times H_i) \atop +(C_V \times V_i) + (C_R \times R_i) \tag{1}$$

where i is each technique, $C(Q)$ is the cost of crop production, in Baht ha^{-1}, and C_W, C_T, C_S, C_L, C_F, C_P, C_H, C_V, and C_R are costs of water management, tillage, seed, labor, fertilizer, insecticide, herbicide, harvest, and land rental, in Baht^{-1} unit, respectively.

In addition, the specific details of the methods used to estimate the costs of each technique are described below.

(1)   MD Technique

The cost of the MD technique was calculated by multiplying the quantity of fuel used for pumping water back into the fields, using the fuel price per unit. The cost of this technique was investigated depending on the distance from the fields and the ownership of the water source by dividing the farms into two groups: (1) those far away from water sources (natural sources or irrigation systems at >100 m or >50 m from the fields, respectively); and (2) farms with their own surface pond or artesian well.

(2)   AS Technique

The use of ammonium sulfate (21-0-0) instead of urea (46-0-0) requires changes in the quantities of the fertilizers used and their costs. The relevant calculations are as follows: (1) 1 kg of urea contains 0.46 kg N; (2) it takes 2.19 kg of ammonium sulfate to replace 1 kg of urea, providing 0.46 kg of N; (3) the amount of ammonium sulfate used, multiplied by its unit price, is equal to the total cost of the ammonium sulfate used.

(3)   SSNM Technique

The cost of the SSNM technique was calculated based on the following steps. Firstly, the amount of each fertilizer to be used was calculated based on the instructions provided by the Land Development Department of Thailand after soil factor analysis. For instance, in the Nong Phra sub-district, Wang Sai Phun district, the soil series is Chiang Rai, suitable for growing photosensitive rice varieties. Suggested fertilizers are 31 kg ha^{-1} of 46-0-0, 71 kg ha^{-1} of 16-20-0, and 37 kg ha^{-1} of 0-0-60, to be applied 7–10 days after sowing or 25–30 days after transplanting, and 31 kg ha^{-1} of 46-0-0, to be applied again during the early flowering phase. After the suitable amounts of all fertilizers were established, the cost of each fertilizer used was calculated by multiplying the quantity by the price per unit. Finally, the total fertilizer cost of the SSNM technique was compared to the fertilizer cost of the BAU case.

2.5.2. Average Abatement Cost (AAC)

The AAC was used to assess the economic potential for the reduction of GHG emissions in this study; AAC refers to the cost of implementing a technique to reduce GHG emissions to an anticipated level. Similar to the GHG emission estimations, AAC was estimated using four scenarios comprising the BAU case and the use of the MD, AS, and SSNM techniques. The AAC (THB kgCO$_2$eq^{-1}) of each technique was calculated by dividing the total abatement cost (THB ha^{-1}) (TAC) by the total abatement potential (kgCO$_2$eq ha^{-1}) (TAP), and each TAC and TAP were obtained by subtracting the cost under the BAU scenario. Indeed, the reduction of GHG emissions is involved with cropping system, mitigation techniques, and farmers' behavior. Therefore, ACC was then presented to the farmers of each farm during their assessments on each mitigation technique. This is because ACC can help the farmers to visualize about being environmentally friendly and reducing production costs.

2.6. Farmers' Assessment and Analysis Tools

After the last crop year (2015/2016) for data collection, the investigation of the farmers' assessment for each farm was taken place in 2017. A multiple criteria evaluation was developed to assess farmers in the qualitative evaluation of the mitigation techniques. In this study, the criteria applied in the multiple criteria evaluation for farmers' assessment on the three mitigation techniques were as defined in Table 1, adapted from Webb et al. [55]. To reduce the bias and uncertainty from the farmers' assessment, the survey was administered via a face-to-face interview in November 2016 and August 2017, with the same 156 farm owners. The farmers were introduced and explained the purposes of the survey. The farmers' assessment was investigated after calculating the AAC for each scenario and each farm, but the farmers were allowed to choose only one suitable technique to implement. A questionnaire was presented to the farmers to evaluate the rating of each mitigation technique. A four-Likert scale was adopted for the evaluation [56]. The rating scale for the farmers' assessment was: '4' = very good, '3' = good, '2' = poor, and, '1' = very poor. We used a four-point scale to interpret the farmers' response because a mid-point is considered as too ambiguous for decision making [57], which was also mentioned in Webb et al. [55]. The scores of each farmer were summed up from the scores of each criterion for the three mitigation techniques. For instance, 78 farmers gave a score of 4 (very good) to the MD technique on the criteria of effectiveness; the total score was 312 (78 $\times$ 4). Moreover, the farmers were asked about their needs for policies and incentives to support their farming.

Table 1. Definitions of the criteria for farmers' assessment (adapted from Webb et al. [55]).

Criteria	Definition
Effective	Evaluates whether or not the mitigation technique reduces GHG emissions
Flexible	Evaluates whether or not the ability of the mitigation technique to enhance opportunity for other cropping systems and places
Economically efficient	Evaluates whether or not implementing the mitigation technique reduces production cost and increases household income
Easy to implement	Evaluations whether a mitigation technique is easy to implement by farmers with technical and managerial ease
Ability to trial	Evaluates whether a mitigation technique can be easily trialed or tested before full implementation
Institutional compatibility	Evaluates whether a mitigation technique is consistent with the current management framework, laws, regulations and will be promoted and supported by the government in the near future

2.7. Estimating the Determinants of Mitigation Techniques and Socio-Economic Variables

Factors that might influence the farmers' decision to adopt or reject the mitigation techniques were examined using the multinomial logistic regression (MNL) model. The MNL model is an extension of logistic regression, which is generally effective when the dependent variable is composed of a polytomous category with multiple choices. Explanatory variables included in the MNL model were defined as two types: dichotomous and continuous variables, as detailed below (Table 2). The model was estimated using the following specification:

$$\begin{aligned} Y = \quad & \beta_0 + \beta_1 AREA + \beta_2 EXP + \beta_3 OWN + \beta_4 SIZE + \beta_5 INC + \beta_6 LIB \\ & + \beta_7 LABOR + \beta_8 MEM + \beta_9 PYIELD + \beta_{10} PGHG + \beta_{11} MEA \\ & + \beta_{12} TRAIN + \beta_{13} DOUB + \beta_{14} TRI + u \end{aligned} \quad (2)$$

where Y is the acceptability of the mitigation technique; *AREA* is the planted area; *EXP* is the experience; *OWN* is the land owner; *SIZE* is the land size; *INC* is the farmer′s income; *LIB* is liability; *LABOR* is the amount of labor; *MEM* is the membership of the environment group; *PYIELD* is the perception of yield; *PGHG* is the perception of GHG emissions; *MEA* represents government measures; *TRAIN* represents attendance at training; *DOUB* is the double cropping system; *TRI* is the triple cropping system; and μ is the error term.

Table 2. Definition and descriptive statistics of variables used in the MNL model.

Variable	Description
Planted area	Dummy, 1 if the farm is located in a rain fed area; 0 irrigated area
Experience	Continuous, rice cultivation experience of farmer (years)
Land owner	Dummy, 1 if the farmer is a land owner; 0 otherwise
Land size	Continuous, size of plantation (ha)
Farmer income	Continuous, farmer income from in-farm and off-farm (THB year^{-1} household^{-1})
Farmer liability	Continuous, farmer liability from formal and informal financial institutions (THB household^{-1})
Number of labor	Continuous, number of laborers in the household (persons)
Membership of environment group	Dummy, 1 if the farmer is the member of an environmental group or institution; 0 otherwise
Perception on yield	Dummy, 1 if the farmer's perception is that the mitigation technique will increase the rice yield; 0 otherwise
Perception on GHG emissions	Dummy, 1 if the farmer thinks that the mitigation technique can reduce GHG emissions; 0 otherwise
Perception on measures	Dummy, 1 if the farmer's perception is that the mitigation technique will be supported by government agencies; 0 otherwise
Attendance in training	Dummy, 1 if the farmer had attended the training about the impact of climate change impact on the environment; 0 otherwise
Double cropping system	Dummy, 1 if the farmer practices as usual the double cropping system; 0 otherwise
Triple cropping system	Dummy, 1 if the farmer practices as usual the triple cropping system; 0 otherwise

3. Results and Discussion

3.1. Cost of Rice Production under BAU and Mitigation Techniques

Marked significant differences in costs between irrigated and rain-fed areas were revealed using the t-test ($p < 0.05$). The average production costs under BAU were 27,521 and 24,240 THB ha^{-1} for irrigated and rain-fed areas, respectively. Using cost structure analysis, the average variable cost was 22,375 THB ha^{-1}, consisting of an average labor cost of 11,918 THB ha^{-1} and an average material cost of 10,456 THB ha^{-1}, while the average fixed cost was 4213 THB ha^{-1}. Furthermore, a lack of laborers and water for planting were the outstanding factors increasing the production costs. The average rice yields were 5.58 and 4.58 tons ha^{-1} for irrigated and rain-fed areas, respectively. The net profit in irrigated areas was higher than that in rain-fed areas, being 34,079 and 32,960 THB ha^{-1}, respectively.

This study found that when implementing the MD technique, the average cost of rice production was 30,100 and 29,662 THB ha^{-1} for irrigated and rain-fed areas, respectively. Rain-fed areas were associated with higher average production costs than irrigated areas, about 2840 THB ha^{-1} or double the increase in costs. Comparing the cost of water source distance, farmers who owned their surface pond or artesian well, implementing MD, would face average costs 1946 THB ha^{-1} higher than those for BAU. Meanwhile, at distances of 100 and 50 m from the water sources, the costs would be 6843 and 5584 THB ha^{-1}, respectively. Consequently, this study reflects that the cost of implementing MD is reduced by 28–35% if farmers own their own surface pond or artesian well for cultivation, while the average cost will be higher with increasing distance to the water source.

To implement the AS technique, the average production costs were 28,985 and 25,998 THB ha^{-1} for irrigated and rain fed-areas, respectively. An interesting point is that organic farmers following the AS technique can reduce their costs by about 645 and 863 THB ha^{-1} for irrigated and rain-fed areas, respectively, due to their lower costs for chemical fertilizer application under the BAU case.

Therefore, if organic farmers switch from using urea to ammonium sulfate, their average costs will be reduced as well. A cost-benefit analysis showed that organic rice farming could generate higher net profits than conventional farming, of about 437 and 289 THB ha^{-1} for irrigated and rain-fed areas, respectively. Consequently, to effectively implement the AS technique, organic fertilizer should be applied in combination to further reduce costs and increase net profit while not affecting rice yields.

For SSNM, the average production costs were 26,450 and 23,354 THB ha^{-1} for irrigated and rain-fed areas, respectively. Following this technique, farmers could achieve reductions in the average production cost compared with BAU of 1068 and 885 THB ha^{-1} for irrigated and rain-fed areas, respectively. The average production costs in irrigated areas were about 182 THB ha^{-1} lower than those in rain-fed areas, as lower amounts of chemical fertilizer were applied under BAU conditions.

Comparing the cost of BAU and using mitigation techniques for both irrigated and rain-fed areas, performing SSNM can reduce the average production costs compared with BAU. However, MD and AS resulted in higher production costs than BAU. Overall, the average production costs were higher in irrigated areas than in rain-fed areas. This result reflects that the average production costs are higher when farmers own more land for growing rice, but this higher average cost tends to decrease when farmers adapt their rice cultivation behavior by adopting the option that has lower costs than BAU, without reducing the rice yields.

3.2. GHG Emissions, Abatement Potential, and AAC Under BAU and Mitigation Techniques

The results of estimates of GHG emissions, abatement potential, and AAC between BAU and the different mitigation techniques are presented in Table 3 and Figures 2 and 3. There were highly significant differences in the first and second cultivations between irrigated and rain-fed areas and for each technique. These results reflect the fact that MD is more appropriate for implementation in irrigated rather than rain-fed areas and more appropriate for the second rice cultivation than for the first cultivation. The AS technique led to a higher abatement potential for the second rice cultivation than for the first one. Meanwhile, SSNM generated a 42.6% higher abatement potential for the second rice cultivation than for the first one, with a 9.8% lower AAC for irrigated than rain-fed areas. However, among all techniques, SSNM was the most appropriate one because its AAC was lower than that for BAU, and it had a 60.2 and 58.1% higher abatement potential than MD and AS, respectively.

Table 3. Average abatement cost (AAC) using different mitigation techniques (Authors own calculation).

	GHG Emissions under BAU (kgCO$_2$eq ha^{-1})	GHG Emissions under Mitigation Technique (kgCO$_2$eq ha^{-1})	Abatement Potential (kgCO$_2$eq ha^{-1})	Abatement Cost (THB ha^{-1})	AAC (THB kgCO$_2$eq^{-1})
MD technique					
1st rice					
Irrigated	3549	3411	138	7372	53
Rain-fed	3214	3089	125	8975	71
2nd rice					
Irrigated	2767	2590	176	7960	45
Rain-fed	2185	2046	139	9663	69
AS technique					
1st rice					
Irrigated	3549	3403	146	3405	23
Rain-fed	3214	3062	151	3002	19
2nd rice					
Irrigated	2767	2618	148	3641	24
Rain-fed	2185	2022	163	3499	21
SSNM technique					
1st rice					
Irrigated	3549	3276	273	−4718	−17
Rain-fed	3214	2888	326	−3747	−11
2nd rice					
Irrigated	2767	2269	497	−6600	−13
Rain-fed	2185	1828	357	−5738	−15

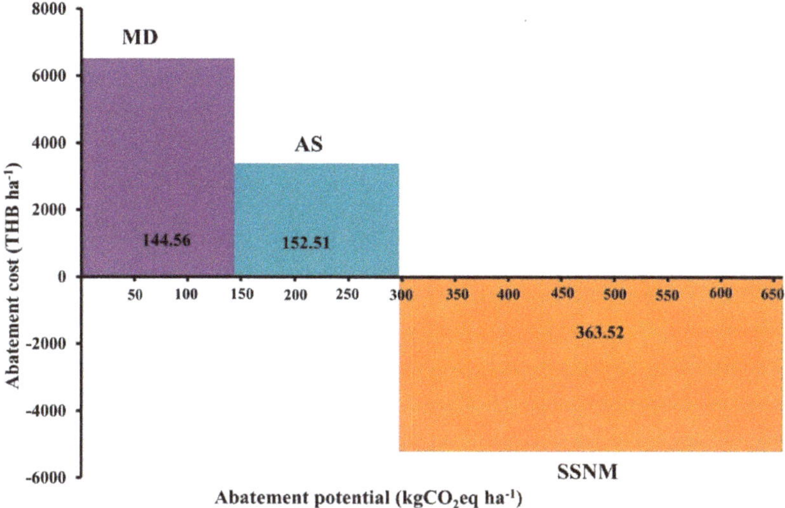

Figure 2. Comparison between abatement cost and abatement potential for each mitigation technique (Authors own calculation).

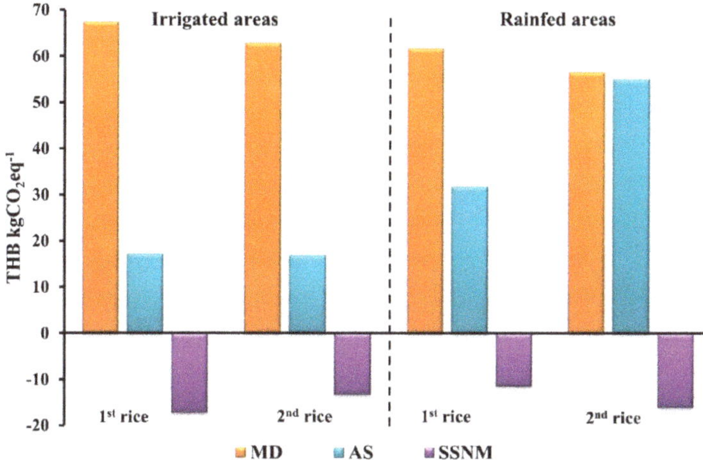

Figure 3. Average abatement cost (AAC) under BAU and using mitigation techniques (Authors own calculation).

3.3. Farmers' Assessment on Mitigation Techniques and Barriers

In the survey, farmers were requested to indicate their opinion on all mitigation techniques. Farmers' assessments across multiple criteria and the total score of each mitigation technique are provided in Table 4. As a result, the SSNM technique was the most favored one and presented the highest score, followed by MD and AS, respectively. The criteria of effectiveness, flexibility, economic efficiency, and institutional compatibility indicated the highest score regarding the SSNM technique. This is in line with Dobermann et al. [58], who reported that the higher benefit for farmers from the implementation of nutrient management strategies can increase the profitability of rice cropping, enhance socio-economic conditions, and mitigate labor shortage. Moreover, efficient nutrient management can also result in environmental benefits through a reduction of chemical fertilizers without a reduction in yield [59]. The criteria "easy to implement" and "ability to trial" were implementing the MD technique because it is easy to drain the water out of the rice field, but farmers need reliable control over irrigation water to implement this technique, otherwise rice yields are impacted. On the other hand, the AS technique obtained the lowest scores for the criteria "economic efficiency", "easy to implement", and "institutional compatibility".

Table 4. Summary of farmers' assessment with multiple criteria evaluation of each mitigation technique (Authors own calculation).

Assessment Criteria	Mitigation Techniques		
	MD	AS	SSNM
Effectiveness	542	393	588
Flexibility	317	446	565
Economic efficiency	376	201	603
Farmer implementability	496	233	468
Ability to trial	510	420	464
Institutional compatibility	495	233	570
Total score	2736	1926	3258

The scale used for scoring is presented in Table 4; green reflects low scores, while red reflects high scores.

The percentage of farmers ranking the mitigation techniques for each criterion, indicating the level of agreement, across the survey is provided in Table 5. The SSNM technique was the technique most favored by the farmers, with 86.5% indicating that they strongly agreed with the highest economic efficiency compared with other mitigation techniques, while only 13.5% of farmers indicated that they strongly agreed that this technique is easy to implement. Indeed, 4.5% of the farmers considered its "ability to trial" as very poor. Similarly, Chinese farmers willing to adopt low-carbon technology when the expenses of required inputs increase less after application [60]. In terms of the MD technique, 50% of the farmers strongly agreed with "effectiveness", followed by "institutional compatibility" (43.6%), "farmer implementability" (41.7%), and "ability to trial" (40.4%). However, 87.8% and 17.9% of farmers considered "flexibility" as poor and "economic efficiency" as very poor, respectively. Further, 32.1 and 10.9% of farmers evaluating the AS technique selected very good in terms of "flexibility" and "ability to trial". On the other hand, 71.2% of the farmers considered "economic efficiency" of the AS technique as very poor.

Table 5. The percentage of farmers showing a score of the level of agreement for each criteria (Authors own calculation).

Criteria/Rank	Mitigation Techniques											
	MD				AS				SSNM			
	1	2	3	4	1	2	3	4	1	2	3	4
Effectiveness	0	2.6	47.4	50.0	10.9	32.1	51.3	5.8	0	0	23.1	76.9
Flexibility	4.5	87.8	7.7	0	3.8	38.5	25.6	32.1	0	2.6	39.1	58.3
Economic efficiency	17.9	41.0	23.1	17.9	71.2	28.8	0	0	0	0	13.5	86.5
Farmer implementability	9.0	5.8	43.6	41.7	50.6	49.4	0	0	0	4.5	82.1	13.5
Ability to trial	1.3	10.9	47.4	40.4	2.6	36.5	50.0	10.9	4.5	16.7	17.9	60.9
Institutional compatibility	0	26.3	30.1	43.6	34.0	54.5	11.5	0	0	0	34.6	65.4

The scale used for scoring is presented in Table 5; green reflects low scores, while red reflects high scores.

When the farmers were asked to select one technique, 58.87% of the respondents were willing to implement SSNM, 29.29% AS, and 11.84% MD. Farmers in irrigated areas were most willing to perform SSNM, followed by AS and MD. In contrast, farmers in rain-fed areas were most willing to operate via SSNM, followed by AS, similar to those in irrigated areas, but no farmers were willing to implement

MD. As a result, we suggest that state policies should encourage SSNM in both irrigated and rain-fed areas as a practice that can result in lower fertilizer use. However, the relative willingness, beliefs, attitudes, and perceptions concerning such choices are indicators of the future likelihood to adopt a certain practice, which have also been described by McCown [61], Morton [62], and Jones et al. [63].

The reasons for the unwillingness to implement MD were water shortage, fear of increased weeds and pests, worries about nutrient losses, potential declines in rice yield, and a perception of MD being time-consuming, labor-consuming, and requiring more investment. Concerning the AS technique, farmers were worried about lower yields when not using urea, as they believe that urea contributes to greater yields, and there was a lack of knowledge about implementing the use of ammonium sulfate. Farmers unwilling to implement SSNM were concerned about yield decrease and felt that SSNM is time-consuming and complex. They also reported a lack of knowledge to support the use of soil analysis and high expenditures on soil analysis as matters of concern.

3.4. Factors Determining Farmers' Decisions

The results of the MNL model are presented in Table 6. The variables that were highly significant in the allocation of the farmers' decisions concerning each mitigation technique were as follows: (i) planted area; (ii) land size; (iii) farmer liability; (iv) farmer's perception of yield; and (v) farmer's perception of GHG emissions. Multicollinearity was checked among independent variables. The variance inflation factor (VIF) for all independent variables ranged from 1.108 to 1.265 (VIF < 5), which means that multicollinearity should not be a serious concern in this regression ($p < 0.01$).

Table 6. Estimated marginal effects of the farmers' decision to use the mitigation technique.

Variable	Mitigation Technique		
	MD	AS	SSNM
Planted area	−0.246 ** (0.0732)	−1.082 *** (0.153)	0.381 *** (0.022)
Experience	0.00384 (0.00492)	0.00376 (0.00348)	0.00743 (0.00315)
Land owner	0.00485 (0.0105)	0.0255 * (0.00503)	0.0466 * (0.0062)
Land size	−1.208 *** (0.0632)	−0.00478 (0.00255)	0.050 * (0.0260)
Farmer income	0.164 ** (0.00478)	0.403 ** (0.00455)	0.365 ** (0.00173)
Farmer liability	−0.411 * (0.00251)	−0.548 *** (0.000751)	0.332 *** (0.000177)
Number of labor	0.0301 (0.0137)	0.00428 (0.00199)	0.0676 (0.00295)
Membership of environment group	0.0446 (0.00662)	0.0507 (0.00227)	0.215 ** (0.00351)
Perception on yield	−0.0643 * (0.0338)	0.0661 (0.0255)	0.332 *** (0.00708)
Perception on GHG emissions	−0.0162 (0.0582)	−0.314 ** (0.0122)	−0.209 *** (0.00314)
Perception on measures	0.00944 (0.0132)	0.0407 (0.00671)	0.00194 (0.0118)
Attendance in training	0.0552 (0.00831)	0.0253 (0.0448)	0.0158 ** (0.00257)
Double cropping system	0.0308 (0.000744)	−0.206 ** (0.00678)	0.0321 (0.0186)
Triple cropping system	−0.0269 (0.00731)	0.0316 (0.0733)	−0.00736 (0.00228)
Constant	122461.72 ** (13562.15)	140939.82 ** (18953.05)	−159005.10 *** (10535.43)
Observations	156	156	156

* $p < 0.1$; ** $p < 0.05$; *** $p < 0.01$; SE in parentheses.

In the area studied, a great number of rice fields are located in rain-fed areas. The negative coefficient for rain-fed areas for MD and AS implies that these techniques are considerably less likely to be implemented in rain-fed areas compared with the irrigated areas, or not implemented at all. The reason is that when implementing MD in rain-fed areas, it is difficult to drain water into rice fields after it has been drained out, resulting in higher costs. Similarly, in terms of the AS technique, the farmers felt unaccustomed to the use of ammonium sulfate fertilizers. If adopting AS, farmers face higher costs as more ammonium sulfate fertilizer is required to maintain the same level of nutrients while possibly achieving lower yields. On the other hand, SSNM has a positive and significant influence when implemented, and it is highly likely that farmers will implement this technique.

Land size is an important factor influencing farmers' decisions in terms of various mitigation techniques. Land size had a negative and significant influence on MD, which probably means that the larger the land, the less likely the farmers are to implement MD. The same is true for AS, which can generate higher production costs in water and chemical fertilizer management. In contrast, farmers who owned more land were interested in SSNM because of its obvious cost savings. However, farmers with large areas of land were also worried about high expenses for soil characteristics analysis.

Of the significant variables, farmer liability had a positive influence favoring SSNM, while having a negative influence towards AS. Therefore, farmers with greater liabilities were interested in low-cost techniques and may reject high-cost techniques.

The effect on rice yield of each mitigation technique was the priority of the farmers. Consequently, farmers' perception of yield was one of the significant variables influencing their decision making. The results show that farmers' perception of yield had a positive and significant influence favoring SSNM. It can be inferred that farmers perceived that implementing SSNM could increase their yields, so they decided to use it.

Farmers' perception of GHG emissions had a negative and significant influence favoring SSNM and AS, meaning that farmers perceived that implementing SSNM and AS techniques would reduce GHG emissions, which was particularly the case for SSNM. Likewise, MD had a negative but non-significant influence, which might be because most farmers still do not have sufficient knowledge about the mitigation potential of each technique. It should be noted that relevant and responsible organizations should encourage and provide knowledge on GHG reduction techniques. Sources of information, including extensions, workshops, and training can enhance the adoption of a certain technology [30]. However, there are several farmers who have less chances for training, probably due to a limitation of time and budget. Therefore, participatory action research should receive more attention both from research-funding organizations and researchers to support collaborations among academicians, local authorities/leaders, and farmers [64]. This would increase the effectiveness of transferring knowledge, the sharing of knowledge and experiences, and could serve as a means to raise awareness about the positive effects of mitigation techniques.

3.5. Prioritizing Incentive Measures for the Adoption of Mitigation Techniques

Understanding farmers' decision-making behavior regarding their current practices is important and must be based on the knowledge of why farmers reject or accept different techniques [65]. Based on the results of the field survey and the in-depth interviews, three incentive measures were important from the point of the view of farmers: (1) cash incentives from governmental agencies to convince farmers to adapt their practices; (2) assistance for cost reduction—seed support and soil property analysis; and (3) support for water system development for agricultural activities—digging ponds and drilling wells near rice fields. The classification of farmers' characteristics for prioritizing supporting measures were identified as follows.

3.5.1. Planted Area

Farmers in irrigated areas rated cash incentive measures as the highest priority, while farmers in rain-fed areas were more concerned about supports for water system development.

3.5.2. Land Size

According to land tenure, farmers could be grouped as: (1) small land owners (1.3–6.5 ha); (2) medium land owners (6.6–11.6 ha); and (3) large land owners (11.7–16.8 ha). Medium land owners rated supports for water system development as the highest priority, while small and large land owners rated assistance with cost reduction as their major concern.

3.5.3. Farmer Income

Farmers could be categorized into three groups based on their income: (1) low-income farmers (52,800–128,000 THB year^{-1} household^{-1}); (2) medium-income farmers (128,001–203,200 THB year^{-1} household^{-1}); and (3) high-income farmers (203,201–278,400 THB year^{-1} household^{-1}). Farmers with medium and high incomes rated support for water system development as the first priority, followed by assistance with cost reduction and cash incentive measures. For farmers with low income, cash incentive measures were most important, because this measure had a direct impact on their income and expenses for implementing GHG mitigation techniques.

3.5.4. Farmer Liability

Regarding the levels of liability, there were three groups of farmers: low liability (58,400–538,933 THB household^{-1}), medium liability (538,934–1,019,467 THB household^{-1}), and high liability (1,019,468–1,500,000 THB household^{-1}). Low liability farmers mainly highlighted support for water system development, while medium liability farmers stressed assistance with cost reduction. High liability farmers highly valued cash incentive measures due to their direct and immediate impact on income. Farmers with low or medium liability gave higher priority to investment in their land (seeds, soil property analysis, and water sources).

3.5.5. Number of Laborers in a Household

According to the number of household members, farms were grouped into low-labor households (1–3 persons) and high-labor households (3–5 persons). Low-labor households made seed support a higher priority than high-labor households. This was because most low-labor households conducted their agricultural activities on smaller areas, so seed support and soil property analysis could greatly help to reduce their production costs. High-labor households prioritized support for water system development, because potential improvements in their water systems could allow them to increase their agricultural activities and gain more income.

3.5.6. Cropping System Pattern

Farmers using a double cropping system preferred support for water system measures, followed by cash incentives and assistance for cost reduction measures. This was because although the farmers' way of making a living in Thailand was based on rice cultivation, these farmers had limited water sources, so they selected crop rotation, which requires less water during the dry season. This could also reduce the cost of water management for agricultural activities. Among farmers using a triple cropping system, assistance for cost reduction measures was the first priority as it reduces the costs of seeds and soil property analysis.

The outstanding point was that cash incentives can be appropriate for low-income farmers or small land owners, who have fewer opportunities to increase their income and need more assistance. These farmers obviously considered subsidies are the priority. Besides, small land owners also placed emphasis on developing their land to be more appropriate for agricultural activities, as their main income relies on their land. On the other hand, high-income farmers and large land owners were aware of other alternatives to increase their income, whether from rice grain or crop rotation. Farmers with medium incomes or medium land owners were more concerned about water system development for agricultural activities than the other groups, because having enough water could lead to greater income and increased crop production efficiency [66]. For farmers with high liabilities, subsidies were of greater concern than for farmers with low or medium liabilities due to their direct and immediate effect on income.

4. Conclusions

Site-specific nutrient management (SSNM) was evaluated as the highest abatement potential (363.52 kgCO$_2$eq ha^{-1}), the negative value of abatement cost (-2565 THB ha^{-1}), and the negative value of the average abatement cost (-14 THB kgCO$_2$eq^{-1}). Based on farmers' assessment to be a mitigation technique for rice cultivation, SSNM reached the highest score for effectiveness, flexibility, economic efficiency, and institutional compatibility. This indicated that SSNM was obviously preferable and presented the highest scores for farmer acceptability, followed by the replacement of urea with ammonium sulfate ((NH$_4$)$_2$SO$_4$) and mid-season drainage. Irrigation systems, land size, farmers' liability, and perception of yield and GHG emissions were found as the main factors affecting the farmers' decision to accept the mitigation techniques. Therefore, incentive measures, such as subsidies or cost-sharing measures can convince farmers to adopt new techniques and enhance their practices. More support of water system development can increase their availability.

Author Contributions: N.A. collected data and preformed life cycle assessment and average abatement cost analyses, wrote, and revised the manuscript. S.S. collected data, analyzed farmers' assessment, and collaborated in discussion. N.P. provided the discussion and advice to this study.

Acknowledgments: This article was supported by the Thailand Research Fund (TRF) (grant number: TRG 5880123), and National Research Council of Thailand (NRCT) (KO-BO-NGO/2557-68). Furthermore, the author would like to thank the reviewers for their helpful comments to improve the manuscript.

Conflicts of Interest: The authors declare no conflict of interest.

References

1. Smith, P.; Martino, D.; Cai, Z. Agriculture. In *Climate Change 2007: Mitigation. Contribution of Working Group III to the Fourth Assessment Report of the Intergovernmental Panel on Climate Change*; Metz, B., Davidson, O.R., Bosch, P.R., Dave, R., Meyer, L.A., Eds.; Cambridge University Press: Cambridge, UK; New York, NY, USA, 2007; pp. 497–540.

2. Intergovernmental Panel on Climate Change (IPCC). *The Supplementary Report to IPCC Scientific Assessment*; Houghton, J.T., Callander, B.A., Varney, S.K., Eds.; Cambridge University Press: Cambridge, UK, 1992.

3. IPCC. *Climate Change 2007: The Physical Science Basis. Contribution of Working Group I to the Fourth Assessment Report of the Intergovernmental Panel on Climate Change*; Solomon, S., Qin, D., Manning, M., Chen, Z., Marquis, M., Averyt, K.B., Tignor, M., Miller, H.L., Eds.; Cambridge University Press: Cambridge, UK; New York, NY, USA, 2007.

4. Adviento-Borbe, M.A.A.; Haddix, M.L.; Binder, D.L.; Walters, D.T.; Dobermann, A. Soil greenhouse gas fluxes and global warming potential in four high-yielding maize systems. *Glob. Chang. Biol.* **2007**, *13*, 1972–1988. [CrossRef]

5. Zou, J.; Huang, Y.; Zheng, X.; Wang, Y. Quantifying direct N_2O emissions in paddy fields during rice growing season in mainland China: Dependence on water regime. *Atmos. Environ.* **2007**, *41*, 8030–8042. [CrossRef]

6. Shang, Q.; Yang, X.; Gao, C.; Wu, P.; Liu, J.; Xu, Y.; Shen, Q.; Zou, J.; Guo, S. Net annual global warming potential and greenhouse gas intensity in Chinese double rice-cropping systems: A 3-year field measurement in long-term fertilizer experiments. *Glob. Chang. Biol.* **2011**, *17*, 2196–2210. [CrossRef]

7. Wang, W.; Dalal, R.; Reeves, S.; Butterbach-Bahl, K.; Kiese, R. Greenhouse gas fluxes from an Australian subtropical cropland under long-term contrasting management regimes. *Glob. Chang. Biol.* **2011**, *17*, 3089–3101. [CrossRef]

8. Yao, Z.; Zheng, X.; Dong, H.; Wang, R.; Mei, B.; Zhu, J. A 3-year record of N_2O and CH_4 emissions from a sandy loam paddy during rice seasons as affected by different nitrogen application rates. *Agric. Ecosyst. Environ.* **2012**, *152*, 1–9. [CrossRef]

9. Van Groenigen, K.J.; van Kessel, C.; Hungate, B.A. Increased greenhouse-gas intensity of rice production under future atmospheric conditions. *Nat. Clim. Chang.* **2013**, *3*, 288–291. [CrossRef]

10. Yagi, K.; Tsuruta, H.; Minami, K. Possible options for mitigating methane emission from rice cultivation. *Nutr. Cycl. Agroecosyst.* **1997**, *49*, 213–220. [CrossRef]

11. Towprayoon, S.; Smakgahn, K.; Poonkaew, S. Mitigation of methane and nitrous oxide emissions from drained irrigated rice fields. *Chemosphere* **2005**, *59*, 1547–1556. [CrossRef] [PubMed]

12. Nayak, D.; Saetnan, E.; Cheng, K.; Wang, W.; Koslowski, F.; Cheng, Y.-F.; Zhu, W.Y.; Wang, J.-K.; Liu, J.-X.; Moran, D.; et al. Management opportunities to mitigate greenhouse gas emission from Chinese agriculture. *Agric. Ecosyst. Environ.* **2015**, *209*, 108–124. [CrossRef]

13. Cai, Z.; Yan, X.; Yan, G.; Xu, H.; Tsuruta, H.; Yagi, K.; Minami, K. Methane and nitrous oxide emissions from rice paddy fields as affected by nitrogen fertilisers and water management. *Plant Soil* **1997**, *196*, 7–14. [CrossRef]

14. Wassmann, R.; Neue, H.U.; Lantin, R.S.; Makarim, K.; Chareonsilp, N.; Buendia, L.V.; Rennenberg, H. Characterization of methane emissions from rice fields in Asia: II. Differences among irrigation, rainfed, and deepwater rice. *Nutr. Cycl. Agroecosyst.* **2000**, *58*, 13–22. [CrossRef]

15. Lu, W.F.; Chen, W.; Duan, B.W.; Guo, W.M.; Lu, Y.; Lantin, R.S.; Wassmann, R.; Neue, H.U. Methane emissions and mitigation options in irrigated rice fields in southern China. *Nutr. Cycl. Agroecosyst.* **2000**, *58*, 65–73. [CrossRef]

16. Yan, X.; Ohara, T.; Akimoto, H. Development of region-specific emission factors and estimation of methane emission from rice fields in the East, Southeast and South Asian countries. *Glob. Chang. Biol.* **2003**, *9*, 237–254. [CrossRef]

17. Zou, J.; Huang, Y.; Jiang, J.; Zheng, X.; Sass, R.L. A 3-year field measurement of methane and nitrous oxide emissions from rice paddies in China: Effects of water regime, crop residue, and fertilizer application. *Glob. Biogeochem. Cycl.* **2005**, *19*, GB2021. [CrossRef]

18. Zheng, X.H.; Wang, M.X.; Wang, Y.S.; Shen, R.X.; Shangguan, X.J.; Heyer, J.; Kögge, M.; Papen, H.; Jin, J.S.; Li, L.T. CH_4 and N_2O emissions from rice paddies in southeast China. *Chin. J. Atmos. Sci.* **1997**, *21*, 167–174.

19. Zheng, X.; Wang, M.; Wang, Y.; Shen, R.; Gou, J.; Li, J.; Jin, J.; Li, L. Impacts of soil moisture on nitrous oxide emission from croplands: A case study on the rice-based agro-ecosystem in South-east China. *Chemosphere Glob. Chang. Sci.* **2000**, *2*, 207–224. [CrossRef]

20. Cai, Z.C.; Xing, G.X.; Shen, G.Y.; Xu, H.; Yan, X.Y.; Tsuruta, H. Measurements of CH_4 and N_2O emissions from rice paddies in Fengqiu, China. *Soil Sci. Plant Nutr.* **1999**, *45*, 1–13. [CrossRef]

21. Towprayoon, S. Greenhouse Gas Mitigation Options from Rice Field. 2004. Available online: https://unfccc.int/files/meetings/workshops/other_meetings/application/vnd.ms-powerpoint/towprayoon.ppt (accessed on 30 June 2014).

22. Wassmann, R.; Lantin, R.S.; Neue, H.U.; Buendia, L.V.; Corton, T.M.; Lu, Y. Characterization of methane emissions from rice fields in Asia. III. Mitigation options and future research needs. *Nutr. Cycl. Agroecosyst.* **2000**, *58*, 23–36. [CrossRef]

23. Sela-Adler, M.; Ronen, Z.; Herut, B.; Antler, G.; Vigderovich, H.; Eckert, W.; Sivan, O. Co-existence of Methanogenesis and Sulfate Reduction with Common Substrates in Sulfate-Rich Estuarine Sediments. *Front. Microbiol.* **2017**, *8*, 766. [CrossRef] [PubMed]

24. Akiyama, H.; Yan, X.; Yagi, K. Evaluation of effectiveness of enhanced-efficiency fertilizers as mitigation options for N_2O and NO emissions from agricultural soils: Meta-analysis. *Glob. Chang. Biol.* **2009**, *16*, 1837–1846. [CrossRef]

25. Edmeades, D.C. *Nitrification and Urease Inhibitors—A Review of the National and International Literature on Their Effects on Nitrate Leaching, Greenhouse Gas Emissions and Ammonia Volatilisation from Temperate Legume-Based Pastoral Systems*; Technical Report 2004/22; Environment Waikato: Hamilton, New Zealand, 2004.

26. Xiang, Y.; Jin, J.Y.; Ping, H.E.; Liang, M.Z. Recent advances on the technologies to increase fertilizer use efficiency. *Agric. Sci. China* **2008**, *7*, 469–479.

27. Dobermann, A.; Cassman, K.G. Plant nutrient management for enhanced productivity in intensive grain production systems of the United States and Asia. *Plant Soil* **2002**, *247*, 153–175. [CrossRef]

28. Bruulsema, T.W.; Witt, C.; Garcia, F.; Li, S.; Rao, T.N.; Chen, F.; Ivanova, S. A global framework for fertilizer BMPs. *Better Crops* **2008**, *92*, 13–15.

29. Iheke, O.R.; Echebiri, R.N. Rural Land tenancy and Resource Use Efficiency of cassava farmer in South Eastern Nigeria. *J. Food Fibre Prod.* **2010**, *3*, 455–465.

30. Arunrat, N.; Wang, C.; Pumijumnong, N.; Sereenonchai, S.; Cai, W. Farmers' intention and decision to adapt to climate change: A case study in the Yom and Nan basins, Phichit province of Thailand. *J. Clean. Prod.* **2017**, *143*, 672–685. [CrossRef]

31. Sani, S.; Chalchisa, T. Farmers' Perception, Impact and Adaptation Strategies to Climate Change among Smallholder Farmers in Sub-Saharan Africa: A Systematic Review. *J. Ressour. Dev. Manag.* **2016**, *26*, 1–8.

32. Krause, M.A.; Deuson, R.R.; Baker, T.G.; Preckel, P.V.; Lowenberg-DeBoer, J.; Reddy, K.C.; Maliki, K. Risk-Sharing Versus Low-cost Credit Systems for International Development. *Am. J. Econ.* **1990**, *72*, 911–922. [CrossRef]

33. Tazeze, A.; Haji, J.; Ketema, M. Climate Change Adaptation Strategies of Smallholder Farmers: The Case of Babilie District, East Harerghe Zone of Oromia Regional State of Ethiopia. *J. Econ. Sustain. Dev.* **2012**, *3*, 1–13.

34. Sasaki, H. Farmer Behaviour, Agricultural Management and Climate Change. Available online: http://www20.iadb.org/intal/catalogo/pe/2012/09774.pdf (accessed on 18 June 2016).

35. Howden, S.M.; Soussana, J.F.; Tubiello, F.N.; Chhetri, N.; Dunlop, M.; Meinke, H. Adapting Agriculture to Climate Change. *Proc. Natl. Acad. Sci. USA* **2007**, *104*, 19691–19696. [CrossRef] [PubMed]

36. McCarl, B.A. Analysis of climate change implications for agriculture and forestry: An interdisciplinary effort. *Clim. Chang.* **2010**, *100*, 119–124. [CrossRef]

37. Grafakos, S.; Pacteau, C.; Delgado, M.; Landauer, M.; Lucon, O.; Driscoll, P.; Wilk, D.; Zambrano, C.; O'Donoghue, S.H.; Roberts, D. Integrating Mitigation and Adaptation as Win-Win Actions. In *Climate Change and Cities: Second Assessment Report of the Urban Climate Change Research Network. Summary for City Leaders, Chapter: 4, Publisher: Urban Climate Change Research Network (UCCRN)*; Rosenzweig, C.W., Solecki, P., Romero-Lankao, S., Mehrotra, S., Dhakal, T., Bowman, S., Ibrahim, A., Eds.; Columbia University: New York, NY, USA, 2015; Volume 6, pp. 1–24.

38. Vanslembrouck, I.; Van Huylenbroeck, G.; Verbeke, W. Determinants of the Willingness of Belgian Farmers to Participate in Agri-environmental Measures. *J. Agric. Econ.* **2002**, *53*, 489–511. [CrossRef]

39. Maddison, D. *The Perception and Adaptation to Climate Change in Africa*; CEEPA Discussion Paper No. 10; Centre for Environmental Economics and Policy in Africa; University of Pretoria: Pretoria, South Africa, 2006.

40. Ahnstrom, J.; Francis, C.; Hockert, J.; Skelton, P.; Bergea, H.; Hallgren, L. Farmers and nature conservation: What is known about attitudes, context factors and actions affecting conservation? *Renew. Agric. Food Syst.* **2009**, *24*, 38–47. [CrossRef]

41. OAE (Office of Agricultural Economics). *Mitigation Abatement Cost of Greenhouse Gases from Rice Production*; Ministry of Natural Resources and Environment: Bangkok, Thailand, 2012; 355p. (In Thai)

42. Lattanzio, R.K. *Life-Cycle Greenhouse Gas Assessment of Coal and Natural Gas in the Power Sector*; Congressional Research Service: Washington, DC, USA, 2015.

43. Gadde, B.; Bonnet, S.; Menke, C.; Garivait, S. Air pollutant emissions from rice straw open field burning in India, Thailand and the Philippines. *Environ. Pollut.* **2009**, *157*, 1554–1558. [CrossRef] [PubMed]

44. Kanokkanjana, K.; Cheewaphongphan, P.; Garivait, S. Black Carbon Emission from Paddy Field Open Burning in Thailand. In *Proceedings of the 2011 2nd International Conference on Environmental Science and Technology, IPCBEE (6), Singapore, 26–28 February 2011*; IACSIT Press: Singapore, 2011.

45. Niu, H.; Cheng, W.; Hu, W.; Pian, W. Characteristics of individual particles in a severe short-period haze episode induced by biomass burning in Beijing. *Atmos. Pollut. Res.* **2016**, *7*, 1072–1081. [CrossRef]

46. Zhang, X.; Lu, Y.; Wang, Q.; Qian, X. A high-resolution inventory of air pollutant emissions from crop residue burning in China. *Atmos. Chem. Phys. Discuss* **2018**. [CrossRef]

47. IPCC. *The Physical Science Basis: Working Group I Contribution to the Fifth Assessment Report of the Intergovernmental Panel on Climate Change*; Cambridge University Press: Cambridge, UK; New York, NY, USA, 2013.

48. Arunrat, N.; Wang, C.; Pumijumnong, N. Alternative cropping systems for greenhouse gases mitigation in rice field: A case study in Phichit province of Thailand. *J. Clean. Prod.* **2016**, *133*, 657–671. [CrossRef]

49. Ecoinvent Centre. *Ecoinvent Database*, version 3.2; Swiss Centre for Life Cycle Inventories: Duebendorf, Switzerland, 2005. Available online: http://www.ecoinvent.org/ (accessed on 25 August 2016).

50. The National Technical Committee on Product Carbon Footprinting. *The National Guideline on Product Carbon Footprint*, 3rd ed.; Amarin Publishing: Bangkok, Thailand, 2011.

51. IPCC. Agriculture, forestry and other land use. In *2006 IPCC Guidelines for National Greenhouse Gas Inventories*; Eggleston, H.S., Buendia, L., Miwa, K., Ngara, T., Tanabe, K., Eds.; Institute for Global Environmental Strategies (IGES): Hayama, Japan, 2006; Volume 4.

52. Maciel, V.G.; Zortea, R.B.; Silva da, W.M.; Cybis, L.F.A.; Einloft, S.; Seferin, M. Life Cycle Inventory for the agricultural stages of soybean production in the state of Rio Grande do Sul, Brazil. *J. Clean. Prod.* **2015**, *93*, 65–74. [CrossRef]

53. EPA. *Emission Factors for Greenhouse Gas Inventories*; United States Environmental Protection Agency: Washington, DC, USA, 2014. Available online: https://www.epa.gov/sites/production/files/2015-07/documents/emission-factors_2014.pdf (accessed on 17 February 2016).

54. Lal, R. Carbon emission from farm operations. *Environ. Int.* **2004**, *30*, 981–990. [CrossRef] [PubMed]

55. Webb, N.P.; Stokes, C.J.; Marshall, N.A. Integrating biophysical and socio-economic evaluations to improve the efficacy of adaptation assessments for agriculture. *Glob. Environ. Chang.* **2013**, *23*, 1164–1177. [CrossRef]

56. Likert, R. A technique for the measurement of attitudes. *Arch. Psychol.* **1931**, *22*, 1–55.

57. Marshall, N.A. Understanding social resilience to climate variability in primary enterprises and industries. *Glob. Environ. Chang.* **2010**, *20*, 36–43. [CrossRef]

58. Dobermann, A.C.; Witt, D.; Dawe, S.; Abdulrachman, H.C.; Gines, R.; Nagarajan, S.; Satawathananont, T.T.; Son, P.S.; Tan, G.H.; Wang, N.V.; et al. Adviento Site-specific nutrient management for intensive rice cropping systems in Asia. *Field Crops Res.* **2002**, *74*, 37–66. [CrossRef]
59. Wang, G.H.; Dobermann, C.; Witt, Q.; Sun, Q.Z.; Fu, R.X. Performance of site-specific nutrient management for irrigated rice in Southeast China. *Agron. J.* **2001**, *93*, 869–878. [CrossRef]
60. Zhu, H.J.; Tian, Z.H. Analysis on the rice farmer's willingness to use low-carbon technologies measures: Based on the survey of the rice-producing areas in Southern China. *J. Agric. Econ.* **2013**, *215*, 62–71.
61. McCown, R.L. New thinking about farmer decision makers. In *The Farmer's Decision*; Hatfield, J.L., Ed.; Soil and Water Conservation Society: Ankeny, Lowa, 2005; pp. 11–44.
62. Morton, L.W. Citizen involvement. In *Pathways for Getting to Better Water Quality: The Citizen Effect*; Morton, L.W., Brown, S.S., Eds.; Springer: New York, NY, USA, 2011.
63. Jones, A.K.; Jones, D.; Edwards-Jones, G.; Cross, P. Informing decision making in agricultural greenhouse gas mitigation policy: A Best–Worst Scaling survey of expert and farmer opinion in the sheep industry. *Environ. Sci. Policy* **2013**, *29*, 46–56. [CrossRef]
64. Abebaw, D.; Belay, K. Factors influencing adoption of high yielding maize varieties in South-western Ethiopia. An application of logit. *Q. J. Int. Agric.* **2001**, *40*, 149–167.
65. Nowak, P. Why farmers adopt production technology. *J. Soil Water Conserv.* **1992**, *47*, 14–16.
66. Richard, T.; Shively, E.G. Technical Change and Productive Efficiency: Irrigated Rice in the Philippines. *Asian Econ. J.* **2007**, *21*, 155–168.

Article

Farmers' Net Income Distribution and Regional Vulnerability to Climate Change: An Empirical Study of Bangladesh

Md. Shah Alamgir [1,2], Jun Furuya [3,*], Shintaro Kobayashi [3], Mostafiz Rubaiya Binte [1] and Md. Abdus Salam [4]

[1] University of Tsukuba, Tsukuba, Ibaraki 305-8577, Japan; salamgir.afb@sau.ac.bd (M.S.A.); ruba_zhumu@yahoo.com (M.R.B.)
[2] Sylhet Agricultural University, Sylhet 3100, Bangladesh
[3] Japan International Research Center for Agricultural Sciences; Tsukuba, Ibaraki 305-8686, Japan; shinkoba@affrc.go.jp
[4] Bangladesh Rice Research Institute, Gazipur 1701, Bangladesh; asalam_36@yahoo.com
* Correspondence: furuya@affrc.go.jp; Tel.: +81-29-838-6304

Received: 14 June 2018; Accepted: 16 July 2018; Published: 23 July 2018

Abstract: Widespread poverty is the most serious threat and social problem that Bangladesh faces. Regional vulnerability to climate change threatens to escalate the magnitude of poverty. It is essential that poverty projections be estimated while bearing in mind the effects of climate change. The main purpose of this paper is to perform an agrarian sub-national regional analysis of climate change vulnerability in Bangladesh under various climate change scenarios and evaluate its potential impact on poverty. This study is relevant to socio-economic research on climate change vulnerability and agriculture risk management and has the potential to contribute new insights to the complex interactions between household income and climate change risks to agricultural communities in Bangladesh and South Asia. This study uses analysis of variance, cluster analysis, decomposition of variance and log-normal distribution to estimate the parameters of income variability that can be used to ascertain vulnerability levels and help us to understand the poverty levels that climate change could potentially generate. It is found that the levels and sources of income vary greatly among regions of Bangladesh. The variance decomposition of income showed that agricultural income in Mymensingh and Rangpur is the main cause of the total income difference among all sources of income. Moreover, a large variance in agricultural income among regions is induced by the gross income from rice production. Additionally, even in the long run the gradual, constant reduction of rice yield due to climate change in Bangladesh is not a severe problem for farmers. However, extreme events such as floods, flash floods, droughts, sea level rise and greenhouse gas emissions, based on Representative concentration pathways (RCPs), could increase the poverty rates in Mymensingh, Rajshahi, Barisal and Khulna—regions that would be greatly affected by unexpected yield losses due to extreme climatic events. Therefore, research into and development of adaptation measures to climate change in regions where farmers are largely dependent on agricultural income are important.

Keywords: income distribution; cost distribution; vulnerable region; adaptation measures; Bangladesh

---

1. Introduction

Bangladesh has experienced severe famines [1–3]. However, heavy investments in agriculture following these famines have given rise to enhanced food production and have caused significant increases in domestic rice production [4,5]. Both the cultivation techniques and cropping patterns relating to rice production have gradually changed in terms of yield potential [6,7]. Despite huge population pressures, the country has reached self-sufficiency in rice production [8–10]. Additionally, Bangladesh's economic situation is improving; as such, it is one among a rather small group of countries that have seen remarkable progress in terms of both economic performance and development indicators [11]. However, poverty remains a critical social concern in this country [6,12,13].

Climate change will have a largely adverse impact on agricultural production in Asia [14]. For particular geographical locations and due to other environmental reasons, Bangladesh is one of the world's most disaster-prone countries [15–18]. Given climate change impacts, natural resource constraints and competing demands, agriculture and food systems continue to face considerable challenges. The livelihoods of the poor who are directly reliant on agriculture already face a profound threat due to the current climate change in Bangladesh [19,20], which could lead to increased pauperization. At the household level, climate change significantly affects food production [21] which in turn influences food prices and directly affects the poverty of low-income household [22,23]. Agricultural income and non-farm income are the most significant factors in poverty reduction among rural people [24–27]. However, Chaudhry and Wimer reported that household income plays a vital role in the social and economic development of a community and income from agriculture might result in increasing per capita income [28].

Agriculture is strongly influenced by weather and climate, which in turn have impacts on agricultural production [29]. Over the last three decades, temperature has been increasing in Bangladesh [30,31] and the average daily temperature is predicted to undergo an increase of 1.0 °C by 2030 and 1.4 °C by 2050 [32,33]. The annual rainfall is also unevenly distributed in some areas of Bangladesh. Rainfall patterns might change with increasing temperature and drought occur in some areas; however, total rainfall sometimes increases and heavy rainfall induces floods in Bangladesh. Increasing temperature also enhances extreme events, such as cyclones in coastal areas and adversely affects rice production [7,30,34–36]. Additionally, climate change is projected to affect agriculture and it is very likely that climate change will induce significant yield reduction in the future due to climate variability in Bangladesh [37–39], with a projected decline of 8–17% in rice production by 2050 [33,40]. In Bangladesh, nearly 80% of the total cropped area is dedicated to rice production, accounting for almost 90% of total grain production [39,41–46]. Agricultural production, farm income and food security are significantly affected by seasonal growing temperatures [47].

Some previous studies have projected the impacts of climate change on food production and national food security [48,49], as well as their impact on agricultural production, by collecting information under drought, rainfall, sea level rise, flood and temperature increases [39,43,50] and the impact of coastal flooding on rice [7,51,52]. However, there have been fewer studies from micro or regional points of view based on integrated household survey data or poverty measurements under yield reductions of crops due to climate change vulnerabilities. Farmers' low incomes are the main reinforcing factors in poverty traps, so this context of research is not sufficient. To consider suitable adaptation technologies and policies for farmers, impact projections in terms of regional characteristics and poverty are needed far more. To alleviate the severity of climate change's impact on farm production and poverty, adaptation strategies, such as new crop varieties, changing planting times, homestead gardening, planting trees and migration, are vital approaches [6]. Furthermore, research that projects climate change's impacts on poverty or that pinpoints especially vulnerable regions and the vulnerability of farm household income under the impact of climate change is still needed [53,54]. Using statistical analysis, the current study attempts to derive an understanding of regional characteristics in terms of income and agriculture and to assess the contributions of different components on the observed total variance of income and cost, with an eye towards determining regional vulnerability to climate change and projecting the potential effects of climate change on poverty in Bangladesh. In this study, we used high-quality plot-level agricultural production data from the nationally representative survey by the International Food Policy Research Institute (IFPRI) (Appendix A.1). We used different analytical techniques to evaluate regional characteristics and to assess the potential climate change impacts on farm production and poverty under newly developed representative concentration pathways (RCPs) and other climate scenarios. The objective of this study was to project the poverty under the impacts of climate change on crop production and to provide possible adaptive measures.

The paper is designed as follows: we draw a review of the related literature concerning climate change, vulnerability and poverty in Section 2; Section 3 is the methodology section, in which we describe the data sources, compilation procedures and the analytical approaches of the data; in Section 4, descriptive statistics and empirical results of the analysis with discussion are presented; and in Section 5, we conclude by emphasizing the future research directions and some policy guidelines.

2. Review of the Literature

The research on climate change scenarios and poverty in terms of regional characteristics is outlined concisely in this section. Climate change is a reality that is occurring and will increasingly affect the poor; moreover, it is a serious threat to poverty eradication [55]. Poor agricultural communities are always disrupted by climate change's impact on household food security and poverty [56,57]; climate change impacts could increase household poverty [55]. Poverty as a dynamic and multidimensional condition is characterized by the interaction of individual and community features, socioeconomic and political issues, environmental processes and historical circumstances. Particularly in less developed countries and regions through several direct and indirect channels, climatic variability and change can worsen poverty [58]. Lade et al. reviewed the socio-ecological relationship in rural development concepts, emphasizing the economic, biophysical and cultural aspects of poverty. This study classified the poverty alleviation strategies and developed multidimensional poverty trap models and it stated that interventions that ignore nature and culture can reinforce poverty [59].

A multi-factor impact analysis framework was developed by Yu et al. [39] and using this framework [50] Ruane et al. provided sub-regional vulnerability analyses and quantified key uncertainties in climate and crop production. Climate change impacts increase under the higher emissions scenarios and agriculture in Bangladesh is severely affected by sea level rise [50]. Over the same period, several attempts have been made regarding climate scenario development in Bangladesh, mainly using Global Climate Models (GCMs) and in some cases Regional Climate Models (RCMs) [60–62]. From these studies, the overall conclusions include increases in temperature and rainfall, different drought seasons and impacts on crop production.

The projected future yield of rice cultivars in 2030 and 2050 in different areas of Bangladesh by DSSAT crop modelling showed that Bagerhat, Dinajpur, Gaibandha, Maulvibazar, Panchagarh, Rangpur, Sirajganj and Thakurgaon districts will have high yield losses due to climate change impacts. Rainfall, temperature and CO_2 affect the yield for *aman* rice in Rangpur and Khulna divisions and for *boro* rice in Rajshahi, Barisal and the southwest region [63]. Changing patterns of rainfall and temperature in different regions of Bangladesh are significantly higher, compared to IPCC predictions. For sustainable adaptation, location-specific management of seed, crop and irrigation is needed [21]. Soil tolerance, flood tolerance and shorter varieties of rice and other crops could be used to adapt to climate change impacts [64]. Climate change is likely to have an adverse effect on rice and wheat production [5] and significant yield reductions in the future due to climate variability [38] are also directly associated with extreme weather events [19]; due to population pressures, future food production is a challenge in maintaining food security in Bangladesh [5]. Food demand changes because of urbanization, population structure, among other factors; however, food supply can change due to extreme climate change impacts on agricultural production in Bangladesh. The combined effects on rice of major climatic variables were checked by Karim et al. and they found that rice yield would decrease by 33% in both 2046–2065 and 2081–2100 for Rangpur, Barisal and the Faridpur region [65].

Total annual income of a farm household depends on farm and non-farm income. Farm income is always unstable due to the dependency of weather and even if farm income is high poverty may occur;

however, higher non-farm income could reduce the poverty [28]. Farm households in Bangladesh are the most prone to the impacts of climatic hazards. Uncertainty is high in farm income and it depends on the wide fluctuations of yields and prices. Unexpected weather can easily damage crop production, rendering farms more vulnerable [66]. In Bangladesh, farmers are fully dependent on weather for their crop production, resulting in lower farm income if extreme climatic events occur. Unexpected yield reductions cause fluctuating farm income and increase food insecurity and poverty. Agriculture is the main source of income of farmers in Bangladesh [8,21] and it might cause per capita income to increase, which in turn could further reduce poverty. The participation of government programs and off-farm income is significantly important in reducing poverty [24].

There has been much research on climate change impacts, adaptations and projections in agriculture. The IPCC's fifth assessment report showed that food production in Asia will vary and decline in many regions under the impact of climate change [37]. Rajendra et al. focused on climate change impacts on farming in northern Thailand, where the vulnerability of farm households persists under the negative impact of climate change [54]. Yamei et al. assessed the adverse effects of future climate on rice yields and provided potential adaptive measures [67]. Nazarenko et al. examined the climate response under a representative concentration pathway (RCP) for the 21st century [68], while there are fewer comprehensive scenarios for the whole country regarding farm income and poverty projections.

In addition, in-depth empirical research on farm income distribution and regional vulnerability to climate change has been lacking. Furthermore, most of the previous studies of climate change impacts on agricultural production have been for specific regions. However, a comprehensive study of climate change impacts comparing the regions of Bangladesh could be enormously significant. One of the motivations of the study is to summarize the farmers' net income scenarios for all of the regions of Bangladesh, assessing the contributions of different components on the observed total variance in income and costs and possible poverty under climate change impacts on agricultural production. Moreover, understanding farmers' local economic situations and coping strategies with climate change impacts could have immense significance for regional point of view. Based on actual farm income, this study evaluates the projected farm income under the scenario that extreme climatic events occur. It then determines the projected poverty to identify vulnerable regions and to suggest appropriate coping and poverty alleviation strategies.

3. Methodology

3.1. Survey Data

In its empirical analysis, this study uses cross-sectional data drawn from nine administrative regions across Bangladesh. These data were derived from the International Food Policy Research Institute (IFPRI), which adopted a multi-stage stratified random sampling method to collect primary data: first a selection of primary sampling units (325 villages) and then a selection of farm households (20 farms) from each primary sampling unit. Randomly selected villages with probability proportional to size (PPS) sampling using the number of households from the Bangladesh population census data in 2001. Randomly selected 20 farm households in each village from the aforementioned national census list. IFPRI researchers designed the Bangladesh Integrated Household Survey (BIHS) (Appendix A.1), the most comprehensive, nationally representative household survey conducted to date. Plot-wise crop production data were collected via semi-structured questionnaire by the IFPRI from 6503 sample farmers across Bangladesh vis-à-vis cultivated crops; the survey period was from 1 December 2010, to 30 November 2011. The original data were collected in a typical agricultural year according to rice production statistics; there was no severe crop loss in the 2010 or 2011 rice years in Bangladesh [69].

3.2. Data Compilation

This study models the poverty rate change under climate change vulnerability in different regions of Bangladesh. Based on the purpose of this study, to analyze the data we applied descriptive, inferential, statistical and multivariate techniques. Plot-wise raw data were compiled in line with the study objectives. We compiled data pertaining to many income sources for each separate household into some important sectors. In addition, for agricultural activities, we also compiled all types of input cost data into some important cost items and output values for each crop. We then compiled and combined them into one data set of households for all 6503 farms. Bangladesh consists of 30 agro-ecological zones (AEZs) that overlap with each other [69,70]. For the convenience of this research, some homogenous agro-ecological zones were combined into the nine administrative regions with their geographical locations. In this manner, we tried to develop nine mutually exclusive regions for our research. To overcome the resulting challenge in consistency under the same impact of climate change in each region [50], we categorized all the sample farmers per the nine administrative zones of Bangladesh, calling each a division (nine different colors indicating the individual divisions) (Figure 1): Barisal (700 sample farmers), Chittagong (300), Comilla (660), Dhaka (1380), Khulna (1020), Mymensingh (600), Rajshahi (580), Rangpur (543) and Sylhet (720).

Figure 1. Map of the objective regions of Bangladesh.

We estimated the costs and incomes associated with 17 major crops produced by farmers in Bangladesh (each is considered an important crop); other crops (such as pulses, oil seeds, spices except for chili and onion, vegetables, leafy vegetables, etc.) and all types of fruits (such as banana, mango, pineapple, jackfruit, papaya, guava, litchi, orange, etc.) were added to another group, "all other crops." The 18 groups are *aus* (Appendix ??), rice local, *aus* rice LIV, *aus* rice HYV, *aman* rice local, *aman* rice LIV, *aman* rice HYV, *aman* rice Hybrid, T *aus* rice HYV, *boro* rice HYV, *boro* rice Hybrid, wheat local, wheat HYV, maize, jute, potato, chili, onion and all other crops.

To estimate per-capita income for farm household members in all nine administrative regions of Bangladesh, this study considers all income sources, including income from agriculture. The basic unit of analysis is each farm, while farming is the only significant source of income among other sources, such as employment, small business and so on, for the family in a one-year period. Net income for the farm household from agriculture was calculated by deducting total input costs from gross income:

$$\pi = \sum_i P_i Y_i - \sum_i \sum_j P_{ij} X_{ij} \tag{1}$$

where π is net income, P_i is price of crop i, Y_i is production of crop i, P_{ij} is price of input j for crop i and X_{ij} is input j for crop i.

This analysis used only the accounting costs to estimate net income from agriculture (Appendix B.1); these costs include the so-called explicit costs actually incurred by the farms and in surveys, farmers reported their own cost data. For this reason, this study regards supply of one's own land and family labor as part of agricultural income. The farm gate price of each crop for each household was used to estimate gross income derived from agricultural crops, livestock and poultry and fish production; additionally, actual input prices were used to estimate the production costs cited by each farmer and in-kind payments by crops are deducted for estimating gross income. For farmers with no information about farm gate prices or input prices for their respective crops, we used the average prices from the region. This study crosschecked the farm gate prices and input prices with data pertaining to the average national retail price data of select commodities in Bangladesh [71] during the aforementioned study period. Farmers used farm gate prices to sell their crops and for this reason, there was some divergence between national retail prices and the farmers' prices. To estimate per-capita income for each member of the farm, this study assumes that all negative returns tend towards zero so that we can calculate shares of income sources.

Income data were collected for each household and these data were used to calculate overall household income. Income was broadly classified into seven major sectors, as follows:

(i)   Agricultural crop income: income from all crop types produced by farmers throughout the year;
(ii)   Income from fish/shrimp farming;
(iii)   Income from livestock and poultry enterprises;
(iv)   Nonagricultural enterprise income: income from nurseries, food processing, fishing, nonagricultural day labor, retail, wholesale, construction, manufacturing, wooden furniture and other businesses;
(v)   Remittances: remittances from within or outside Bangladesh, with the persons who sent the remittances excluded from their respective households;
(vi)   Employment: both formal and informal employment, income from self-employed and/or owned businesses that are not agricultural, income received from relatives and friends not presently living with the household and so on; and
(vii)   Other income: income received from land rent or property rent, income from life and nonlife insurance, profit from shares, gratuities, or retirement benefits, income from lotteries or prizes, interest received from banks, charity assistance, other cash receipts and/or other in-kind receipts.

These seven sectors of household income were used to determine the actual income and income sector shares, both of which reflect income distributions significantly.

3.3. Analytical Approach

This study used four types of statistical analysis.

3.3.1. Analysis of Variance (ANOVA)

After dividing farm households into the nine aforementioned regions, we conducted single-factor analysis of variance (ANOVA) to examine differences among the farm households of the nine regions in Bangladesh in terms of mean per-capita income.

3.3.2. Cluster Analysis

The cluster analysis (CA) technique was used to determine the main and dominant income sources in Bangladesh's various regions. Environmental (i.e., topographical) divergence is a common phenomenon in Bangladesh and it diversifies farm production, although farm households within a certain region do tend to be similar. Ward's hierarchical method and the partitioning method can be used to determine the most appropriate clusters regarding the main income sources in each region. A dendrogram—a graphical representation of the hierarchy of nested cluster explanations—is a manifestation of Ward's method and it provides clues for finding the preferable number of clusters regarding income sources.

3.3.3. Decomposition of Variances

To understand the interregional differences and to assess the contributions of different components to the observed total variance of input cost and income, different crop production data are used [72–75]. These data include per hectare crop yields, prices and all costs at the farm level and we decompose the variances in net cost and net income into different factors using the following relations.

$$V(X \pm Y) \; = \; V(X) + V(Y) \pm 2\text{Cov}(X, Y) \tag{2}$$

where X and Y are stochastic variables, such as the costs of inputs or incomes from different sectors; $V(\cdot)$ is variance and $\text{Cov}(\cdot)$ is covariance.

3.3.4. Projections: Log-Normal Distributions

There are different types of probability distributions studied in probability theory. Lognormal distribution is one of the most important one and was established long ago [76–78]. Lognormal distribution is a type of a continuous distribution. It is a probability distribution in which the logarithm of the random variable is distributed normally. This distribution is closely related to the normal distribution. Lognormal distribution is very commonly used in the social sciences, economics and finance [79].

Arata [80] pointed out that the income distribution among individuals is very important and is one of the main themes in economics. Income distribution is widely understood to be well described by a log-normal distribution.

Lognormal distribution has two parameters: mean (μ) and standard deviation (σ). If x is distributed log-normally with parameters μ and σ, then $\log(x)$ is distributed normally with mean μ and standard deviation σ. The log-normal distribution is applicable when the quantity of interest must be positive since $\log(x)$ exists only when x is positive. A positive random variable X is log-normally distributed if the logarithm of X is normally distributed.

$$ln(X) \sim N\left(\mu, \sigma^2\right) \tag{3}$$

Let Φ and φ be, respectively, the cumulative probability distribution function and the probability density function of the $N(0, 1)$ distribution.

The probability density function of the log-normal distribution is;

$$f(x|\mu, \sigma) \; = \; \frac{1}{x\sigma\sqrt{2\pi}} exp\left\{\frac{-(lnx - \mu)^2}{2\sigma^2}\right\}; \; x > 0 \tag{4}$$

If we substitute a poverty line into x and integrate the probability density function up to x, we can obtain a poverty rate. The poverty line, which is estimated by world Bank, is inserted into the equation [12,67].

We estimate the incomes of all sample families on the assumption of climate change impacts and draw the distribution of the estimated incomes, assuming that the distribution follows log normal distribution. To draw log normal distribution, we must find the mean and standard deviation of $ln(x)$ (Appendix B.2). From the actual per-capita income of household members in the study areas, we obtain the actual distribution of per-capita income using the lognormal distribution. Next, we project the crop yield loss from the assumption of the literature reviews and we estimate the projected per-capita income. From projected per-capita income using lognormal distribution, we obtain the estimated distribution of per-capita income. By simulating these two distributions, we find the poverty rate graph.

4. Results and Discussion

4.1. Comparison of Income Levels Among Regions

Agricultural income is a key driver in reducing poverty in Bangladesh, where it accounted for 90% of all poverty alleviation between 2005 and 2010 [81]. In terms of employment, Bangladesh's economy is primarily dependent on agriculture. Approximately 85% of the population is directly or indirectly attached to the agriculture sector [38,69].

Agriculture continues to be the main source of income in the sample households in all regions (Table 1) and this result is consistent with Hossain and Silva (2013) [5]. However, in all regions, nonagricultural profit and employment are important income sources and these results are consistent with Bangladesh Economic Review [45]. The amount of remittances varies by region: that in Sylhet is not the highest nationally but the people there do consider remittances to be the main income source in the region. The agricultural income is higher in Rajshahi than in other regions and the per capita income of this region per the study sample is US$ 423.6 (Table 2). Diversification of agricultural crops results in this region having highest income from agriculture.

Table 1. Each income sector's share in total household income (%), by region.

	B	CH	CO	D	K	M	RJ	RN	S	BD
Agril. crops	12.71	8.14	5.50	13.55	19.43	20.15	18.72	21.41	9.03	14.32
Main crops	6.08	2.89	2.34	8.25	10.81	11.44	11.72	14.84	6.15	8.36
Other crops	6.63	5.25	3.16	5.30	8.62	8.71	7.00	6.58	2.87	5.96
Fish	9.23	1.54	0.57	2.18	7.93	6.06	2.87	1.14	3.16	3.96
Livestock	2.19	1.17	1.48	3.60	6.15	5.12	4.43	3.10	1.80	3.47
Non-ag. profit	20.76	19.25	14.13	21.22	18.09	17.66	19.61	14.88	20.05	18.80
Remittance	11.04	24.99	41.48	15.68	7.64	9.11	4.48	7.58	17.77	15.22
Employment	38.91	44.35	30.80	41.10	38.52	39.04	38.83	50.54	44.02	40.10
Other income	5.16	0.55	6.04	2.66	2.23	2.86	11.06	1.35	4.18	4.12
Total	100	100	100	100	100	100	100	100	100	100

B = Barisal, CH = Chittagong, CO = Comilla, D = Dhaka, K = Khulna, M = Mymensingh, RJ = Rajshahi, RN = Rangpur, S = Sylhet, BD = Bangladesh, Main crops = *Aus*, *Aman* and *Boro* rice and other crops = Wheat, maize, jute, potato, chili, onion and so on.

Table 2. Mean, median and standard deviation of per-capita income (US$/yr), by region.

	B	CH	CO	D	K	M	RJ	RN	S	BD
Mean	308.93	336.75	378.35	362.17	369.84	307.63	423.63	308.76	301.63	327.55
Median	289.93	217.83	246.25	242.87	254.11	215.04	283.14	226.99	204.82	232.94
SD	314.75	418.11	314.22	403.66	382.81	278.08	372.71	246.61	301.02	348.64
PR	0.51	0.48	0.46	0.46	0.42	0.51	0.33	0.47	0.49	0.46

B = Barisal, CH = Chittagong, CO = Comilla, D = Dhaka, K = Khulna, M = Mymensingh, RJ = Rajshahi, RN = Rangpur, S = Sylhet, SD = Standard deviation and PR = Poverty rate.

Table 1 shows significant differences in main income sources among farmers in various regions in Bangladesh. Employment is the predominant income source in most regions, followed by nonagricultural profits and agriculture. The share of agriculture in total income varies by region. Among Bangladeshi farming households, the employment share is 40.10%, although the overall share of agriculture in total income is 14.32%. Rangpur has the highest share of agricultural income in total annual income (21.41%), followed by the Mymensingh region (20.15%). Comilla's share of remittances in total annual income was highest (41.48% of total income); in comparison, the share generated by agricultural crops in Comilla was only 5.50%. Currently, overseas workers are more often from the Comilla region than other regions in Bangladesh, with a significant proportion of them sending remittances, becoming a vital source of income in the Comilla region. Rice and other crops were the main sources of income among the sampled farm households in the study areas (Appendix C). Incomes from maize and potato appear to be growing but their respective shares remain small. There are regional land conditions and climate differences among Bangladesh's regions, so wheat, maize, onion and potato production is not familiar to all farmers. Consequently, farmers in all areas of Bangladesh tend to focus on rice cultivation.

Table 2 shows descriptive statistics of income status by region. Poverty rates were estimated by applying the poverty line and the purchasing power parity from the World Bank [22] to log-normal income distributions. The findings presented in Table 2 indicate differences in mean, median and standard deviation of net incomes among the nine regions in Bangladesh; using these findings, one can pinpoint relatively rich and poor regions.

In terms of mean net income, incomes of sampled farm households in Rajshahi are the highest, while those of Barisal, Mymensingh, Rangpur and Sylhet are lower. As some farmers had negative or zero per-capita income, the standard deviation is relatively large in certain regions. The highest standard deviation value is found in Chittagong (US\$ 418.1), reflecting a large income gap among the farmers there.

The highest poverty rate (i.e., 0.51) was found in Mymensingh and Barisal (Table 2), while the lowest (i.e., 0.33) was in Rajshahi; overall, the country's upper poverty rate is 0.46. The rates in Chittagong and Sylhet were also relatively low (i.e., 0.49). The officially estimated upper poverty rate and national average poverty rate are both in the vicinity of 0.35 [12,82], which makes sense because the original data were collected from rural, farming-engaged people and excluded affluent or single urban people.

Among regions where the poverty rates were high, Barisal, Mymensingh and Sylhet had the lower mean incomes. In contrast, Chittagong had the highest standard deviation, compared to the other regions. In the regions of Barisal, Mymensingh and Sylhet, it appeared that the mean income level was low; however, in the other regions, the mean income was large. These results show that these low-income regions are vulnerable regions and should be the targets of farmers' support policies.

From results of Table 2, this study found that there are differences in mean, median and standard deviation of net incomes among the nine regions in Bangladesh and for validation of this difference, we perform ANOVA and report the results in Table 3. Analysis of variance (ANOVA) is a statistical test designed to examine means across more than two groups by comparing variances, based upon the variability in each sample and in the combined samples. We analyzed the variance within and between the sample farmers to determine the significance of any differences in per capita income of farm household members among the regions of Bangladesh. The results of the overall F test in the ANOVA summary shows the results regarding the variability of means between groups and within groups. As indicated, the overall F test is significant (i.e., p-value < 0.05), indicating that means between groups are not equal and it is statistically concluded that there have been significant differences among the regions in terms of mean per-capita income.

Table 3. ANOVA mean differences across regions.

Source of Variation	SS	df	MS	F	p-Value	F Crit
Between groups	6.31×10^{10}	9	7.01×10^{9}	4.757462	2.39×10^{-6}	1.880604
Within groups	1.91×10^{13}	12,996	1.47×10^{9}			
Total	1.92×10^{13}	13,005				

The first column in ANOVA provides us with the sum of squares between and within the groups and for the total sample farmers. The total sum of squares represents the complete variance on the dependent variable for the total sample. The second column represents the degrees of freedom, $(n - 1)$. The total degrees of freedom represent $13,006 - 1 = 13,005$; degrees of freedom between groups equals the number of groups minus one $(10 - 1 = 9)$. The within groups degrees of freedom equals $13,005 - 9 = 12,996$. The third (mean square) column contains the estimates of variability between and within the groups. The mean square estimate is equal to the sum of the squares divided by the degrees of freedom. The between groups mean square is 7.01×10^9; the within-groups mean square is 1.47×10^9. The fourth column, the F ratio, is calculated by dividing the mean square between groups by the mean square within the groups. The F ratio should be one if the null hypothesis is true, while both mean square estimates are equal. However, as shown in Table 3, larger F values (4.757462) imply that the means of the per capita income groups are greatly different from each other, compared to the variation in the individual sample farmers in each group. The next column is the significance level (p-value) and it indicates that the value of F ratio is sufficiently large to reject the null hypothesis. The significance level is 2.39×10^{-6}, which is less than 0.05. Therefore, the mean per capita incomes of sample households among the regions of the country were significantly different in the study year.

4.2. Regional Characteristics on Income Source

This section intends to classify regions of Bangladesh to determine the regional characteristics of income sources in each administrative region. Sectoral income shares from Table 1 are analyzed by cluster analysis and are shown in Figure 2. Here, a dendrogram depicts the income source relationships among the regions. The horizontal axis of the dendrogram (in Figure 2) represents the distance or dissimilarity between clusters and the vertical axis represents the objects (regions) of clusters. From the cluster analysis, this study attempted to find the similarity and clustering with the dendrogram, which visually displays a certain cluster shape. Regions that are close to each other (have small dissimilarities) are linked near the right side of the plot. In Figure 2, we note that Khulna and Mymensingh are very similar compared to the regions that link up near the left side, which are very different. For example, Comilla appears to be quite different from any of the other regions. The number of clusters formed at a particular cluster cutoff value can be quickly determined from this plot by drawing a vertical line at this value and counting the number of lines that the vertical line intersects. In this study, we can see that, if we draw a vertical line at the value of 18.0, four clusters will result. One cluster contains four regions, one contains three regions and two clusters each contain only one region, as shown in Figure 2, in which Barisal, Mymensingh, Khulna and Rajshahi are more alike than resembling Rangpur. In addition, Chittagong, Dhaka and Sylhet are more alike than resembling Comilla.

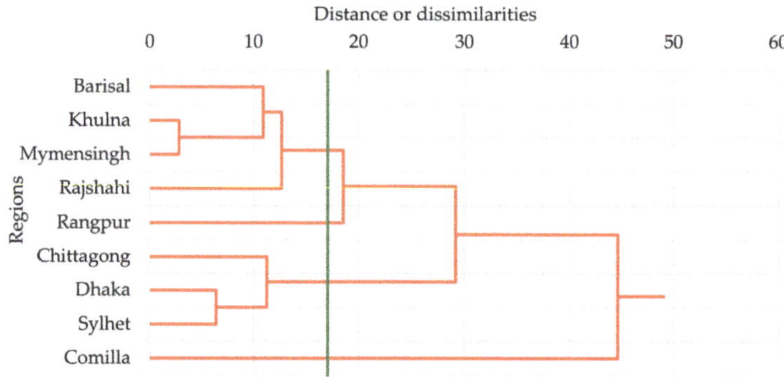

Figure 2. Dendrogram showing clusters for main income sources, by region.

Table 4 summarizes regional characteristics of income sources. Clusters 1 and 2 are largely dependent on agriculture. Clusters 3 and 4 are not largely dependent on agriculture. This result indicates the importance of agricultural research for clusters 1 and 2.

Table 4. Cluster characteristics of main income sources, by region.

Cluster	Region	Main Income Source	Distinction
1	Barisal, Mymensingh, Khulna, Rajshahi	Agricultural. crops, non-agricultural profit, employment	
2	Rangpur		Dominant Employment
3	Chittagong, Dhaka, Sylhet	Non-agricultural profit, remittance, employment	
4	Comilla		Dominant Remittance

Using the dendrogram in Figure 3 (agricultural crop share in total agricultural income analyzed by cluster analysis), four clusters were determined (Table 5) as the clusters suitable for representing agricultural crop income sources among the regions. We followed the same procedure for this dendrogram (Figure 3) that we followed in Figure 2.

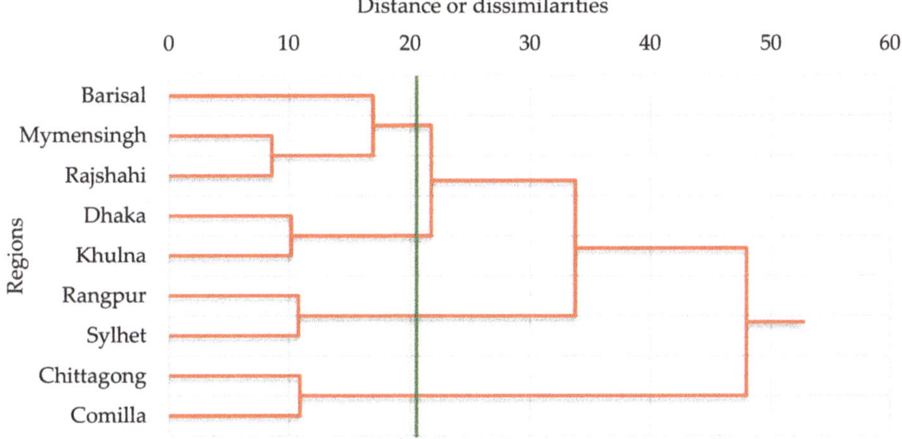

Figure 3. Dendrogram showing clusters for agricultural income sources, by region.

Table 5. Cluster characteristics of agricultural income sources, by region.

Cluster	Region	Main Income Source	Distinction
1	Barisal, Mymensingh, Rajshahi	Rice, other crops	
2	Rangpur, Sylhet		Dominant rice
3	Chittagong, Comilla		Dominant other crops
4	Dhaka, Khulna	Rice, jute, chili, onion, other crops	

The selected clusters show significant differences among the regions. Rice and other crops were identified as the main agricultural income sources of clusters 1–3, whereas rice, jute, chili, onion and other crops were those of cluster 4. The selected clusters produced the significant differences among the regions. In addition, rice predominated in cluster 2, while other crops predominated in cluster 3. These findings imply, for example, that rice is the main agricultural income source in Rangpur and Sylhet, while other crops are those in Chittagong and Comilla.

4.3. Reasons for Broad Income Distribution within a Region

To grasp the diversity of income for sampled farm households, the income can be decomposed into seven broad components, such as Agriculture, Fish, Livestock and poultry, Nonagricultural enterprise profit, Remittance, Other income and Employment income, in each region. We applied decomposition of variances and the results are shown in Table 6. The decomposition of variances is useful in evaluating how much each source of income contributes to total income variation of farm households. The decomposed variance share was derived from annual per capita income from the seven aforementioned broad income source sectors. Across Bangladesh, differences in remittances, other income and employment are important factors that all contribute the largest share of variation in total income. If a family can find good employment both inside and outside its region, it can become relatively wealthy, although income share from employment does not significantly more contribute in all regions (Table 6).

Table 6. Share of broad income components (%) in total income variation, by region.

	B	CH	CO	D	K	M	RJ	RN	S	BD
V(b)	6.57	1.67	1.94	4.19	8.18	13.87	3.18	20.59	2.49	4.79
V(c)	20.03	0.19	0.03	1.57	35.73	8.17	1.11	0.23	1.98	6.42
V(d)	1.08	0.18	0.17	0.87	1.78	4.58	2.81	0.98	1.05	1.54
V(e)	17.39	13.64	6.33	16.50	13.47	11.90	5.09	7.84	19.73	11.63
V(f)	8.70	40.78	54.36	10.94	10.22	12.99	1.61	30.23	29.95	17.78
V(g)	4.84	0.05	14.76	1.16	0.61	2.38	69.70	0.37	2.82	21.63
V(h)	19.44	27.29	11.61	44.54	17.17	25.26	7.16	38.32	21.01	22.05
2*Cov(e,h)	21.95	15.22	10.81	20.22	12.85	14.22	7.32		20.96	14.16
2*Cov(b,c)								1.43		
2*Cov(c,h)							2.03			
2*Cov(f,g)		0.99								
2*Cov(c,e)						6.63				
Total	100	100	100	100	100	100	100	100	100	100

B = Barisal, CH = Chittagong, CO = Comilla, D = Dhaka, K = Khulna, M = Mymensingh, RJ = Rajshahi, RN = Rangpur, S = Sylhet and BD = Bangladesh; b = Agriculture, c = Fish, d = Livestock and poultry, e = Nonagricultural enterprise profit, f = Remittance, g = Other income and h = Employment income.

We found in Table 6 that agriculture is one of the main contributors to income differences in Mymensingh and Rangpur regions. Figure 4 shows total income distribution by income sources for the whole country, of which 22% of income inequality of total income is explained by inequality of employment income, while 13.87% and 20.59% of income inequality of total income explained by agriculture in Mynemnsingh and Rangpur respectively (Figures 5 and 6). Furthermore, this result indicates that remittance is the most important sector inducing income disparity in Comilla, compared to employment in Dhaka and Rangpur. In addition, other income sources are significant sources of income to confirm the total income disparity in Rajshahi. This finding likely explains that the income inequality of total income makes the larger contribution of inequality in agricultural income for crop farm households in Bangladesh.

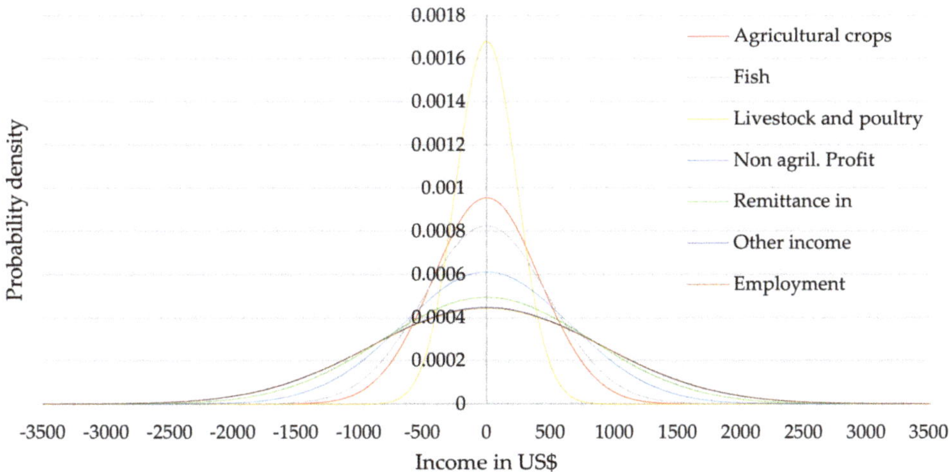

Figure 4. Distribution of total income for farm households in Bangladesh by income sources.

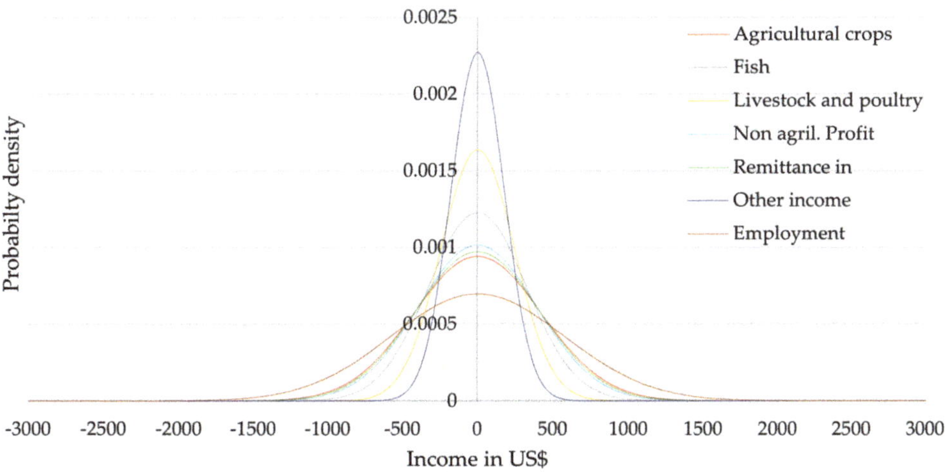

Figure 5. Distribution of total income (US$) for farm households in Mymensingh by income sources.

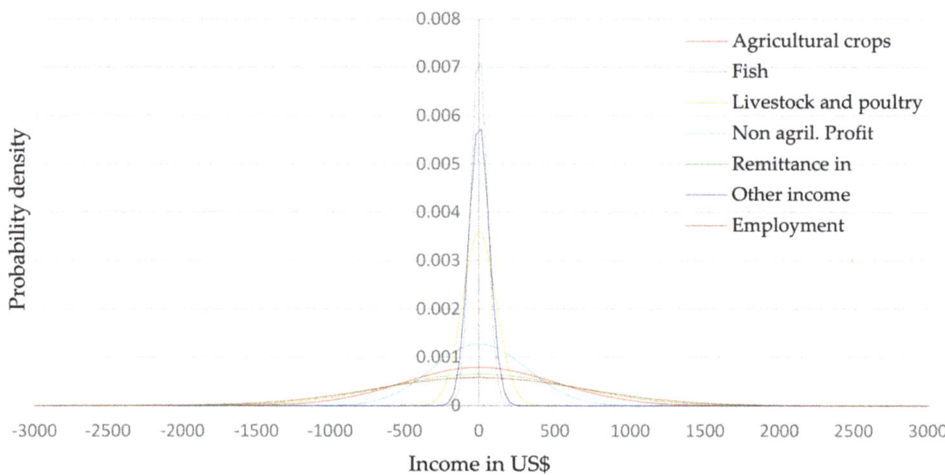

Figure 6. Distribution of total income (US$) for farm households in Rangpur by income sources.

4.4. Factors in Agricultural Income Differences

The main factors of agricultural income differences are shown in Table 7 obtained by the decomposed variance method. We estimate the variance component shares of crops for all farms across nine regions. From Table 6, we identify that agriculture is one of the main reasons for income differences in Mymensingh, Rangpur, Barisal, Khulna and Rajshahi. The empirical estimates of Table 7 indicate that the main variation in agricultural income comes from *aman* HYV (g) and *boro* HYV (j) rice. However, the results also display the contributions of other crop income to total agricultural income variation.

Table 7. Shares of crop income (%) in total agricultural income variation, by region.

	B	CH	CO	D	K	M	RJ	RN	S	BD
V(b)	0.35	0.07	0.03	0.15	0.10	0.00	0.01	0.00	0.36	0.11
V(c)	0.08	0.04	0.03	0.00	0.00	0.06	0.06	0.01	0.04	0.04
V(d)	0.64	0.43	0.01	0.02	1.54	0.06	0.13	0.13	1.06	0.53
V(e)	5.23	0.00	0.36	0.36	0.53	0.50	0.50	0.15	2.06	1.02
V(f)	0.47	0.02	0.16	0.02	0.07	0.06	0.01	0.15	0.00	0.10
V(g)	8.95	7.67	1.12	1.63	10.15	3.84	7.64	12.95	7.88	8.50
V(h)	0.02	0.00	0.00	0.00	0.09	0.09	0.05	0.11	0.00	0.06
V(i)	0.70	0.00	0.06	0.01	0.06	0.00	0.00	0.36	0.16	0.14
V(j)	6.36	4.32	8.13	34.03	17.72	20.89	17.72	14.03	48.26	25.30
V(k)	2.49	2.13	1.26	5.71	3.88	0.69	3.56	3.40	17.82	5.03
V(l)	0.00	0.00	0.00	0.00	0.00	0.00	0.00	0.00	0.00	0.00
V(m)	0.00	0.00	0.01	0.04	0.15	0.00	0.23	0.18	0.00	0.11
V(n)	0.00	0.00	0.27	0.07	0.10	0.00	0.53	0.65	0.00	0.28
V(o)	0.26	0.00	4.28	4.74	2.46	0.04	0.91	0.93	0.14	2.38
V(p)	0.49	0.04	20.77	0.35	0.03	0.08	1.78	6.48	0.16	2.68
V(q)	1.65	0.90	0.81	11.56	12.40	0.98	0.17	0.49	0.08	6.00
V(r)	0.00	0.00	0.00	6.51	0.54	0.00	0.63	0.02	0.00	1.91
V(s)	67.37	75.85	43.55	29.35	44.77	62.62	16.16	24.67	21.98	44.00
2*Cov(o,r)				5.43	0.85		0.81			1.79
2*Cov(g,j)		5.75				9.73	11.64	13.34		
2*Cov(g,k)		2.79			0.37		4.55	7.94		
2*Cov(g,p)						0.02	3.58	11.66		
2*Cov(o,p)			18.45			0.34	6.19	2.33		
2*Cov(g,s)							9.54			
2*Cov(j,s)							13.61			
2*Cov(d,j)	4.95		0.72		4.20					
Total	100	100	100	100	100	100	100	100	100	100

B = Barisal, CH = Chittagong, CO = Comilla, D = Dhaka, K = Khulna, M = Mymensingh, RJ = Rajshahi, RN = Rangpur, S = Sylhet, BD = Bangladesh; b = *Aus* rice local, c = *Aus* rice LIV, d = *Aus* rice HYV, e = *Aman* rice Local, f = *Aman* rice LIV, g = *Aman* rice HYV, h = *Aman* rice Hybrid, i = T *Aus* rice HYV, j = *Boro* rice HYV, k = *Boro* rice Hybrid, l = Wheat Local, m = Wheat HYV, n = Maize, o = Jute, p = Potato, q = Chili, r = Onion, s = All other crops.

Rice is the leading crop in Bangladesh, accounting for more than 90% of total cereal production covering 75% of Bangladesh's total cropped area [45,69]. For Mymensingh and Rangpur, variances in both *aman* HYV and *boro* HYV rice are high. For other regions, variances in *boro* HYV are high.

All other crops(s) are among the main causes (44% variance share) of income differences for all of Bangladesh since all types of pulses, oil seeds, spices, vegetable, leafy vegetables and fruits are included in the group of "all other crops." Moreover, all other crops(s) explain the larger contribution to total agricultural income variation because, in some regions, vegetables and fruits, among others, excluding rice, are important agricultural income sources.

The distribution of crop income among total agricultural income for the whole country is shown in Figure 7, which follows in Figures 8 and 9 for Mymensingh and Rangpur, respectively, with selected crops mainly produced by farmers in these regions. We found that *boro* rice has the widest variation in both the region and the highest inequality of total agricultural income, explained by the inequality of *boro* HYV income.

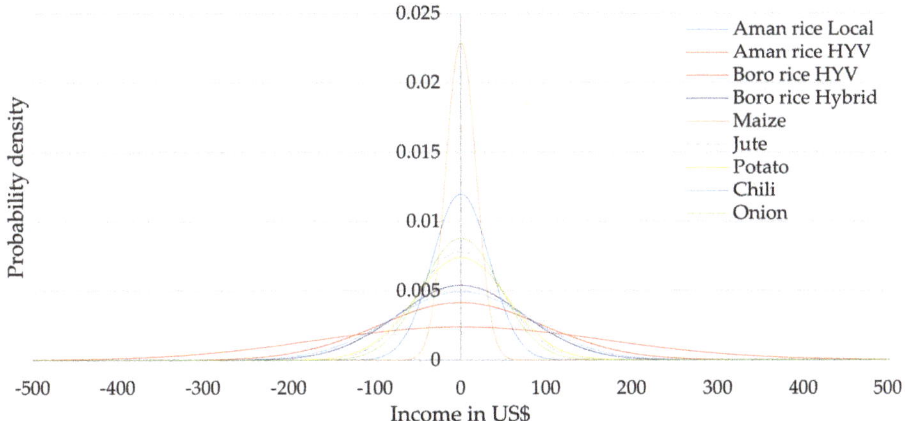

Figure 7. Distribution of agricultural income for farm households in Bangladesh by crop income.

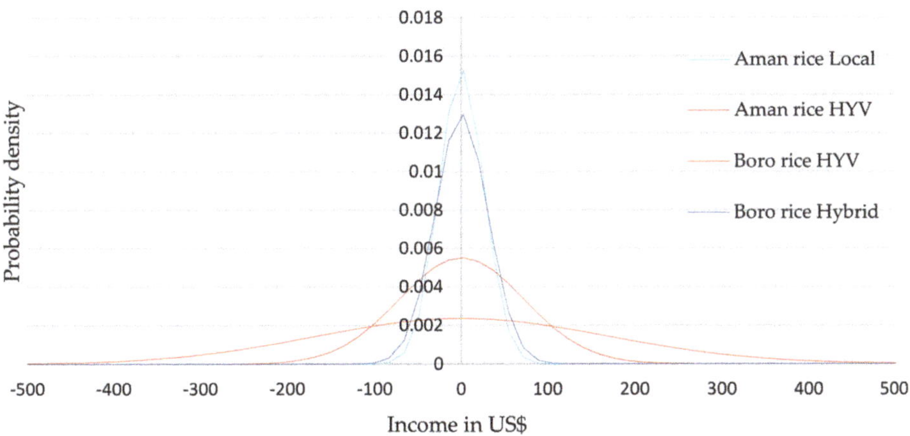

Figure 8. Distribution of agricultural income for farm households in Mymensingh by crop income.

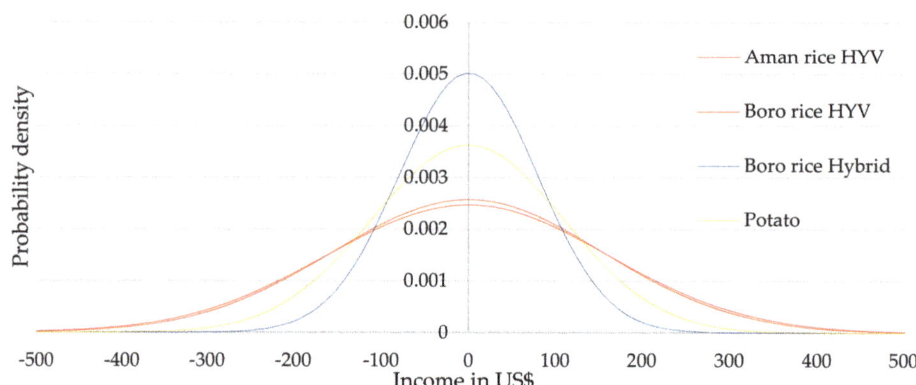

Figure 9. Distribution of agricultural income (US$) for farm households in Rangpur by crop income.

4.5. Factors Contributing to Variations in Income from Aman HYV and Boro HYV Rice Production

According to the results of Table 7, it is important to determine the factor causing the net income differences in *aman* HYV production. From decomposed variance of gross income and gross cost, we find in Table 8 that gross income is the main factor in net income difference, indicating that, although farmers in same region cultivated *aman* HYV rice, their gross incomes were different.

Table 8. Decomposed variances share (%) of GI and GC for *aman* HYV rice, by region.

	B	CH	CO	D	K	M	RJ	RN	S	BD
V(GI)	75.31	74.34	98.38	53.87	76.53	57.17	66.88	74.25	45.49	69.45
V(GC)	80.97	33.57	35.80	91.18	36.13	49.23	55.56	30.27	55.10	45.67
−2*Cov(GI, GC)	−56.27	−7.91	−34.18	−45.06	−12.66	−6.39	−22.44	−4.52	−0.59	−15.11
Total	100	100	100	100	100	100	100	100	100	100

B = Barisal, CH = Chittagong, CO = Comilla, D = Dhaka, K = Khulna, M = Mymensingh, RJ = Rajshahi, RN = Rangpur, S = Sylhet and BD = Bangladesh; GI = Gross income; and GC = Gross cost.

These gross income differences mainly induce the net income disparities in Comilla, Khulna, Chittagong and Rangpur, while gross cost induces the income disparities in Dhaka and Barisal for *aman* HYV rice. Additionally, gross cost also contributes to the total net income disparity of *aman* HYV rice production. To determine the variance in gross cost for *aman* HYV rice production, we estimate the variance component shares of all costs contributing to gross cost and present them in Table 9.

The results show the factors responsible for large variations in cost from *aman* HYV rice production. As shown in Table 9, variances in seed (c) shows in third row, chemical fertilizer (g) in row seven and hired labor costs (k) in row eleven, are high across all regions. In Dhaka, the highest 80% of inequality of gross cost for *aman* HYV rice production is explained by the inequality of hired labor cost (k), while in Barisal, the highest 25% inequality of gross cost is explained by inequality of seed cost. These costs were the main factors inducing the income differences in *aman* HYV rice production. This result indicates the importance of farming knowledge and easy input access to rice cultivation.

Table 9. Decomposed variances share (%) of costs for *aman* HYV rice production, by region.

	B	CH	CO	D	K	M	RJ	RN	S	BD
V(b)	3.64	3.73	3.79	0.97	3.66	5.50	3.72	8.79	4.32	3.24
V(c)	25.01	1.87	24.54	1.47	3.55	5.56	3.12	6.78	3.81	5.15
V(d)	0.53	1.79	1.04	1.32	8.33	2.04	4.15	6.70	0.67	3.69
V(e)	0.07	0.18	0.19	0.08	0.41	0.64	0.77	0.64	0.23	0.33
V(f)	0.54	0.48	0.28	0.07	0.65	0.10	0.65	0.54	0.14	0.35
V(g)	5.32	9.73	6.27	1.54	12.74	6.72	7.57	7.05	3.38	6.42
V(h)	0.98	0.06	0.01	0.04	0.30	2.76	0.05	0.57	1.42	0.50
V(i)	9.49	2.29	1.88	0.35	4.25	1.29	1.31	2.70	1.62	2.10
V(j)	3.47	0.58	1.62	0.10	0.44	0.70	0.15	0.26	3.04	0.69
V(k)	15.16	39.90	45.37	80.58	37.61	70.65	40.88	58.04	74.50	59.53
2*Cov(f,g)	1.72	2.37	1.33	0.33	2.14	0.77	3.05	1.26		1.41
2*Cov(i,f)	2.07		0.59	0.13			1.17	1.03	0.41	0.54
2*Cov(i,g)	11.50		3.88	0.77	5.69	3.26	4.29	4.69	1.94	3.32
2*Cov(k,g)	5.46	20.32		8.55	19.47		18.35			12.74
2*Cov(c,j)	15.04							0.95	4.52	
2*Cov(k,f)		3.79		2.04			4.82			
2*Cov(k,i)		1.90	9.21	1.67	0.75		5.94			
2*Cov(c,k)		11.0								
Total	100	100	100	100	100	100	100	100	100	100

B = Barisal; CH = Chittagong; CO = Comilla; D = Dhaka; K = Khulna; M = Mymensingh; RJ = Rajshahi; RN = Rangpur; S = Sylhet; and BD = Bangladesh; b = Rental cost of land; c = Seed cost; d = Irrigation cost; e = Manure/compost cost; f = Pesticide cost; g = Chemical fertilizer cost; h = Draft animal cost for land preparation; i = Rental cost for tools and machinery; j = Threshing cost; and k = Hired labor cost.

In Table 7, we note that *boro* HYV also had an influence on agricultural income. It is essential to determine the factors affecting the net income variation for *boro* HYV rice cultivation. Table 10 summarizes the decomposed variance of gross income and gross cost from *boro* HYV rice production and shows that gross income is the main factor in net income differences for *boro* HYV rice production, except for in Chittagong and Sylhet. However, gross cost also contributes to the total net income disparity of *boro* HYV rice production.

Next, we want to know which costs are the main factors in income differences in *boro* HYV rice production. To know the variance in gross costs for *boro* HYV rice production, we estimate the variance component shares of all cost expenditures contributing to gross cost and present them in Table 11. We found that the variances in seed (c) shows in third row, irrigation (d) in row four, chemical fertilizer (g) in row seven and hired labor cost (k) in row eleven, are high in all regions, indicating that adaptation strategies, such as low input costs, have priorities for the large gross income variances of *boro* rice cultivation.

Table 10. Decomposed variance share (%) of gross income and cost of *boro* HYV rice, by region.

	B	CH	CO	D	K	M	RJ	RN	S	BD
V(GI)	101.34	46.75	264.6	62.73	79.59	70.15	69.81	80.61	67.68	91.68
V(GC)	43.86	79.49	97.26	41.17	40.46	47.38	60.96	28.25	84.98	54.04
−2*Cov (GI, GC)	−45.20	−26.24	−261.9	−3.90	−20.05	−17.53	−30.77	−8.86	−52.66	−45.72
Total	100	100	100	100	100	100	100	100	100	100

B = Barisal, CH = Chittagong, CO = Comilla, D = Dhaka, K = Khulna, M = Mymensingh, RJ = Rajshahi, RN = Rangpur, S = Sylhet and BD = Bangladesh; GI = Gross income and GC = Gross cost.

These input costs were made the net income differences in this rice production for sample farmers. Based on the findings in Table 11, it is also important to note that, in Chittagong region, the variance in hired labor cost (k) is highest (69.84%) while it is lowest in Comilla region (27.25%). This result implies that 69.84% of inequality of gross cost is elucidated by the inequality of hired labor cost in Chittagong region. As shown in the fourth row, irrigation cost (d) contributes a significant share of the variation of gross cost; the highest 22.93% of inequality of gross cost is explained by the inequality of irrigation cost in Dhaka, compared to the lowest in Chittagong. This result implies that reduction of input cost variances will ensure the low net income differences for this rice production. Farm households are not entirely self-sufficient regard the labor supply for their farming. In peak times of agricultural production, such as transplanting, weeding and harvesting, hired labor demand occurs. However, the labor supply is low in Chittagong due to hill tract areas of Bangladesh [69], resulting in the higher costs of labor.

Table 11. Decomposed variance share (%) of costs for *boro* HYV rice production, by region.

	B	CH	CO	D	K	M	RJ	RN	S	BD
V(b)	2.87	0.66	0.50	1.88	2.66	4.11	1.32	5.32	2.63	2.27
V(c)	4.10	0.71	2.21	3.67	4.78	2.72	1.73	4.34	2.20	3.61
V(d)	8.89	2.70	4.06	22.93	22.39	22.42	10.70	16.00	7.57	18.01
V(e)	0.24	0.05	1.10	0.31	0.76	0.88	0.33	2.56	0.12	0.80
V(f)	0.89	0.09	0.18	0.16	0.48	0.33	0.31	0.60	0.07	0.33
V(g)	7.71	3.31	1.98	6.71	14.76	12.82	4.71	13.54	3.23	8.21
V(h)	0.04	0.03	0.00	0.05	0.79	10.08	0.13	0.38	2.04	1.16
V(i)	2.42	0.89	1.01	0.93	1.47	1.09	0.47	1.68	1.12	1.23
V(j)	0.98	0.20	0.15	1.08	0.75	2.24	0.24	0.39	0.18	0.78
V(k)	38.05	69.84	27.25	42.04	38.45	31.49	51.04	38.17	65.10	51.51
2*Cov(f,g)	3.91	0.73	0.66	0.90	2.15		1.49	3.46	0.50	1.55
2*Cov(d,g)	4.98		1.18				4.35			
2*Cov(f,i)	1.07	1.15	2.62	0.39	0.52		0.52	0.97	0.26	0.61
2*Cov(g,i)	4.68	2.70	1.99	2.87	5.47	3.76	2.14	5.69	1.99	3.43
2*Cov(g,k)	11.72	14.45	6.27	11.25			10.64		11.72	
2*Cov(i,k)	7.46		6.84	4.83	4.58	8.05	3.89			5.90
2*Cov(e,i)		2.50	9.58					1.25	0.22	0.60
2*Cov(f,k)			5.34				5.99			
2*Cov(e,g)			1.50					4.90	0.44	
2*Cov(e,f)			7.04					0.76	0.63	
2*Cov(d,k)			8.70							
2*Cov(e,k)			9.85							
Total	100	100	100	100	100	100	100	100	100	100

B = Barisal, CH = Chittagong, CO = Comilla, D = Dhaka, K = Khulna, M = Mymensingh, RJ = Rajshahi, RN = Rangpur, S = Sylhet and BD = Bangladesh; b = Rental cost of land, c = Seed cost, d = Irrigation cost, e = Manure/compost cost, f = Pesticide cost, g = Chemical fertilizer cost, h = Draft animal cost for land preparation, i = Rental cost for tools and machinery, j = Threshing cost and k = Hired labor cost.

4.6. Future Projections

Production levels in agriculture, fishery and livestock raising are projected to change due to climate change [39,83]. We therefore sought to project the impact of rice yield change on the state of poverty in Bangladesh. If rice is a commercial crop, a price hike due to any damage from climate change could increase Bangladeshi farmers' living standards. However, rice remains a subsistence crop among most Bangladeshi farmers; therefore, we assume that rice yield reduction will lead to a rice consumption reduction.

The effects of climate change on rice yields, as has been estimated and shown by International Food Policy Research Institute [37], are such that, without adaptation to climate change impacts, *aman* HYV and *boro* HYV rice yields will decline by 3.5% and 10.2%, respectively, in Bangladesh. According to the Geophysical Fluid Dynamics Laboratory (GFDL) scenarios, if temperature changes by 4.0 °C, then 17% decline in overall rice will occur in Bangladesh [84].

According to this projection, we assumed that, due to climate change effects on *boro* HYV and *aman* HYV, rice yields will be reduced by 10% and 4%, respectively, as well as a 17% reduction in overall rice among the sample households. We applied log-normal distribution to project the poverty rate due to income reduction by yield loss on the effects of climate change.

Figure 10 shows the annual per-capita income (actual and projected) in US$ of the sample households across Bangladesh. In general, one can see from this figure that the sample population density (i.e., probability density) mostly lies within the low annual per-capita income range, which is less than the poverty line. Additionally, the probability density of the low-income range increases in the projected income distribution when one considers rice yield loss due to climate change.

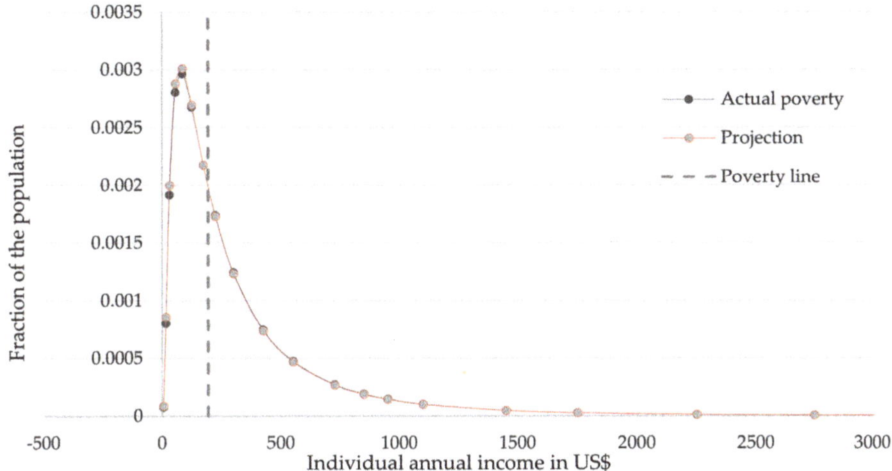

Figure 10. Annual per-capita income (US$) distribution of Bangladesh (17% loss of rice).

From the decomposed variance share of income sources in Table 9, we found that agriculture was the main reason for income differences in Mymensingh and Rangpur. Now, we can examine the effects of climate change on rice production (10% and 17% losses) in these two regions by log-normal distribution.

We analyzed and found that constant reduction of rice yield (10% loss) by climate change in Bangladesh is not such a severe problem for farmers. Because the change in net per-capita income is very small, there is not a dramatic change of poverty rate. However, if unexpected extreme events, such as floods, flash floods, droughts and sea level rise, occur in specific areas of Bangladesh, they create a more vulnerable situation for the farmers' livelihood. In addition, the probability density of low-income range increases (Figures 11 and 12) in both Mymensingh and Rangpur districts, where rice income decreases due to climate change.

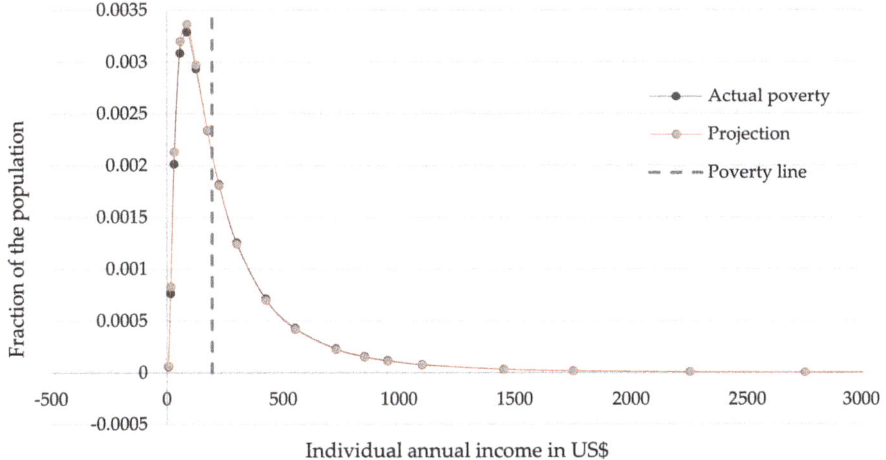

Figure 11. Annual per-capita income (US$) distribution of Mymensingh (17% loss of rice).

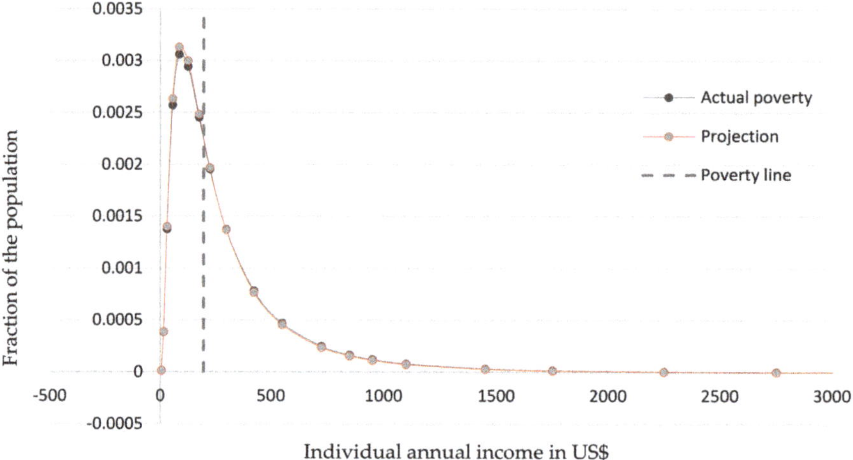

Figure 12. Annual per-capita income (US$) distribution of Rangpur (17% loss of rice).

We also applied the same analysis in Figures 10–12 to all of the regions and Table 12 shows the results of the poverty rate after income changes due to assumed yield losses of *aman* HYV, *boro* HYV rice and overall rice.

Table 12. Change in poverty rate following a loss of rice yield due to climate change.

		B	CH	CO	D	K	M	RJ	RN	S	BD
	Actual	0.507	0.484	0.446	0.455	0.415	0.496	0.323	0.462	0.484	0.454
	Projected	0.508	0.491	0.447	0.458	0.417	0.502	0.330	0.466	0.487	0.457
10% loss	Change	0.001	0.007	0.001	0.003	0.002	0.006	0.007	0.004	0.003	0.003
	Increase (%)	0.197	1.446	0.224	0.659	0.482	1.210	2.167	0.866	0.620	0.661
	Projected	0.513	0.494	0.449	0.460	0.422	0.511	0.335	0.473	0.490	0.461
17% loss	Change	0.006	0.010	0.003	0.005	0.007	0.015	0.012	0.011	0.006	0.007
	Increase (%)	1.183	2.066	0.673	1.099	1.687	3.024	3.715	2.381	1.240	1.542

B = Barisal, CH = Chittagong, CO = Comilla, D = Dhaka, K = Khulna, M = Mymensingh, RJ = Rajshahi, RN = Rangpur, S = Sylhet and BD = Bangladesh.

The estimated results suggest that rice yield loss would reduce the annual per-capita income of the sample farm households and increase the poverty rate in various regions across Bangladesh. It was found that the highest poverty rate increase (3.024%) would occur in Mymensingh, Rajshahi (3.715%) and Rangpur (2.381%). Rajshahi and Rangpur are in northwestern Bangladesh and are prone to drought; climate change would affect rice production specifically in the summer, when *boro* rice is being produced. Mymensingh is affected by floods, flash floods and heavy rainfall each year, owing to the effects of climate change on *aman* and *boro* harvests.

Climate Change Impact Scenario

Extreme events, such as floods, droughts and changes in seasonal rainfall patterns, negatively impact crop yields in vulnerable areas [85–87]. In Bangladesh, the rural poverty rate would be exacerbated [88] as a result of the impacts of extreme events on the yield of rice crop and increases in food prices and the cost of living [89,90]. The impacts of climate change on poverty would be heterogeneous among countries [91]. Due to the impact of climate change, rice production would decrease and some rice exporting countries, such as Indonesia, the Philippines and Thailand, would benefit from global food price rises and reduced poverty, while Bangladesh would experience a net increase in poverty of approximately 15% by 2030 [89,91].

Climate change refers to changes in climate attributed directly as temperature, precipitation, CO_2 concentrations and solar radiation or indirectly as river floods, flash floods and sea level rise that alter the composition of the global atmosphere, as well as to natural climate variability observed over comparable time periods [33,50].

Temperature Increase

Temperature is an important factor for *boro* rice production and the maximum temperature is always more vulnerable with a negative impact on rice yields. In Bangladesh, seasonal temperature suddenly fluctuates, causing drastically declines in the yield of *boro* rice. *Boro* rice yields decrease by a maximum of 18.7% due to an increase in minimum temperature of 2.0 °C–4.0 °C and by 36.0% for 2.0 °C–4.0 °C maximum temperature increases in different location of Bangladesh in 2008 [92]. According to the Intergovernmental Panel on Climate Change (IPCC), SRES emissions scenarios and climate models being considered, global mean surface temperature is projected to rise in the range of 1.8 to 4.0 °C by 2100 [93]. Following the previous assessment, the IPCC concludes in their fifth assessment report (AR5) that it will be difficult to adapt with large-scale warming of approximately 4°C or more, which will increase the likelihood of severe, pervasive and irreversible impacts [91,94,95].

According to the previous projection of temperature fluctuations in Bangladesh, we assume that, due to the maximum and minimum temperature fluctuations, in the future, the overall rice production will decrease by approximately 17% of the sample farmers and results are shown in Table 12. The table shows that maximum 3.7% poverty will increase in Rajshahi and second highest (3.0%) in Mymensingh region and this implies that it is important to adaptation strategies for Rajshahi and Mymensingh for high temperature.

Rainfall Decreases (Drought)

Inadequate rainfall leads to greater drought frequency and intensity, while increased evaporation increases the chance of complete crop failure [96,97]. Drought is the most widespread and damaging of all environmental stresses [35,98]. In South and Southeast Asia, including some states of India, severe drought affects rain-fed rice and yield, with losses as high as 40% and the total area affected measuring 23 million hectares, amounting to $800 million [99]. Bangladesh experienced severe drought in different years and locations in the districts of the northwestern border [100]. Erratic rainfall and drought reduce crop production by 30% and 40%, respectively [84]. *Boro* rice production will decrease due to rainfall in winter [92]. This study noted that, with 5-mm and 10-mm rainfall reductions in the future, *boro* rice will decrease by a maximum of 16.6% and 24.2%, respectively, in the winter. Drought caused 25% to 30% crop reduction in the northwestern part of Bangladesh based on from 2008 [101]. Due to the high rainfall variability and dryness, the northwestern region is the most drought-prone area in Bangladesh [102,103]. Rajshahi, Chapai-Nawabganj, Naogaon, Natore, Bogra, Joypurhat, Dinajpur and Kustia districts are drought prone areas in Bangladesh because of their moisture-retention capacity and infiltration rate characteristics [104].

According to the previous projection of drought, we assume that, if rainfall decreases and drought occur in the future, the overall rice production will decrease by approximately 20% of the sample farmers in northwestern districts of Bangladesh. By using log-normal distribution, we project the poverty rate due to income reduction by yield loss because of drought.

Table 13 shows the results of the poverty rate (Figure 13) after income changes due to assumed yield losses of overall rice by drought in the northwestern region in Bangladesh, while the Dinajpur (10.175% poverty increase), Rajshahi (5.670% poverty increase) and Naogaon (11.245% poverty increase) districts are most vulnerable to poverty. Dependency on agriculture with high variability of annual rainfall has made the northwestern regions highly susceptible to droughts and high poverty rates, compared to other parts of the country. Conservation of water could play an important role in reducing the impact of drought and alleviating poverty in this area [103].

Table 13. Poverty rate in drought-prone districts on rainfall decrease.

	BG	CN	DI	KU	NG	NT	RJ	JT
Actual	0.242	0.354	0.285	0.447	0.249	0.448	0.388	0.268
Projected	0.263	0.361	0.314	0.452	0.277	0.452	0.410	0.282
Change	0.021	0.007	0.029	0.005	0.028	0.004	0.022	0.014
Increase (%)	8.678	1.977	10.175	1.119	11.245	0.893	5.670	5.224

BG = Bogra, CN = Chapai-Nawabganj, DI = Dinajpur, KU = Kustia, NG = Naogaon NT = Natore, RJ = Rajshahi and JT = Joypurhatr.

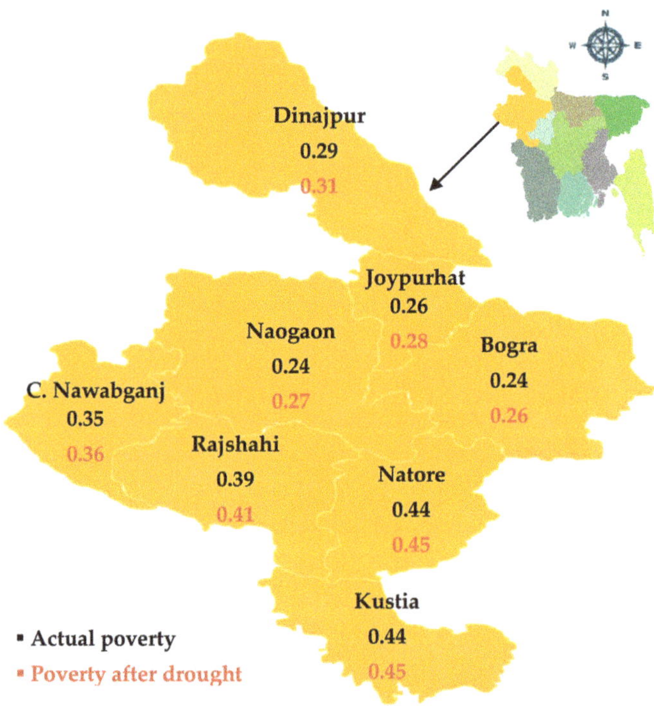

Figure 13. Changing poverty rates caused by drought in northwestern regions.

Flood

From the GBM basins, the monsoonal discharge of water causes seasonal floods and affects most of the areas of Bangladesh, with extent varying by year [50]. Floods occur almost every year and in 1998, floods covered almost 70% of total land area in Bangladesh, causing the maximum damage by floods in Bangladesh [105]. According to the IPCC's fourth assessment report, the intensity and frequency of floods and cyclones will increase in the near future [33]. Moreover, the IPCC's fifth assessment report (AR5) predicts that greater risks of flooding will increase on the regional scale [91,94–99]. In addition, extreme flood events will reduce crop production by 80% in Bangladesh [37,84].

Mymensingh, Sylhet, Dhaka, Comilla, some parts of Rangpur and Khulna regions are the mainly river-flooded areas in Bangladesh [50]. We assume that, if extreme floods, as in 1998 (the magnitude of the 1998 flood was the maximum in Bangladesh), occur, farm production will decrease by 80% in the flood-prone regions of Bangladesh. By log-normal distribution we project the poverty rate due to income reduction by yield loss due to the effects of extreme floods. The results are shown in Table 14.

Table 14. Poverty rate due to yield loss by flood in Bangladesh.

	CO	D	K	M	RN	S
Actual	0.446	0.455	0.415	0.496	0.462	0.484
Projected	0.465	0.502	0.479	0.554	0.529	0.519
Change	0.019	0.047	0.064	0.058	0.067	0.035
Increase (%)	4.260	10.330	15.422	11.694	14.502	7.231

CO = Comilla, D = Dhaka, K = Khulna, M = Mymensingh, RN = Rangpur and S = Sylhet.

The estimated results in Table 14 suggest that rice yield loss would reduce the annual per-capita income of the sample farm households and increase the poverty rate in various regions across Bangladesh (Figure 14). It was found that the highest poverty rate increases would occur in Rangpur (14.502%) and Khulna (15.422%). This result implies that coping strategies to highly flood affected areas of crops loss should have priority.

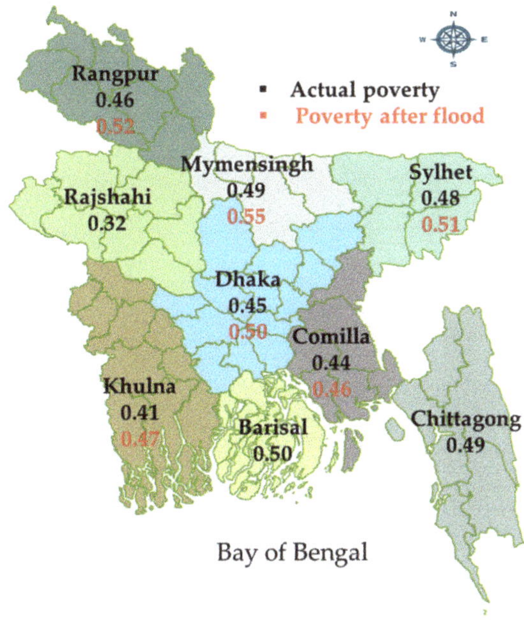

Figure 14. Changing poverty rates caused by floods in different regions.

Flash Floods

The northeastern parts of Bangladesh—mostly Sunamganj, Kishorganj, Netrokona, Sylhet, Habiganj and Maulvibazar—are prone to flash floods during the months of April to November and these areas are covered by many haors, where water remains stagnant [106]. Farmers of these districts produced *boro* rice in almost 80% of their land, while only approximately 10% of the area is covered by transplanted *aman* production [107]. In 2017, flash floods affected these areas and damaged almost 90% (maximum) of *boro* rice [108]. According to this scenario, we assumed that if in the future this extreme event occurs in haor areas, *boro* rice yields will be reduced by a maximum of 90% of the sample households. We applied log-normal distribution to project the poverty rate due to income reduction by yield loss due to the effects of flash floods on *boro* rice yields by a maximum of 90%.

Table 15 shows the results of the poverty rate after incomes changed due to assumed yield loss of *boro* rice in flash flood regions in Bangladesh, while Kishorganj district is most vulnerable to poverty

(19.214% increase) if flash floods occur (Figure 15). The projected results are treated as flash flood to be changed the poverty in northern-eastern parts of Bangladesh and this region are vulnerable on flash flood. Therefore, ex-ante coping strategies are important to the damages of flash flood.

Table 15. Poverty rate in flash flood region in Bangladesh.

	HB	KI	MV	NT	SU	SY	TH
Actual	0.354	0.458	0.624	0.585	0.511	0.427	0.354
Projected	0.381	0.546	0.637	0.628	0.550	0.452	0.381
Change	0.027	0.088	0.013	0.043	0.039	0.025	0.027
Increase (%)	7.627	19.214	2.083	7.350	7.632	5.855	7.627

HB = Habiganj, KI = Kishorganj, MV = Maulvibazar, NT = Netrokona SU = Sunamganj, SY = Sylhet and TH = Total Haor.

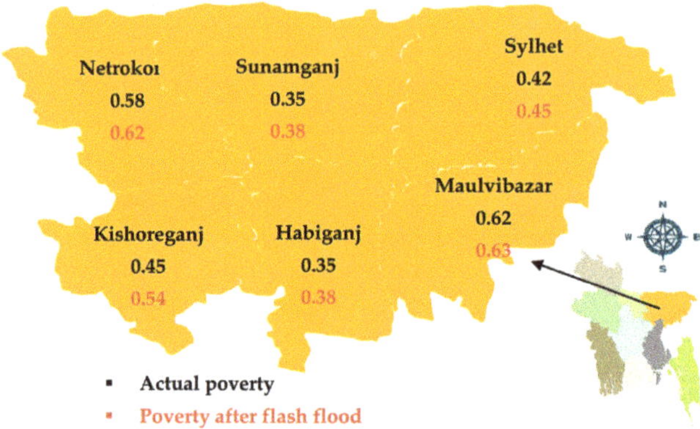

Figure 15. Changing poverty rate caused by flash floods in northeastern regions.

Sea Level Rise

Approximately 80% of the land of Bangladesh is flatlands, while 20% is 1 m or less above sea level, which is the coastal area (southern 19 districts beside the Bay of Bengal) and particularly vulnerable to sea level rise [109]. The coastal area covers approximately 20% of the country (including 19 districts beside the Bay of Bengal), which is approximately 30% of the net cultivable area and 25.7% of the population of Bangladesh [110,111]. Sea level rise will directly result in increased coastal flooding, which will increase in the event of storm surges. IPCC's fourth assessment report [33] reports that a 1-m sea level rise will displace approximately 14,800,000 people by inundating a 29,846-sq. km. coastal area [112]. Nicholls and Leatherman in 1995 [113] predicted that a 1-m sea level rise would result in a 16% of national rice production loss in Bangladesh [114].

In terms of number of people affected with respect to sea level rise, Bangladesh has been rated as the third most vulnerable country in the world. By 2050, approximately 33 million people would be suffering from surging, assuming a sea level rise of 27 cm. A full 18% of the total land area in Bangladesh would submerge with a 1-m rise in sea level [115]. Based on the IPCC fifth annual report (AR5), across all representative concentration pathways (RCPs), global mean temperature (°C) is projected to rise by 0.3 to 4.8 °C by the late-21st century and global mean sea level (m) is projected to increase by 0.26 to 0.82 m [91]. The Global Circulation Model (GCM) predicts an average temperature increase of 1.0 °C by 2030, 1.4 °C by 2050 and 2.4 °C by 2100; the study revealed that the sea level will rise by 14 cm, 32 cm and 62 cm, respectively. A rise in temperature would cause significant decreases in production of 28 % and 68 % for rice and wheat, respectively [84].

According to this scenario, we assumed that, due to sea level rise in the southern part of Bangladesh, *boro* rice yields will be reduced by 30% of the sample households. We applied log-normal distribution to project the poverty rate due to income reduction with yield loss based on the effects of sea level rise.

Table 16 shows the results of the poverty rate after income changes due to assumed yield loss of rice in coastal regions due to sea level rise, while Khulna district is the most vulnerable to poverty and poverty will increase by 6.752% (Figure 16). Changing continuous sea level rise in the coastal region result in no significant loss reduction for rice.

Table 16. Poverty rate in sea level rise regions in Bangladesh.

	SK	KH	BT	PR	JL	BG	BS	PT	BL	LK	NK	FN	CT	CX
Actual	0.599	0.295	0.363	0.388	0.640	0.532	0.419	0.628	0.491	0.529	0.438	0.481	0.505	0.462
Projected	0.609	0.315	0.370	0.390	0.650	0.545	0.431	0.636	0.493	0.533	0.440	0.487	0.515	0.464
Change	0.010	0.020	0.007	0.002	0.011	0.013	0.013	0.008	0.002	0.004	0.002	0.007	0.010	0.002
Increase (%)	1.688	6.752	1.924	0.527	1.674	2.388	3.081	1.255	0.491	0.770	0.410	1.361	1.901	0.367

SK = Satkhira, KH = Khulna, BT = Bagerhat, PR = Pirozpur, JL = Jhalakati, BG = Barguna, BS = Barisal, PT = Patuakhali, BL = Bhola, LK = Lakshmipur, NK = Noakhali, FN = Feni, CT = Chittagong and CX = Cox's Bazaar.

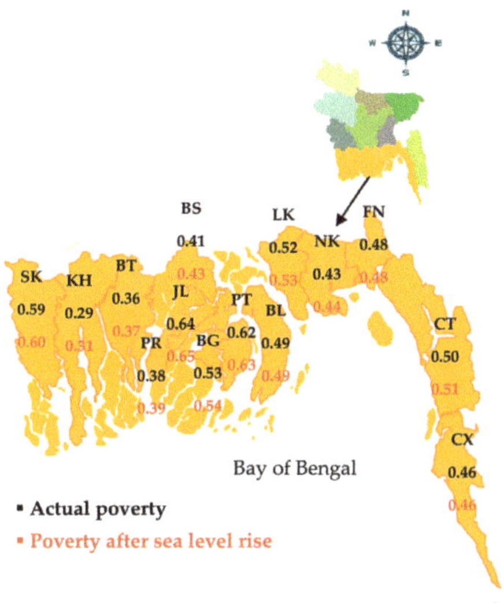

Figure 16. Changing poverty rate caused by sea level rise in southern regions.

Representative Concentration Pathways (RCPs)

In assessing future climate change, the fifth assessment report (AR5) of the IPCC selected four RCPs, –RCP 2.6, RCP 4.5, RCP 6.0 and RCP 8.5 [91], with RCP 4.5 and RCP 8.5 covering both medium and extreme scenarios. These four RCPs describe four probable climate futures depending on how much greenhouse gasses are emitted over the next 85 years.

According to the IPCC's fifth annual report (AR5), across all representative concentration pathways (RCPs), global mean temperature (°C) is projected to rise by 0.3 to 4.8 °C by the late-21st century [68]. Increasing temperatures will increase the number of growing days over time. Heat stress is a major issue for crop production and reduces yields.

Climate change will certainly continue in coming decades and affect agricultural production. Yamei Li et al. worked on simulating total climate change impacts on rice production under RCP scenarios and projected that average rice yields during the 2020s, 2050s and 2080s would decrease by 12.3%, 17.2% and 24.5% under RCP 4.5 and by 14.7%, 27.5% and 47.1% under RCP 8.5, respectively [67].

According to this scenario, we assumed that, due to total climate change impacts, rice yields would be reduced by a maximum of 47% based on RCP 8.5 among the sample households. We applied log-normal distribution to project the poverty rate due to income reduction by yield loss. Table 17 shows that, under RCP 4.5 and RCP 8.5, the poverty rate will increase in all of the regions because of rice income reductions.

Additional increases in average poverty occur in Rajshahi, Mymensingh, Rangpur, Khulna and Sylhet region under both RCP 4.5 and RCP 8.5 with variations in the total climate change impacts on rice production. The yield of rice is predicted to decrease more under RCP 8.5 than RCP 4.5, resulting in per-capita income decreases. Under RCP 8.5, this study predicts a maximum increase in poverty of 10.526% in Rajshahi and the lowest of 3.139% in Comilla (Table 17). It is possible that our predicted rice yield declines by RCP scenario and relatively drought prone areas, such as Rajshahi, will be more vulnerable (Figure 17). The results from our drought scenarios are comparable to the results for RCP 8.5 and it is consistent that Rajshahi region is more vulnerable under climate change impacts. In both scenarios, our predicted yield decline and resulting per-capita income decline increase poverty. Climate change forces a decline in rice yield [116], suggesting that the predicted decreases in heat stress yield can be mostly attributed to an increased drought tolerant variety.

Table 17. Changes in poverty rates following a loss of rice yield due to RCPS.

		B	CH	CO	D	K	M	RJ	RN	S	BD
25% loss of rice under RCP 4.5	Actual	0.507	0.484	0.446	0.455	0.415	0.496	0.323	0.462	0.484	0.454
	Projected	0.516	0.490	0.455	0.462	0.424	0.510	0.345	0.471	0.497	0.463
	Change	0.009	0.006	0.009	0.007	0.009	0.014	0.022	0.009	0.013	0.009
	Increase (%)	1.775	1.240	2.018	1.538	2.169	2.823	6.811	1.948	2.686	1.982
47% loss of rice under RCP 8.5	Projected	0.524	0.500	0.460	0.470	0.438	0.526	0.357	0.488	0.507	0.474
	Change	0.017	0.016	0.014	0.015	0.023	0.030	0.034	0.026	0.023	0.020
	Increase (%)	3.353	3.306	3.139	3.297	5.542	6.048	10.526	5.628	4.752	4.405

B = Barisal, CH = Chittagong, CO = Comilla, D = Dhaka, K = Khulna, M = Mymensingh, RJ = Rajshahi, RN = Rangpur, S = Sylhet and BD = Bangladesh.

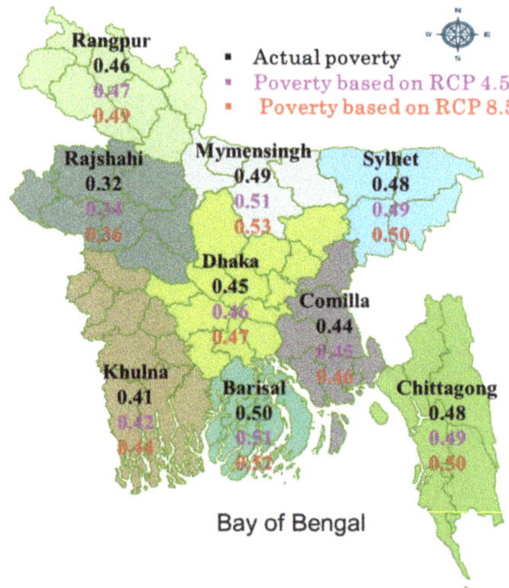

Figure 17. Changing poverty rate caused by total climate change impact based on RCP 4.5 and 8.5.

5. Conclusions

This paper has focused on the agrarian sub-national regional analysis of climate change vulnerability in Bangladesh under various climate change scenarios and its potential impact on poverty. It has drawn some significant evidence of regional vulnerability to climate change from regional characteristics, per-capita income, total income disparity, cost of production and poverty, based on statistical analysis of farm survey data. Our findings indicated that some regions are vulnerable to climate change impact on agricultural production among the administrative regions of Bangladesh, where coping strategies and techniques are important.

Bangladeshi farmers are producing crops, although there is much uncertainty due to associated risks of climate change. The results of our study show that, from the income shares of income source sectors, farmers in Mymensingh and Rangpur are largely dependent on agriculture. Of these regions, Mymensingh is one of the regions with the highest poverty rates. The income share in income sources revealed that income category shares across the various regions of Bangladesh are far from uniform. Income share comparisons and cluster analysis classified the regions into three groups as follows. (1) In some regions, namely Rajshahi, Khulna and Dhaka, income from agriculture is important and these regions receive relatively high income. (2) In other regions, namely Mymensingh, Rangpur and Barisal, agriculture income is important but the regions receive relatively low income. (3) The other regions, which are Comilla, Chittagong and Sylhet, are not strongly dependent on agriculture and Comilla region strongly relies on income from remittances. The principal targets of agricultural research for poverty reduction are considered to be in group (2).

Variance decomposition of income showed that agricultural income in Mymensingh and Rangpur is the main cause of income differences. Moreover, large variances in agricultural income in the regions are induced by gross incomes from rice production, indicating that rice yield can have large impacts on income levels. Therefore, research and development and technical support for farmers to realize high and stable rice yields in these regions are important.

This paper used modelling to predict crop yield changes by different aspects of climate change under droughts, floods, flash floods, sea level rise and RCP scenarios. We account for some uncertainty in crop yields and the resulting reduction in per-capita income of farm households. The proposed lognormal distribution projected the poverty rate and examined the vulnerable regions. The key is to understand the future projections of poverty rates on assumptions of *boro* HYV and *aman* HYV rice yield decreases on each farm due the climate change impacts and climate volatility subjecting the poor to poverty rate increases in different regions. Current climate change impacts are not the same in different regions; in particular, different extreme climatic events in specific regions often result in irreversible losses. One of the examples of the interventions of climatic events is that dependency on agriculture with high variability in annual rainfall has render the northwestern parts highly vulnerable to droughts and has increased the high poverty rates, compared to other parts of the country. Extreme floods can increase the poverty rates in Rangpur, Mymensingh and Khulna regions. Kishorganj district is the most vulnerable on poverty (8.8% increase) if sudden flash floods occur in the northeastern part of the country. Due to sea level rise, coastal areas will face poverty.

Strategies and techniques to cope with climate change for regions where small-scale farmers are largely dependent on agriculture are important challenges. Among the negative consequences of climate change impacts, subsistence farmers are suffering more from vulnerabilities such as extreme poverty or hunger. However, adaptation techniques in agriculture are a vital tool to avoid the adverse impacts of climate change [117]. Given the complex nature of droughts, floods, flash floods and sea level rise as phenomena, the development of drought-tolerant, short-maturing and salt-tolerant varieties is critically important.

More generally, our results are focused on farm income and poverty, including regional vulnerability due to climate change impacts on agricultural production. In recent years, climate change impacts have played a vital role in increasing the poverty rate and income variability among farm households in Bangladesh. Extreme environmental hazards are faced by farmers in this country and their net farm production decreases drastically, increasing the poverty rate while changes in weather conditions are a less severe problem for farmers due to their involvement in other income activities. We actually performed this study focusing on revealing the comprehensive impact of climate change on farm production and the crops are that the most important for per capita income differences across the country and that enhance the poverty rate, using the covariance and lognormal distribution methods.

This study has attempted to bridge the gap between academic research and professional practices in the context of potential climate change impacts on crop production and poverty. Because of the relatively large sample size, compilation and manipulation of the data were challenging. With the assessment of poverty and regional vulnerability due to climate changes, it is hoped that the study in general will assist in guiding authorities in terms of interventions aimed at climate change risk reduction in Bangladesh. Therefore, we believe that this research will help to reveal the mechanisms behind the per capita income differences and projected poverty rates of farm households based on different climate change impact scenarios across Bangladesh. Future work might also be more micro level for policy making to test root-level poverty and to further evaluate the impact of climate change on different crops and it should include the model for poverty determinants to confirm the relationships studied and their adaptations.

Author Contributions: M.S.A. conceived the research, compilation and analyze the data, drafted, edited and revised the manuscript; J.F. modified the methodology of the research, checked, edited and revised the manuscript; S.K. designed, compilation and analyze the data, edited and revised the manuscript; M.R.B. checked the statistical tools and maps of the objective regions; and M.A.S. helps to compilation of data and first draft of the manuscript.

Funding: This research received no external funding.

Acknowledgments: The authors are grateful to three anonymous reviewers for their constructive comments and suggestions. We would like to thank the International Food Policy Research Institute (IFPRI) for providing us with the primary data. We also acknowledge the support from JIRCAS under the project "Climate Change Measures in Agricultural Systems" and University of Tsukuba.

Appendix A

Appendix A.1

In this study, we used the primary data from Bangladesh Integrated Household Survey (BIHS 2011–2012) by IFPRI, https://dataverse.harvard.edu/dataset.xhtml?persistentId=hdl:1902.1/21266.

Appendix A.2

"aus" is former rainy season, *"aman"* is rainy season and *"boro"* is dry season irrigated rice.

Appendix B

Appendix B.1

$$Net\ accounting\ cost_i = \sum_i \sum_j P_{ij} X_{ij}$$

$$
\begin{aligned}
&= c_{i,\ Rental\ cost\ of\ land} + c_{i,\ seed\ cost} + c_{i,\ irrigation\ cost} \\
&+ c_{i,\ manure\ or\ compost\ cost} + c_{i,\ pesticides\ cost} + c_{i,\ fertilizer\ cost} \\
&+ c_{i,\ draft\ animal\ cost} + c_{i,\ machinery\ cost} + c_{i,threshing\ cost} + c_{i,\ hired\ labor\ cost}
\end{aligned}
\tag{A1}
$$

$$Production\ value_i = \sum_i P_i Y_i \tag{A2}$$

$$
\begin{aligned}
Gross\ income_i &= \sum_i P_i Y_i - Inkind\ payment_i \\
&= \sum_i P_i Y_i - \left(C_{i,\ irrigation\ cost\ paid\ by\ crop} + C_{i,\ labor\ cost\ paid\ by\ crop}\right)
\end{aligned}
\tag{A3}
$$

$$Net\ income(\pi)_i = Gross\ income_i - Net\ accounting\ cost_i \tag{A4}$$

Appendix B.2

We estimate per-capita incomes (US$) of all sample families on assumption of climate change impacts and draw the distribution of the estimated incomes assuming that the distribution follows log normal distribution. To draw log normal distribution, we have to find mean and standard deviation of $ln(x)$. Firstly, we divide the per capita income in different class and make the average (x) of each class and we find the frequency of household (n) in each per-capita income class. Then we find the log of average per-capita class, log (x); and multiplied by the frequency of household in each class, $n * log (x)$. Next average,

$$u = \frac{\sum n\{\log(x)\}}{\sum n} \tag{A5}$$

Then we estimate, $\log(x) - u$, $\{\log(x) - u\}^2$ and $n\{\log(x) - \mu\}^2$
Next standard deviation,

$$\sigma = \sqrt{\frac{\sum n \{\log(x) - u\}^2}{\sum n}} \tag{A6}$$

Returns the lognormal distribution of x, where $\ln (x)$ is normally distributed with parameters Mean and Standard deviation. Use this function to analyze data that has been logarithmically transformed.

$$f_X(x) = \frac{1}{dx} Pr(X \le x) = \frac{1}{dx} Pr(lnX \le lnx) = \frac{1}{dx}\Phi\left(\frac{lnx-\mu}{\sigma}\right) = \varphi\left(\frac{lnx-\mu}{\sigma}\right)\frac{1}{dx}\left(\frac{lnx-\mu}{\sigma}\right)$$
$$= \varphi\left(\frac{lnx-\mu}{\sigma}\right)\frac{1}{\sigma x} = \frac{1}{x}\cdot\frac{1}{\sigma\sqrt{2\pi}}exp\left(-\frac{(lnx-\mu)^2}{2\sigma^2}\right) \tag{A7}$$

Syntax: LOGNORM.DIST(x, mean, standard deviation and cumulative)

Appendix C

Table A1. Household income (US$/yr.) from different sources, by region.

	B	CH	CO	D	K	M	RJ	RN	S	BD
Agril. crops	159.35	124.17	82.83	194.67	273.63	225.23	322.78	246.71	131.77	200.28
Main crops	76.23	44.11	35.22	118.52	152.25	127.87	202.10	170.95	89.86	116.89
Other crops	83.13	80.06	47.61	76.16	121.39	97.36	120.69	75.76	41.92	83.39
Fish	115.70	23.47	8.54	31.34	111.73	67.72	49.43	13.14	46.17	55.45
Livestock	27.43	17.81	22.35	51.76	86.61	57.25	76.48	35.67	26.20	48.60
Non-Ag. profit	260.29	293.63	212.95	304.83	254.71	197.39	338.22	171.49	292.70	262.92
Remittance	138.41	381.12	624.89	225.28	107.64	101.84	77.30	87.37	259.51	212.90
Employment	487.70	676.42	464.06	590.46	542.42	436.33	669.77	582.29	642.59	560.94
Other income	64.65	8.41	90.96	38.22	31.36	32.01	190.70	15.53	60.98	57.61
Total	1253.53	1525.04	1506.60	1436.53	1408.12	1117.77	1724.70	1152.23	1459.92	1398.71

B = Barisal, CH = Chittagong, CO = Comilla, D = Dhaka, K = Khulna, M = Mymensingh, RJ = Rajshahi, RN = Rangpur, S = Sylhet, BD = Bangladesh, Main crops = *Aus*, *Aman* and *Boro* rice and other crops = Wheat, Maize, Jute, Potato, Chili, Onion etc.

Table A2. Each agricultural crop's share in total net agricultural income (%), by region.

Crops	B	CH	CO	D	K	M	RJ	RN	S	BD
Rice	45.51	33.66	32.99	37.39	43.52	55.62	51.27	57.72	67.05	47.22
Aus	6.37	2.89	1.51	0.64	3.03	0.84	1.11	1.39	5.19	2.24
Aman	24.36	17.83	6.42	5.22	15.55	15.37	17.27	22.12	18.45	14.96
Boro	14.78	12.95	25.06	31.54	24.95	39.42	32.89	34.21	43.41	30.02
Wheat	0.00	0.00	0.19	0.22	0.70	0.07	1.32	0.96	0.00	0.48
Maize	0.00	0.00	0.84	0.30	0.26	0.00	1.40	2.01	0.00	0.56
Jute	0.61	0.00	3.03	10.53	5.85	0.44	2.80	2.96	0.11	4.37
Potato	0.66	0.37	5.49	0.53	0.18	0.36	4.04	4.68	1.00	1.62
Chili	1.82	2.17	2.69	6.85	5.72	1.54	0.67	1.20	0.53	3.40
Onion	0.00	0.00	0.01	5.79	1.01	0.00	1.81	0.32	0.00	1.70
Other crops	51.39	63.80	54.77	38.38	42.76	41.96	36.67	30.16	31.31	40.65
Total	100	100	100	100	100	100	100	100	100	100

B = Barisal, CH = Chittagong, CO = Comilla, D = Dhaka, K = Khulna, M = Mymensingh, RJ = Rajshahi, RN = Rangpur, S = Sylhet and BD = Bangladesh.

Table A3. Costs and income (US$/ha) associated with aman and boro HYV rice production, by region.

	Aman HYV										*Boro* HYV									
	B	CH	CO	D	K	M	RJ	RN	S	BD	B	CH	CO	D	K	M	RJ	RN	S	BD
b	53.77	74.83	76.13	53.84	30.12	45.34	46.93	38.08	57.28	47.08	64.59	82.41	75.69	50.94	32.39	45.13	49.55	42.71	54.49	49.14
c	64.29	38.14	72.11	79.90	64.27	33.96	45.27	30.77	45.13	47.80	60.51	46.47	70.82	71.22	66.36	42.49	73.53	40.41	43.01	58.24
d	1.33	4.58	8.04	34.37	11.12	27.00	37.48	12.43	5.65	19.52	63.70	60.16	135.28	165.63	114.87	122.83	116.16	93.95	61.48	113.42
e	1.19	1.54	2.87	1.55	1.73	2.81	7.00	2.45	3.84	3.22	2.40	5.36	9.24	4.22	10.59	8.17	8.65	25.41	1.92	7.98
f	5.98	11.33	8.48	6.22	3.34	9.16	9.36	9.31	4.81	7.49	14.01	14.25	13.96	7.34	9.24	13.41	11.12	13.73	3.65	9.72
g	26.33	45.58	60.39	50.65	40.65	63.05	49.46	50.75	27.61	47.88	59.67	92.28	92.46	90.84	97.05	106.66	73.24	107.18	45.80	84.34
h	9.08	0.61	0.43	0.67	2.57	3.96	1.61	1.84	6.60	3.22	1.55	1.58	0.30	1.02	2.82	5.72	2.55	2.06	5.54	3.05
i	26.59	43.06	37.71	33.86	25.06	25.65	22.36	26.46	31.04	27.43	42.48	46.83	36.75	33.65	28.94	26.73	21.83	30.41	25.77	29.51
j	17.58	17.31	9.34	6.11	9.51	8.45	4.04	3.36	5.89	7.64	15.92	29.29	14.55	16.23	19.54	10.05	5.96	9.59	4.27	11.99
k	85.80	155.19	133.77	171.81	113.27	115.80	134.27	106.25	107.67	120.55	152.40	305.40	237.84	242.40	151.19	157.81	190.60	125.47	227.20	192.16
TC	291.93	392.18	409.27	438.98	301.63	335.18	357.78	281.70	295.53	331.82	477.24	684.02	686.89	683.49	533.00	539.01	553.19	490.92	473.14	559.55
TP kg/ha	3573	3655	1913	3131	2515	2776	3650	3500	2572	3023	4659	4821	5136	6181	5122	4950	6025	5733	4218	5304
GI	734.65	710.58	387.39	614.66	477.69	577.30	661.75	669.42	476.78	585.58	841.58	964.00	999.64	1169.99	1009.64	1082.65	1115.88	1115.55	749.11	1023.34
GI-TC	442.72	318.40	−21.88	175.69	176.07	242.12	303.96	387.72	181.25	253.75	364.35	279.98	312.75	486.49	476.65	543.65	562.69	624.64	275.96	463.80

B = Barisal, CH = Chittagong, CO = Comilla, D = Dhaka, K = Khulna, M = Mymensingh, RJ = Rajshahi, RN = Rangpur, S = Sylhet and BD = Bangladesh; b = Rental cost of land, c = Seed cost, d = Irrigation cost, e = Manure/compost cost, f = Pesticide cost, g = Chemical fertilizer cost, h = Draft animal cost for land preparation, i = Rental cost for tools and machinery, j = Threshing cost, k = Hired labor cost, TC = Total cost, TP = Total production and GI = Gross income.

References

1. Schendel, W.V. *A History of Bangladesh*; Cambridge University Press: Cambridge, UK, 2009.
2. Ravallion, M. The Performance of Rice Markets in Bangladesh during the 1974 Famine. *Econ. J.* **1985**, *95*, 15–29. [CrossRef]
3. Gisella, K. *Long Run Impacts of Famine Exposure: A Study of the 1974–1975 Bangladesh Famine*; University of Colorado: Denver, CO, USA, 2012.
4. Dorosh, P.A.; Rashid, S. *Bangladesh Rice Trade and Price Stabilization*; IFPRI Discussion Paper 01209; International Food Policy Research Institute: Washington, DC, USA, 2012; 2p.
5. Hossain, A.; Silva, J.A.T. Wheat and rice, the epicenter of food security in Bangladesh. *Songklanakarin J. Sci. Technol.* **2013**, *35*, 261–274.
6. Alam, G.M.; Alam, K.; Mushtaq, S. Climate change perceptions and local adaptation strategies of hazard-prone rural households in Bangladesh. *Clim. Risk Manag.* **2017**, *17*, 52–63. [CrossRef]
7. Alamgir, M.S.; Furuya, J.; Kobayashi, S. Determinants of Early Cropping of Rice in Bangladesh: An Assessment as a Strategy of Avoiding Cyclone Risk. *Jpn. J. Agric. Econ.* **2017**, *19*. [CrossRef]
8. Israt, J.S.; Misuzu, T.N.; Mana, K.N.; Mohammad, S.H.; Yoshiaki, I. Rice Cultivation in Bangladesh: Present Scenario, Problems, and Prospects. *J. Int. Cooper Agric. Dev.* **2016**, *14*, 20–29.
9. Bell, A.R.; Bryan, E.; Ringler, C.; Ahmed, A. Rice productivity in Bangladesh: What are the benefits of irrigation? *Land Use Policy* **2015**, *48*, 1–12. [CrossRef]
10. Chen, Y.; Lu, C. A Comparative Analysis on Food Security in Bangladesh, India and Myanmar. *Sustainability* **2018**, *10*, 405. [CrossRef]
11. World Bank. *World Development Report 2013: Jobs*; World Bank: Washington, DC, USA, 2012; p. 197.
12. Poverty and Inequality in Bangladesh. *Journey towards Progress (2014–2015)*; Macroeconomic Wing, Finance division, Ministry of Finance, Government of the People's Republic of Bangladesh: Dhaka, Bangladesh, 2015.
13. Sulaiman, M.; Misha, F. *Comparative Cost–Benefit Analysis of Programs for the Ultra-Poor in Bangladesh, Bangladesh Priorities*; Copenhagen Consensus Center: Copenhagen, Denmark, 2016; 29p.
14. Hijioka, Y.; Lin, E.; Pereira, J.J.; Corlett, R.T.; Cui, X.; Insarov, G.E.; Lasco, R.D.; Lindgren, E.; Surjan, A. *Climate Change 2014: Impacts, Adaptation, and Vulnerability. Part B: Regional Aspects. Contribution of Working Group II to the Fifth Assessment Report of the Intergovernmental Panel on Climate Change*; Barros, V.R., Field, C.B., Dokken, D.J., Mastrandrea, M.D., Mach, K.J., Bilir, T.E., Chatterjee, M., Ebi, K.L., Estrada, Y.O., Genova, R.C., et al., Eds.; Cambridge University Press: Cambridge,UK; New York, NY, USA, 2014; pp. 1327–1370.
15. Choudhury, A.M. Managing Natural Disasters in Bangladesh. Presented at the Dhaka Meet on Sustainable Development in Bangladesh: Achievements, Opportunities, and Challenges at Rio10, Bangladesh Unnayan Parishad, Dhaka, Bangladesh, 16–18 March 2002.
16. Shimi, A.C.; Parvin, G.R.; Biswas, C.; Shaw, R. Impact and adaptation to flood—A focus on water supply, sanitation and health problems of rural community in Bangladesh. *Disaster Prev. Manag.* **2010**, *19*, 298–313. [CrossRef]
17. World Bank. *Natural Disaster Hotspots: A Global Risk Analysis*; Disaster Risk Management Series. No. 5; World Bank: Washington, DC, USA, 2005.
18. World Bank. *Disaster Risk Management in South Asia: A Regional Overview*; World Bank: Washington, DC, USA, 2012.
19. Wassmann, R.; Jagadish, S.V.K.; Sumfleth, K.; Pathak, H.; Howell, G.; Ismail, A.; Serraj, R.; Redona, E.; Singh, R.K.; Heuer, S. Regional vulnerability of climate change impacts on Asian rice production and scope for adaptation. *Adv. Agron.* **2009**, *102*, 91–133.
20. World Bank. *Economics of Adaptation to Climate Change: Bangladesh*; World Bank: Washington, DC, USA, 2010; 79p.
21. Titumir, R.A.M.; Basak, J.K. Effects of climate change on crop production and climate adaptive Techniques for Agriculture in Bangladesh. *Soc. Sci. Rev.* **2012**, *29*, 215–232.
22. Ferreira, F.H.G.; Chen, S.; Dabalen, A.; Dikanov, Y.; Hamadeh, N.; Jolliffe, D.; Narayan, A.; Prydz, E.B.; Revenga, A.; Sangraula, P.; et al. *A Global Count of Extreme Poor in 2012: Data Issues, Methodologies and Initial Results*; World Bank: Washington, DC, USA, 2015.

23. Hasan, S.A. The impact of the 2005–2010 rice price increase on consumption in rural Bangladesh. *Agric. Econ.* **2016**, *47*, 423–433. [CrossRef]

24. El-Osta, H.S.; Morehart, M.J. Determinants of Poverty among U.S. Farm Households. *J. Agric. Appl. Econ.* **2008**, *40*, 1–20. [CrossRef]

25. Mat, S.H.C.; Jalil, A.Z.A.; Harun, M. Does Non-Farm Income Improve the Poverty and Income Inequality among Agricultural Household in Rural Kedah? *Proc. Econ. Financ.* **2012**, *1*, 269–275. [CrossRef]

26. Yamano, T.; Sserunkuuma, D.; Otsuka, K.; Omiat, G.; Ainembabazi, J.H.; Shimamura, Y. The 2003 REPEAT Survey in Uganda: Results, Foundation for Advanced Studies on International Development (FASID) and Makerere University. 2004. Available online: http://www3.grips.ac.jp/~globalcoe/j/data/repeat/REPEATinUgandaReport.pdf (accessed on 17 May 2018).

27. Janvry, A.D.; Sadoulet, E.; Zhu, N. *The Role of Non-Farm Incomes in Reducing Rural Poverty and Inequality in China*; CUDARE Working Papers; University of California: Berkeley, CA, USA, 2005; pp. 1–31.

28. Chaudhry, A.; Wimer, C. Poverty is Not Just an Indicator: The Relationship Between Income, Poverty, and Child Well-Being. *Acad. Pediatr.* **2016**, *16*, S23–S29. [CrossRef] [PubMed]

29. Gornall, J.; Richard, B.; Eleanor, B.; Robin, C.; Joanne, C.; Kate, W.; Andrew, W. Implications of climate change for agricultural productivity in the early twenty-first century. *Philos. Trans. R. Soc. B* **2010**, *365*, 2973–2989. [CrossRef] [PubMed]

30. GOB (Government of Bangladesh); UNDP (United Nations Development Program). *The Probable Impacts of Climate Change on Poverty and Economic Growth and Options of Coping with Adverse Effects of Climate Change in Bangladesh*; Policy Study: Dhaka, Bangladesh, 2009.

31. Sarker, M.A.R.; Alam, K.; Gow, J. Exploring the relationship between climate change and rice yield in Bangladesh: An analysis of time series data. *Agric. Syst.* **2012**, *112*, 11–16. [CrossRef]

32. FAO (Food and Agriculture Organization). *Livelihood Adaptation to Climate Variability and Changes in Drought-Prone Areas of Bangladesh*; Food and Agriculture Organization: Rome, Italy, 2006.

33. IPCC (Intergovernmental Panel on Climate Change). *Climate Change 2007: Impacts, Adaptation and Vulnerability: Contribution of Working Group II to the Fourth Assessment Report of the Intergovernmental Panel on Climate Change*; Cambridge University Press: Cambridge, UK, 2007.

34. Alauddin, M.; Hossain, M. *Environment and Agriculture in a Developing Economy: Problems and Prospects for Bangladesh*; Edward Elgar: London, UK, 2001.

35. FAO (Food and Agriculture Organization). Climate Change and Food Security: Risks and Responses. 2016. Available online: http://www.fao.org/3/a-i5188e.pdf (accessed on 27 April 2018).

36. UNDP. *Fighting Climate Change: Human Solidarity in a Divided World*; Human Development Report; Oxford University Press: Oxford, UK, 2008.

37. IFPRI (International Food Policy Research Institute). *Agriculture and Adaptation in Bangladesh, Current and Projected Impacts of Climate Change*; Discussion Paper 01281; International Food Policy Research Institute: Washington, DC, USA, 2013.

38. Islam, M.B.; Ali, M.Y.; Amin, M.; Zaman, S.M. *Climatic Variations: Farming Systems and Livelihoods in the High Barind Tract and Coastal Areas of Bangladesh*; Lal, R., Sivakumar, M., Faiz, S., Mustafizur, Rahman, A., Islam, K., Eds.; Climate Change and Food Security in South Asia; Springer: Dordrecht, The Netherlands, 2011.

39. Yu, W.H.; Alam, M.; Hassan, A.; Khan, A.S.; Ruane, A.C.; Rosenzweig, C.; Major, D.C.; Thurlow, J. *Climate Change Risk and Food Security in Bangladesh*; EarthScan: New York, NY, USA, 2010.

40. BBS (Bangladesh Bureau of Statistics). *Compendium of Environment Statistics of Bangladesh*; Government of the People's Republic of Bangladesh: Dhaka, Bangladesh, 2005.

41. Alauddin, M.; Tisdell, C. Trends and projections of Bangladeshi food production: An alternative viewpoint. *Food Policy* **1987**, *12*, 318–331. [CrossRef]

42. Alauddin, M.; Tisdell, C. *The 'Green Revolution' and Economic Development: The Process and Its Impact in Bangladesh*; Macmillan: London, UK, 1991.

43. Asaduzzaman, M.; Ringler, C.; Thurlow, J.; Alam, S. *Investing in Crop Agriculture in Bangladesh for Higher Growth and Productivity, and Adaptation to Climate Change*; Bangladesh Food Security Investment Forum: Dhaka, Bangladesh, 2010.

44. BBS. *Yearbook of Agricultural Statistics of Bangladesh*; Government of the People's Republic of Bangladesh: Dhaka, Bangladesh, 2009.

45. BER. *Bangladesh Economic Review, Government of the People's Republic of Bangladesh*; Ministry of Finance: Dhaka, Bangladesh, 2017.

46. Karim, Z.; Hussain, S.G.; Ahmed, M. Assessing Impact of climate Variation on food grain production in Bangladesh. *J. Water Air Soil Pollut.* **1996**, *92*, 53–62.

47. Battisti, D.S.; Naylor, R.L. Historical warnings of future food insecurity with unprecedented seasonal heat. *Science* **2009**, *323*, 240–244. [CrossRef] [PubMed]

48. Kobayashi, S.; Furuya, J. Comparison of climate change impacts on food security of Bangladesh. *Stud. Reg. Sci.* **2011**, *41*, 419–433. [CrossRef]

49. Salam, M.A.; Furuya, J.; Alamgir, M.S.; Kobayashi, S. *Policy Adaptation Cost for Mitigation of Price Variation of Rice under Climate Change in Bangladesh*; Center for Environmental Information Science: Tokyo, Japan, 2016; Volume 30, pp. 1–7.

50. Ruane, A.C.; Major, D.C.; Yu, W.H.; Alam, M.; Hussain, S.G.; Khan, A.S.; Rosenzweig, C. Multi-factor impact analysis of agricultural production in Bangladesh with climate change. *Glob. Environ. Chang.* **2013**, *23*, 338–350. [CrossRef]

51. Ali, A. Climate change impacts and adaptation assessment in Bangladesh. *Clim. Res.* **1999**, *12*, 109–116. [CrossRef]

52. Sarwar, M.G.M. Impacts of Sea Level Rise on the Coastal Zone of Bangladesh. Master's Thesis, Lund University, Lund, Sweden, 2005.

53. Abid, M.; Schilling, J.; Scheffran, J.; Zulfiqar, F. Climate change vulnerability, adaptation and risk perceptions at farm level in Punjab, Pakistan. *Sci. Total Environ.* **2016**, *547*, 447–460. [CrossRef] [PubMed]

54. Rajendra, P.S.; Nuanwan, C.; Sunsanee, A. Adaptation to Climate Change by Rural Ethnic Communities of Northern Thailand. *Climate* **2017**, *5*, 57. [CrossRef]

55. African Development Bank; Asian Development Bank; Department for International Development, United Kingdom; Directorate-General for Development, European Commission; Federal Ministry for Economic Cooperation and Development, Germany; Ministry of Foreign Affairs—Development Cooperation, The Netherlands; Organization for Economic Cooperation and Development; United Nations Development Programme; United Nations Environment Programme; and The World Bank: Poverty and Climate Change, Reducing the Vulnerability of the Poor through Adaptation. 2002. Available online: http://www.oecd.org/env/cc/2502872.pdf (accessed on 19 May 2018).

56. Ali, A.; Erenstein, O. Assessing farmer use of climate change adaptation practices and impacts on food security and poverty in Pakistan. *Clim. Risk Manag.* **2017**, *16*, 183–194. [CrossRef]

57. Menikea, L.M.C.S.; Arachchi, K.A.G.P. Adaptation to Climate Change by Smallholder Farmers in Rural Communities: Evidence from Sri Lanka. *Proc. Food Sci.* **2016**, *6*, 288–292. [CrossRef]

58. Leichenko, R.; Silva, J.A. Climate change and poverty: Vulnerability, impacts, and alleviation strategies. *WIREs Clim. Chang.* **2014**, *5*, 539–556. [CrossRef]

59. Lade, S.J.; Haider, L.J.; Engstrom, G.; Schulter, M. Resilience offers escape from trapped thinking on poverty alleviation. *Sci. Adv.* **2017**, *3*, e1603043. [CrossRef] [PubMed]

60. Ahmed, A.U.; Alam, M. *Development of Climate Change Scenarios with General Circulation Models*; Huq, S., Karim, Z., Asaduzzaman, M., Mahtabs, F., Eds.; Vulnerability and Adaptation to Climate Change for Bangladesh; Kluwer Academic Publishers: Dordrecht, The Netherlands, 1998; pp. 13–20.

61. Agrawala, S.; Ota, T.; Ahmed, A.U.; Smith, J.; van Aalst, M. *Development and Climate Change in Bangladesh: Focus on Coastal Flooding and the Sundarbans*; Organisation for Economic Co-Operation and Development: Paris, France, 2003.

62. Tanner, T.M.; Hassan, A.; Islam, K.M.N.; Conway, D.; Mechler, R.; Ahmed, A.U.; Alam, M. *ORCHID: Piloting Climate Risk Screening in DFID Bangladesh*; Detailed Research Report; Institute of Development Studies, University of Sussex: Brighton, UK, 2007.

63. CEGIS (Center for Environmental and Geographic Information Services). *Vulnerability to Climate Induced Drought: Scenario and Impacts*; Center for Environmental and Geographic Information Services: Dhaka, Bangladesh, 2013.

64. Sikder, R.; Xiaoying, J. Climate Change Impact and Agriculture in Bangladesh. *J. Environ. Earth Sci.* **2014**, *4*, 35–40.

65. Karim, M.R.; Ishikawa, M.; Ikeda, M.; Islam, M.T. Climate Change model predicts 33% rice yield decrease in 2100 in Bangladesh. *Agron. Sustain. Dev.* **2012**, *32*, 821–830. [CrossRef]

66. Key, N.; Prager, D.; Burns, C. *Farm Household Income Volatility: An Analysis Using Panel Data from a National Survey*; Economic Research Report, No.26; United States Department of Agriculture, Economic Research Service: Washington, DC, USA, 2017.
67. Li, Y.; Wu, W.; Ge, Q.; Zhou, Y.; Xu, C. Simulating Climate Change Impacts and Adaptive Measures for Rice Cultivation in Hunan Province, China. *Am. Metrol. Soc.* **2016**, 1359–1376. [CrossRef]
68. Nazarenko, L.; Schmidt, G.A.; Miller, R.L.; Tausnev, N.; Kelley, M.; Ruedy, R.; Russell, G.L.; Aleinov, I.; Bauer, M.; Bauer, S.; et al. Future climate change under RCP emission scenarios with GISS ModelE2. *J. Adv. Model. Earth Syst.* **2015**, *7*, 244–267. [CrossRef]
69. BBS. *Yearbook of Agricultural Statistics*; Government of the People's Republic of Bangladesh: Dhaka, Bangladesh, 2016.
70. Quddus, M.A. Crop production growth in different agro-ecological zones of Bangladesh. *J. Bangladesh Agric. Univ.* **2009**, *7*, 351–360. [CrossRef]
71. Department of Agricultural Marketing. *The Average Nationwide Retail Price of Selected Commodities in Bangladesh from 1 December 2010 to 30 November 2011*; Government of the People's Republic of Bangladesh, Ministry of Agriculture: Dhaka, Bangladesh, 2017.
72. Schmit, T.M.; Boisvert, R.N.; Tauer, L.W. Measuring the Financial Risks of New York Dairy Producers. *J. Dairy Sci.* **2001**, *84*, 411–420. [CrossRef]
73. Chang, H.-H.; Schmit, T.M.; Boisvert, R.N.; Tauer, L.W. *Quantifying Sources of Dairy Farm Business Risk and Implications for Risk Management Strategies*; Working Paper WP 2007-11, July 2007; Department of Applied Economics and Management, Cornell University: Ithaca, NY, USA, 2007.
74. Wolf, C.A.; Black, J.R.; Hadrich, J.C. Upper Midwest dairy farm revenue variation and insurance implications. *Agric. Financ. Rev.* **2009**, *69*, 346–358. [CrossRef]
75. Benni, E.; Finfer, R. Where is the risk? Price, yield and cost risk in swiss crop production. In Proceedings of the Selected Paper, IAAE, Triennial Conference, Foz do Iguacu, Brazil, 18–24 August 2012.
76. Galton, F. The geometric mean, in vital and social statistics. *Proc. R. Soc.* **1879**, *29*, 365–367. [CrossRef]
77. McAlister, D. The law of the geometric mean. *Proc. R. Soc.* **1879**, *29*, 367–376. [CrossRef]
78. Gaddum, J.H. Log normal distributions. *Nature* **1945**, *156*, 746–747. [CrossRef]
79. Eckhard, L.; Werner, A.S.; Markus, A. Log-normal Distributions across the Sciences: Keys and Clues. *BioScience* **2001**, *51*, 341–352.
80. Arata, Y. *Income Distribution among Individuals: The Effects of Economic Interactions*; RIETI Discussion Paper Series 13-E-042; RIETI: Tokyo, Japan, 2013.
81. World Bank. *Bangladesh: Growing the Economy through Advances in Agriculture*; World Bank: Washington, DC, USA, 2016.
82. World Bank. *Bangladesh-Household Income and Expenditure Survey: Key Findings and Results 2010*; World Bank: Washington, DC, USA, 2011; pp. 153–154.
83. OECD. *Trend and Agriculture Directorate, Agriculture and Climate Change*; OECD Publishing: Paris, France, 2015.
84. UNDP. *Policy Study on the Probable Impacts of Climate Change on Poverty and Economic Growth and the Options of Coping with Adverse Effect of Climate Change in Bangladesh*; General Economics Division, Planning Commission, Government of the People's Republic of Bangladesh & UNDP Bangladesh: Dhaka, Bangladesh, 2009.
85. Dawe, D.; Moya, P.; Valencia, S. Institutional, policy and farmer responses to drought: El Niño events and rice in the Philippines. *Disasters* **2008**, *33*, 291–307. [CrossRef] [PubMed]
86. Douglas, I. Climate change, flooding and food security in south Asia. *Food Secur.* **2009**, *1*, 127–136. [CrossRef]
87. Kelkar, U.; Narula, K.K.; Sharma, V.P.; Chandna, U. Vulnerability and adaptation to climate variability and water stress in Uttarakhand State. *India Glob. Environ. Chang.* **2008**, *18*, 564–574. [CrossRef]
88. Skoufias, E.; Rabassa, M.; Olivieri, S. *The Poverty Impacts of Climate Change: A Review of the Evidence. Policy Research Working Paper 5622, Poverty Reduction and Equity Unit, Poverty Reduction and Economic Management Network*; The World Bank: Washington, DC, USA, 2010; 35p.
89. Hertel, T.W.; Burke, M.B.; Lobell, D.B. The poverty implications of climate induced crop yield changes by 2030. *Glob. Environ. Chang.* **2010**, *20*, 577–585. [CrossRef]

90. Rosegrant, M.W. Impacts of climate change on food security and livelihoods. In *Food Security and Climate Change in Dry Areas: Proceedings of an International Conference, Amman, Jordan, 1–4 February 2010*; Solh, M., Saxena, M.C., Eds.; International Center for Agricultural Research in the Dry Areas (ICARDA): Aleppo, Syria, 2011; pp. 24–26.

91. IPCC. *Climate Change 2013: The Physical Science Basis. Contribution of Working Group I to the Fifth Assessment Report of the Intergovernmental Panel on Climate Change*; Stocker, T.F., Qin, D., Plattner, G.-K., Tignor, M., Allen, S.K., Boschung, J., Nauels, A., Xia, Y., Bex, V., Midgley, P.M., Eds.; Cambridge University Press: Cambridge, UK; New York, NY, USA, 2013; 1535p. [CrossRef]

92. Basak, J.K.; Ali, M.A.; Islam, M.N.; Rashid, M.A. Assessment of the effect of climate change on boro rice production in Bangladesh using DSSAT model. *J. Civ. Eng.* **2010**, *38*, 95–108.

93. IPCC. *Climate Change 2007: Projections of Future Changes in Climate: Contribution of Working Group I to the Fourth Assessment Report of the Intergovernmental Panel on Climate Change*; Cambridge University Press: Cambridge, UK, 2007.

94. IPCC. *Climate Change 2014: Impacts, Adaptation, and Vulnerability. Contribution of Working Group II to the Fifth Assessment Report of the Intergovernmental Panel on Climate Change*; Field, C.B., Barros, V.R., Dokken, D.J., Mach, K.J., Mastrandrea, M.D., Bilir, T.E., Chatterjee, M., Ebi, K.L., Estrada, Y.O., Genova, R.C., Eds.; Cambridge University Press: Cambridge, UK; New York, NY, USA, 2014.

95. IPCC. *Climate Change 2014: Mitigation of Climate Change. Contribution of Working Group III to the Fifth Assessment Report of the Intergovernmental Panel on Climate Change*; Edenhofer, O., Pichs-Madruga, R., Sokona, Y., Farahani, E., Kadner, S., Seyboth, K., Adler, A., Baum, I., Brunner, S., Eickemeier, P., Eds.; Cambridge University Press: Cambridge, UK; New York, NY, USA, 2014.

96. Liu, S.; Mo, X.; Lin, Z.; Xu, Y.; Ji, J.; Wen, G.; Richey, J. Crop yield responses to climate change in the Huang-Huai-Hai Plain of China. *Agric. Water Manag.* **2010**, *97*, 1195–1209. [CrossRef]

97. Reid, S.; Smit, B.; Caldwell, W.; Belliveau, S. Vulnerability and adaptation to climate risks in Ontario agriculture. *Mitig. Adapt. Strateg. Glob. Chang.* **2007**, *12*, 609–637. [CrossRef]

98. Rogelj, J.; Perette, M.; Menon, A.; Schleussner, C.F.; Bondeau, A.; Svirejeva-Hopkins, A.; Schewe, J.; Frieler, K.; Warszawski, L.; Rocha, M. *Turn Down the Heat: Climate Extremes, Regional Impacts, and the Case for Resilience*; World Bank: Washington, DC, USA, 2013.

99. IRRI. Climate Change-Ready Rice. Available online: http://irri.org/our-impact/tackling-climate-change/developing-drought-tolerant-rice (accessed on 11 January 2018).

100. Khatun, M. Climate Change and Migration in Bangladesh: Golden Bengal to Land of Disasters. *Bangladesh e-J. Soc.* **2013**, *10*, 64–79.

101. Rahman, A.; Alam, M.; Alam, S.S.; Uzzaman, M.R.; Rashid, M.; Rabbani, G. *Risks, Vulnerability, and Adaptation in Bangladesh*; Human Development Report 2007/08, Human Development Report Office Occasional Paper, 2007/13; UNDP: New York, NY, USA, 2008.

102. Paul, B.K. Coping mechanism practiced by drought victims (1994/5) in North Bengal, Bangladesh. *Appl. Geogr.* **1998**, *18*, 355–373. [CrossRef]

103. Shahid, S.; Behrawan, H. Drought risk assessment in the western part of Bangladesh. *Nat. Hazards* **2008**, *46*, 391–413. [CrossRef]

104. Habiba, U.; Shaw, R.; Takeuchi, Y. Chapter 2 socioeconomic impact of droughts in Bangladesh, Droughts in Asian Monsoon Region Community. *Environ. Disaster Risk Manag.* **2011**, *8*, 25–48.

105. Rahman, A.; Roytman, L.; Krakauer, N.Y.; Nizamuddin, M.; Goldberg, M. Use of Vegetation Health Data for Estimation of *Aus* Rice Yield in Bangladesh. *Sensors* **2009**, *9*, 2968–2975. [CrossRef] [PubMed]

106. Alam, M.S.; Quayum, M.A.; Islam, M.A. Crop Production in the Haor Areas of Bangladesh: Insights from Farm Level Survey. *Agriculturists* **2010**, *8*, 88–97. [CrossRef]

107. Huda, M.K. Experience with modern and hybrid rice varieties in haor ecosystem: Emerging Technologies for Sustainable Rice Production. In *Twentieth National Workshop on Rice Research and Extension in Bangladesh*; Bangladesh Rice Research Institute: Gazipur, Bangladesh, 2004; pp. 19–21.

108. Needs Assessment Report. *Flooding in the North-East Bangladesh*; Rapid Repose Team, Shifting the Power, Christian Aid and NAHAB: Dhaka, Bangladesh, 2017; pp. 1–8.

109. Litchfield William Alex. Climate Change Induced Extreme Weather Events & Sea Level Rise in Bangladesh Leading to Migration and Conflict, ICE Case Studies, Number 229; 2010. Available online: http://mandalaprojects.com/ice/ice-cases/bangladesh.htm (accessed on 17 January 2018).

110. BBS. *Bangladesh Population and Housing Census*; Bangladesh Bureau of Statistics: Dhaka, Bangladesh, 2011.
111. Haque, S.A. Salinity Problems and Crop Production in Coastal Regions of Bangladesh. *Pakistan J. Bot.* **2006**, *38*, 1359–1365.
112. Akter, T. Climate Change and Flow of Environmental Displacement in Bangladesh. Unnayan Onneshan-The Innovators. 2009. Available online: www.unnayan.org (accessed on 30 April 2018).
113. Nicholls, R.J.; Leatherman, S.P. The implications of accelerated sea-level rise and developing countries: A discussion. *J. Coast. Res.* **1995**, 303–323.
114. Hossain, M.A. Global Warming Induced Sea Level Rise on Soil, Land and Crop Production Loss in Bangladesh. In Proceedings of the 19th World Congress of Soil Science, Soil Solutions for a Changing World, Brisbane, Australia, 1–6 August 2010.
115. Minar, M.H.; Hossain, M.B.; Shamsuddin, M.D. Climate Change and Coastal Zone of Bangladesh: Vulnerability, Resilience and Adaptability. *Middle-East J. Sci. Res.* **2013**, *13*, 114–120. [CrossRef]
116. Xu, H.; Twine, T.E.; Girvetz, E. Climate change and maize yield in Iowa. *PLoS ONE* **2016**, *11*, e0156083. [CrossRef] [PubMed]
117. Kabir, H.M.; Ahmed, Z.; Khan, R. Impact of climate change on food security in Bangladesh. *J. Pet. Environ. Biotechnol.* **2016**, *7*, 306. [CrossRef]

Article

The Value of Tactical Adaptation to El Niño–Southern Oscillation for East Australian Wheat

Bangyou Zheng [1], Scott Chapman [1,2] and Karine Chenu [3,*

[1]   CSIRO Agriculture and Food, Queensland Bioscience Precinct, 306 Carmody Road,
     St. Lucia, QLD 4067, Australia; bangyou.zheng@csiro.au (B.Z.); scott.chapman@csiro.au (S.C.)
[2]   School of Agriculture and Food Sciences, The University of Queensland, Building 8117A NRSM,
     Gatton, QLD 4343, Australia
[3]   The University of Queensland, Queensland Alliance for Agriculture and Food Innovation (QAAFI),
     203 Tor Street, Toowoomba, QLD 4350, Australia
*   Correspondence: Karine.Chenu@uq.edu.au; Tel.: +61-(0)7-4688-1357

Received: 6 August 2018; Accepted: 8 September 2018; Published: 11 September 2018

Abstract: El Niño–Southern Oscillation strongly influences rainfall and temperature patterns in Eastern Australia, with major impacts on frost, heat, and drought stresses, and potential consequences for wheat production. Wheat phenology is a key factor to adapt to the risk of frost, heat, and drought stresses in the Australian wheatbelt. This study explores broad and specific options to adapt wheat cropping systems to El Niño–Southern Oscillation, and more specifically, to the Southern Oscillation Index (SOI) phases ahead of the season (i.e., April forecast) in Eastern Australia, when wheat producers make their most crucial management decisions. Crop model simulations were performed for commercially-grown wheat varieties, as well as for virtual genotypes representing possible combinations of phenology alleles that are currently present in the Australian wheat germplasm pool. Different adaptation strategies were tested at the site level, across Eastern Australia, for a wide range of sowing dates and nitrogen applications over long-term historical weather records (1900–2016). The results highlight that a fixed adaptation system, with genotype maturities, sowing time, and nitrogen application adapted to each location would greatly increase wheat productivity compared to sowing a mid-maturity genotype, mid-season, using current practices for nitrogen applications. Tactical adaptation of both genotype and management to the different SOI phases and to different levels of initial Plant Available Water ('PAW & SOI adaptation') resulted in further yield improvement. Site long-term increases in yield and gross margin were up to 1.15 t·ha^{-1} and AU\$ 223.0 ha^{-1} for fixed adaptation (0.78 t·ha^{-1} and AU\$ 153 ha^{-1} on average across the whole region), and up to an extra 0.26 t·ha^{-1} and AU\$ 63.9 ha^{-1} for tactical adaptation. For the whole eastern region, these results correspond to an annual AU\$ 440 M increase for the fixed adaptation, and an extra AU\$ 188 M for the PAW & SOI tactical adaptation. The benefits of PAW & SOI tactical adaptation could be useful for growers to adjust farm management practices according to pre-sowing seasonal conditions and the seasonal climate forecast.

Keywords: ENSO; Southern Oscillation Index; SOI; El Niño; La Niña; soil water; environment type; climate adaptation; management practices; crop model; APSIM

---

1. Introduction

In Australia, the wheat industry is challenged by complex genotype x environment x management (GxExM) interactions [1,2], due in part to the high spatial and temporal variability of the Australian climate (e.g., [3]). In the eastern part of the continent, annual variations in temperature and rainfall that are influenced by El Niño–Southern Oscillation (ENSO) [1,4,5] affect frost, heat, and drought stress patterns [5–7], and ultimately, wheat production [1,3,5]. Drought and warmer temperatures, but also greater frost risk due to the clear night sky, are generally associated with the onset of El Niño episodes [5,8,9], and limit grain yield [10–13]. Stronger ENSO climate oscillations are expected in the near future, as climate forecasts project more frequent extreme El-Niño and La-Niña conditions [14,15].

As the major driver of inter-annual climate variability in Eastern Australian [4,5,16], ENSO is a quasiperiodic climate pattern that occurs across the tropical Pacific Ocean every 3–8 years. It is caused by variations in the surface temperature of the tropical eastern Pacific Ocean, and the air surface pressure in the tropical western Pacific [17]. The Southern Oscillation Index (SOI), as measured by surface pressure anomaly difference between Tahiti and Darwin, has been used to investigate ENSO effects on crops. Five SOI phases have been defined through grouping all sequential two-month pairs of the SOI into five clusters, using principal component analysis and a cluster analysis [18]. Hammer et al. [1] found that using the 5-phase SOI classification (based on SOI values for the current and previous month) could significantly increase wheat profits (up to 20%) and decrease failure risk (up to 35% less risk) in Goondiwindi, South-Eastern Queensland, Australia, through adapting wheat cultivars and nitrogen fertiliser.

Strategies for yield improvement include breeding new cultivars and adapting management practices to the target population of environments [19]. Climate forecasting offers new opportunities in terms of agricultural planning and operation [4]. In the Australian broad-acre dryland wheat production area, most major decisions occur prior to sowing. Producers can potentially react to early indicators of upcoming rainfall and temperature. Early estimation of SOI phases can thus help farmers adjust management practices such as which cultivar to sow, when to sow, and what nitrogen fertilisation to apply [5,20,21].

In Eastern Australia, wheat crops rely heavily on soil-stored plant available water (PAW) [6,22]. An appropriate combination of sowing data, variety maturity, and pre-sowing PAW is crucial to allow flowering and grain filling to occur with minimal stress, in particular frost, heat, and drought stress, and thus, to maximise yield potential [6,7,23–25]. In this context, crop modelling can assist farmers to adapt their practices to specific SOI phases through adequate choice of maturity type and sowing date, in order to get extra benefit and increased profit [26].

The aims of this paper were to determine the values of (i) fixed adaptation (no distinction between the years) and (ii) adaptations to specific pre-sowing plant available water (PAW) and/or SOI phase. In this study, adaptation strategies were defined in terms of sowing, maturity type, and nitrogen fertilisation, to target the greatest long-term productivity at each site. The APSIM crop model [27], together with a phenology model [28], frost impact module [12] and heat impact module [10], were used to predict flowering time and yield of wheat, and search for the best long-term adaptation strategies.

2. Materials and Methods

2.1. Climatic Data

Fifteen weather stations representing local pedo-climatic conditions from the East Australian wheatbelt [7,22] were selected (see Chenu et al., 2013 [22] for more details) to compare adaptation options to Southern Oscillation Index (SOI) phases (Figure 1, Table 1).

The SOI, which corresponds to differences in sea level pressure between Tahiti and Darwin, has been classified in five phases [18]: 'consistently negative' (I), 'consistently positive' (II), 'rapidly falling' (III), 'rapidly rising' (IV) and 'consistently near zero' (V). These phases were grouped into three classes: consistently negative and rapidly falling (SOI phases I & III), consistently positive and rapidly rising (SOI phases II & IV), and consistently near zero (SOI phase V), as suggested by Potgieter et al. [3]. Weather records from 1900 to 2016 (117 years) were extracted from the Australian weather database (SILO Patched Point Dataset [29]; http://www.longpaddock.qld.gov.au/silo/). The SOI phase classification was sourced from Seasonal Climate Outlook in the Long Paddock (http://www.longpaddock.qld.gov.au/). In this study, the SOI phases were classified using SOI values from March–April to look at the effects for a pre-season indicator. From 1900 to 2016, 36 years had been classified as SOI phases I & III, 45 years as SOI phases II & IV and 34 years as SOI phase IV.

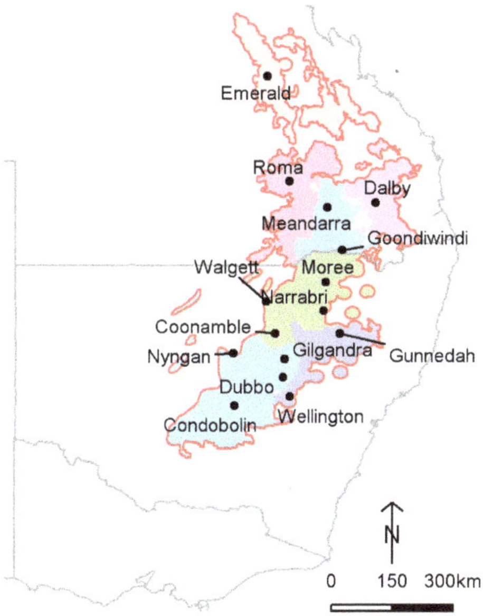

Figure 1. Map of the seven regions of the East Australian wheatbelt, with 15 sites chosen to represent those regions. Details on the locations can be found in Chenu et al., 2013 [22].

Table 1. Regions, locations, soil nitrogen at sowing and nitrogen fertilisation (in the baseline simulations), minimum pant available water (PAW) at sowing chosen to represent the East Australian wheatbelt. Initial and applied nitrogen (N) is indicated by 'x-y-z-a': x, initial N present in the soil at sowing; y, N applied at sowing as urea; z and a, N applied as nitrate at the stages 'beginning of stem elongation' and 'mid-stem elongation', respectively.

Region	Location	Lat.	Long.	Nitrogen (kg ha^{-1})	Minimum PAW at Sowing (mm)
Central Queensland	Emerald	−23.53	148.16	30-50-0-0	80
Eastern Darling Downs	Dalby	−27.18	151.26	30-130-0-0	80
Eastern NSW	Gunnedah	−30.98	150.25	50-70-60 *-0	80
	Wellington	−32.80	148.80	50-50-50 †-0	50
Northern NSW	Moree	−29.48	149.84	30-80-0-0	80
	Walgett	−30.04	148.12	30-80-0-0	80
	Narrabri	−30.32	149.78	30-130-0-0	80
	Coonamble	−30.98	148.38	50-70-60 †-0	50
Southern West Queensland	Roma	−26.57	148.79	30-50-0-0	80
Western Darling Downs	Meandarra	−27.32	149.88	30-80-0-0	80
	Goondiwindi	−28.55	150.31	30-80-0-0	80
Western NSW	Nyngan	−31.55	147.2	50-60-60 †-0	80
	Gilgandra	−31.71	148.66	50-50-50 †-0	50
	Dubbo	−32.24	148.61	50-50-50 †-0	50
	Condobolin	−33.07	147.23	50-60-60 †-0	80

* >80 mm of rainfall from sowing to the stage "end of tillering–beginning of stem elongation". † >100 mm of rainfall from sowing to the stage "end of tillering–beginning of stem elongation".

Monthly temperature and cumulated rainfall were calculated as the average for each month from 1900 to 2016. Daily minimum and maximum temperatures were used to determine occurrences of frost and heat events.

2.2. Crop Simulations and Gross Margins

Wheat yield (dry weight without moisture content) was simulated for the 15 sites (Figure 1, Table 1) from 1900 to 2016. The simulations were performed with the APSIM 7.5 model [27,30], which has been widely tested for wheat across Eastern Australia (e.g., Chenu et al., 2011; Holzworth et al., 2014; Christopher et al., 2016 [6,27,31]; http://www.apsim.info/APSIM.Validation/Main.aspx), and a wheat-phenology gene-based module [28], a heat impact module [10], and a frost impact module with a frost-stress threshold of $-2\,^{\circ}$C [12].

For each site and year, the simulations were begun with a summer fallow starting from 1 November with a soil containing 20% of its potential available soil water capacity (PAWC). Wheat crops were sown at two-day intervals within a fixed sowing window from the 1 April to 30 June for all 15 sites, when the soil held enough plant available water (PAW) at sowing (Table 1). Soil nitrogen and surface organic matter were reset at sowing. The base nitrogen fertilisation was chosen to reflect local farming practices, and therefore, varied with site and seasonal rainfall, as defined in Chenu et al., 2013 [22] (Table 1). Plants were grown at a density of 100 plants per m^2. Seasons with not enough soil water on 1 April (i.e., when management options were chosen for the tactical adaptation scenarios, see below) were excluded from the analysis.

Different management strategies were tested with a range of sowing dates (sowing every two days from 1 April to 30 June) and nitrogen applications. An extra 0 to 140 kg·ha^{-1} (at 20 kg·ha^{-1} interval) of nitrogen was applied to the base simulations. Nitrogen fertilisation was applied at the same stage(s) as in the base simulations (i.e., local farming practices) with the same proportions, i.e., at sowing and/or 'beginning of stem elongation' depending on the seasonal opportunities.

Simulations were performed for 208 genotypes including commercial varieties and virtual genotypes that could potentially be bred based on the flowering alleles present in the Australian germplasm pool (see Zheng et al., 2013 for details [28]). Virtual genotypes were created including all combinations of *VRN-A1*, *VRN-B1*, *VRN-D1*, and *PPD-D1* genes (two alleles for each gene), and the full range of values of additional thermal time requirement from floral to flowering (from 425 to 1025 $^{\circ}$Cd [28]). Genotypes with the same phenology (from different allelic combinations) were disregarded, so that a total of 156 genotypes unique for their phenology were considered. Overall, the selected genotypes had APSIM parameters ranging from 0 to 1.2 for the photoperiod sensitivity (0.6 for the reference genotype Janz), 0.9 to 1.7 for the vernalisation sensitivity (0.9 for Janz), and 425 to 1025 $^{\circ}$Cd for the additional thermal time requirement from floral to flowering (675 $^{\circ}$Cd for Janz).

Odd and even years were first simulated separately as some crops matured after 1 November (date of the simulation initialisation). Odd- and even-year simulations were then merged together. Overall, 800 thousand simulations were run through the CSIRO HTCondor service using ClusterRun platform with the runs being completed in less than 4 h [32].

The gross margin was estimated for each simulation based on wheat and nitrogen prices. Other costs of wheat production were ignored, as only variations in gross margin were considered in this study (i.e., only the fertilisation costs varied among the tested management options). The wheat and nitrogen (as urea) prices were sourced from Australian Commodity Statistics [33], and calculated as median values from 2003 to 2012 (i.e., AU\$ 269 and AU\$ 547 per tonne for wheat and urea, respectively). Variation in grain quality was not considered in this study as the APSIM-wheat model is currently not able to accurately simulate changes in wheat protein content. Increase in gross margin at each site was multiplied with the planting area of the considered region (averaged data from 1975 to 2000, 2004 and 2006; source: Australian Bureau of Statistics), and all regional values were summed to obtain the total increase in gross margin for the Eastern region. Hence, gross margin estimations did not account for changes in fertilizer prices, changes in planting area, changes in wheat prices related to the harvested grain quality, nor changes in wheat prices related to fluctuations in the domestic and/or global market.

2.3. Fixed and Tactical Adaptation Options

Different scenarios exploring the GxExM interactions were evaluated to test their values for different pre-sowing levels of soil water and/or for the three different SOI classes at each studied site. The acceptable range of soil PAW (from minimum required PAW at sowing (Table 1) to the PAWC) calculated at 1 April was divided into three groups (0–33%, 33–67%, 67–100%) to represent Low, Median and High pre-sowing water levels. The 156 genotypes unique for their phenology were considered, as well as sowing dates from 1 April to 30 June and different nitrogen fertilisation options (see previous section).

To provide a reference against conventional practice, a baseline scenario was defined. In this scenario, the reference cultivar Janz was simulated from 1900 to 2016 for a sowing at 21 May using standard farmer practices for fertilisation (Table 1) [22]. A strategic 'fixed scenario' that considers the best management and best genotype for all years (in terms of highest average yield) at each location ('Fixed adaptation') was used as benchmark for best long-term practices (1900–2016). To investigate the potential tactical advantages of adapting, whereby a grower would modify planting decisions based on the SOI phases and/or soil PAW prior to sowing, scenarios with optimised genotypes and management practices specific to SOI classes and/or PAW groups were defined (at each location) through maximizing the average yield for crops grown within each SOI class and/or PAW group. For each location, the tactical adaptation scenarios consisted in either (1) the 'PAW' scenario, which considered the best overall genotype and management within situations from each PAW group, (2) the 'SOI' scenario, which considered the best overall genotype and management within situations from each SOI class, or (3) the 'PAW & SOI' scenario where both genotype and management were optimised for each combination of PAW group × SOI class. For the different scenarios, yield differences and changes in gross margin compared to the baseline and fixed adaptation scenarios were calculated for each year.

3. Results and Discussion

3.1. SOI Impacts on Seasonal Temperature and Rainfall

In the Eastern wheatbelt, SOI phases from the end of summer (calculated in March–April) were typically associated with temperature and cumulated rainfall recorded during the 'summer' fallow preceding the wheat crop (November to April). Higher temperatures were recorded for years with consistently negative SOI (phase I), and to a lesser extent, for years with rapidly falling SOI (III), while lower temperatures occurred in years with consistently positive SOI (II), and to a lesser extent, in years with rapidly increasing SOI (IV) (data not shown, Figure 2A and Figures S1–S3). For instance, the temperature in years from SOI phase I was up to 1.5 °C higher than the 'all years' data in February and March in Emerald, Roma and Gunnedah (data not shown). By contrast, summer rainfall tended to be lowest for SOI phases I & III, and highest for SOI phases II & IV (Figure 2B and Figure S7). As wheat crops in the Eastern wheatbelt heavily rely on soil-stored plant available water (PAW) [6,22], this implies that differences observed in summer rainfall for the different SOI phases are likely to impact crop water-stress pattern and yield, and also the type of genotype and management best suited for specific adaptation.

The impacts of SOI phases (calculated in March–April) for the upcoming 'winter' season (May to October) were weaker than climate variations observed during the previous 'summer' (Figure 2 and Figures S1–S8). In 'winter', differences in temperatures were forecasted for most sites with a tendency for greater differences in the northern sites (e.g., Emerald, Roma, and Meandarra), with highest temperatures for SOI phases I & III, and lowest for SOI phases II & IV (Figure 2B and Figures S4–S6). In any case, monthly temperatures of any SOI phases differed by less than 0.5 °C compared to 'all years' (data not shown, Figure 2A and Figures S4–S6). The impact of SOI phases on rainfall was only visible for a few sites, and mainly for higher rainfall in SOI phase IV years (data not shown, Figure 2B and Figure S8). As found in previous studies (e.g., [34]), ENSO had a substantial impact in northern sites, while a relatively weak impact in southern sites (data not presented for sites south of Condobolin).

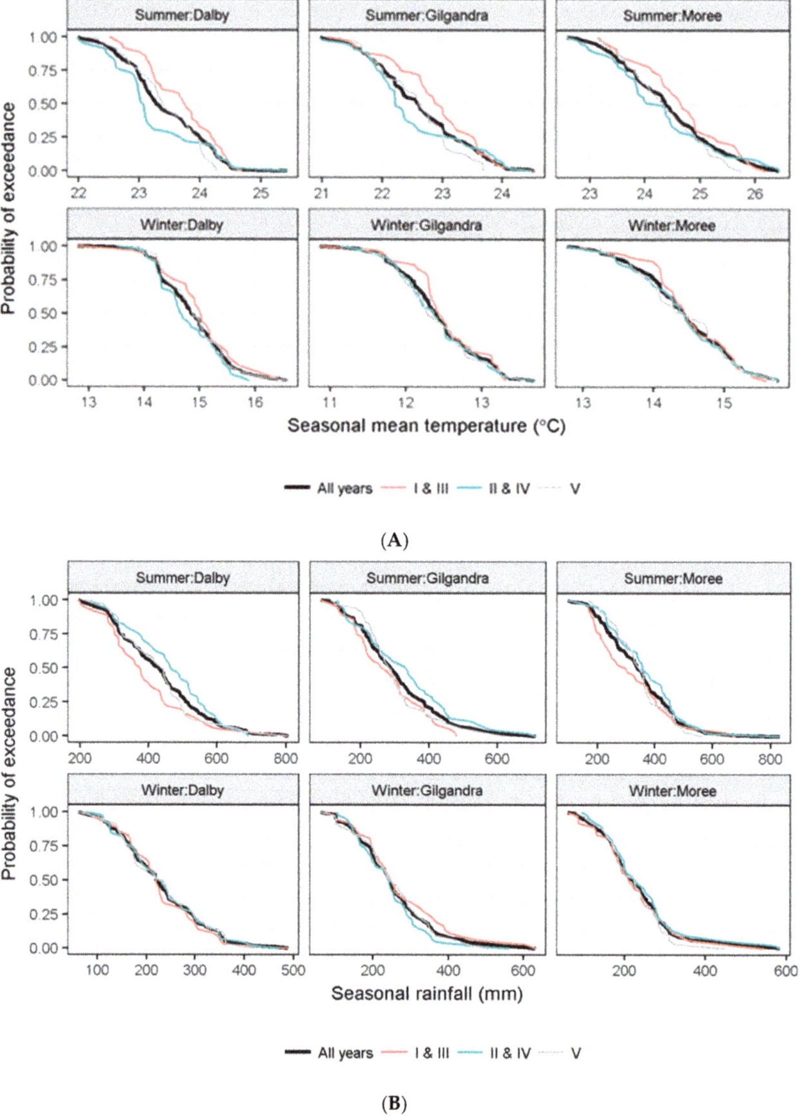

Figure 2. Cumulative probability distributions (probability of exceedance) of 'summer' and 'winter' average temperature (**A**) and cumulated rainfall (**B**) for the three SOI classes, singly (SOI phases I & III, SOI phases II & IV, and SOI phase V) and combined ('all years') for 1900–2016 at three sites in the Eastern wheatbelt. The three SOI classes correspond to SOI consistently negative and rapidly failing (phases I & III), SOI consistently positive and rapidly rising (phases II & IV), and SOI consistently near zero (phase V). SOI phases were determined in March–April, prior to sowing. The 'summer' data were recorded from the previous November to April, while 'winter' data are for the up-coming May to October period. See Figures S1–S8 for average, minimum and maximum temperature, and cumulated rainfall at other sites for both the 'summer' and 'winter' periods.

3.2. ENSO Impacts the Frequency of Occurrence of Frost and Heat Events around Flowering

Extreme temperatures can greatly decrease yield by affecting reproductive organs or impacting grain filling [11,12]. Australian wheat farmers manage their crops to minimise the risk of frost, heat, and drought by targeting the flowering time into an optimum window [7,25]. The last frost day with a 10% risk of frost tended to be earlier in SOI phases II & IV, and delayed in SOI phases I & III mostly in the eastern sites (Figure S9). By contrast, the first heat day with a 30% risk of heat tended to be earlier in SOI phases I & III, and delayed in SOI phases II & IV. Hence, in terms of temperature, the low-risk flowering window tended to last longer for SOI phases II & IV, while it tended to be reduced for SOI phases I & III. Using the three ENSO phases (i.e., El Niño, La Niña and Neutral), Alexander and Hayman [35] found similar trends for distribution and tails of last frost day in 15 sites across the Australian wheatbelt.

3.3. Variations in Yield across SOI Phases

For the reference cultivar and management (i.e., baseline simulations: Janz sown 21 May with farmer fertilisation practices), long-term average yield ranged from 0.93 to 2.71 t·ha^{-1} across sites and averaged 2.06 t·ha^{-1} among the 15 studied sites (Figure 3; 'all years'). Greatest yields were achieved in SOI phases II & IV, with long-term average yield ranging from 0.98 to 2.70 t·ha^{-1} across locations and averaging 2.10 t·ha^{-1} for Eastern Australia. By contrast, long-term average yield in the SOI phases I & III were commonly lower, ranging from 0.90 to 2.70 t·ha^{-1} across locations, and averaging 2.00 t·ha^{-1} for Eastern Australia. Strong links between wheat yield with ENSO were also found in the Eastern wheatbelt in other studies [18,36]. While the early study of Rimmington et al. [37] suggested little impact of ENSO types on wheat yields in Southern and Western Australia, more recent studies found wheat yields to be affected by ENSO in Southern and South-eastern Australia [38,39].

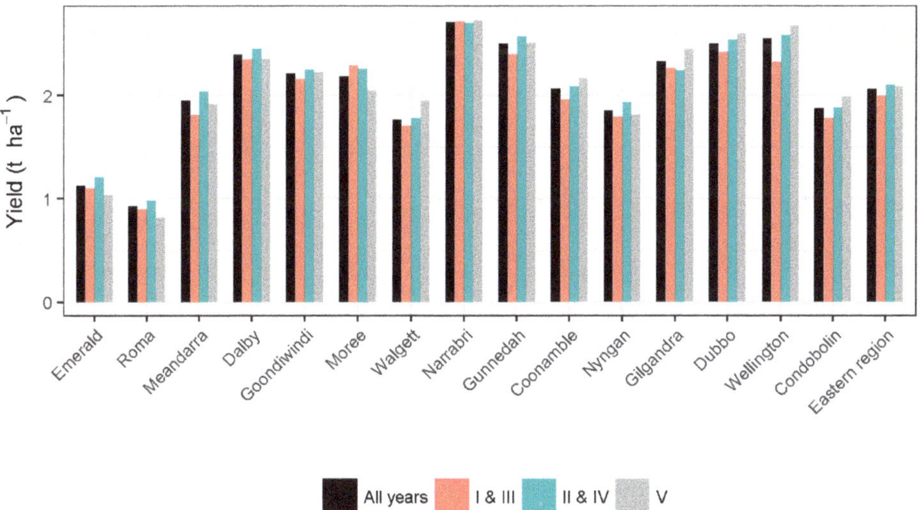

Figure 3. Simulated average yield in the baseline scenario for all years (1900–2016) and for years from each of the three SOI classes at 15 sites across the Eastern wheatbelt and for the whole Eastern wheatbelt region. Baseline simulations corresponded to a standard farmer practice (a medium-season cultivar Janz was sown at semi-optimum sowing date (21 May) with current fertilisation practice). The three SOI classes correspond to SOI consistently negative and rapidly failing (SOI phases I & III), SOI consistently positive and rapidly rising (SOI phases II & IV), and SOI consistently near zero (SOI phase V). SOI phases were determined in March–April, prior to sowing.

3.4. Optimising Genotype and Management across All Years Results in Consistent Yield Improvement and Higher Gross Margins

A large number of adaptation strategies were simulated, combining a wide range of genotypes (with all potential phenology range for Australian wheat) and diverse management practices (a broad range of sowing dates and nitrogen fertilisation options). These strategies were first applied to optimise average yield for a site across all years ('fixed adaptation') by selecting the top yielding genotype × management combination (Figure 4, Figure 5 and Figure S10).

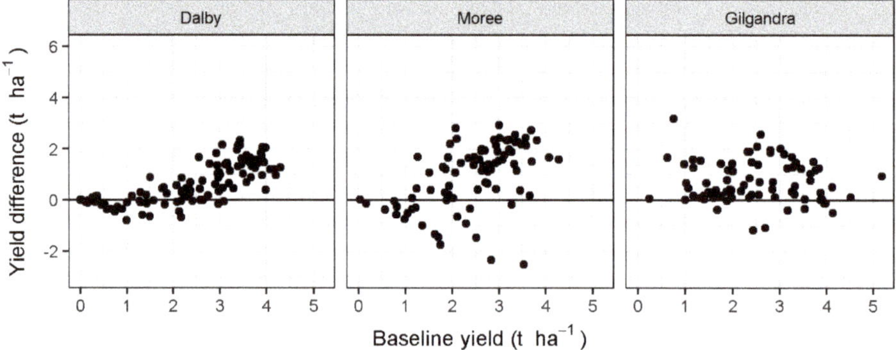

Figure 4. Yield advantage of fixed adaptation over the baseline scenario. The yield difference is calculated for each year. The baseline corresponds to simulated yield for a medium-season cultivar Janz sown at semi-optimum sowing date (21 May) with current fertilisation practice. See Figure S10 for other sites.

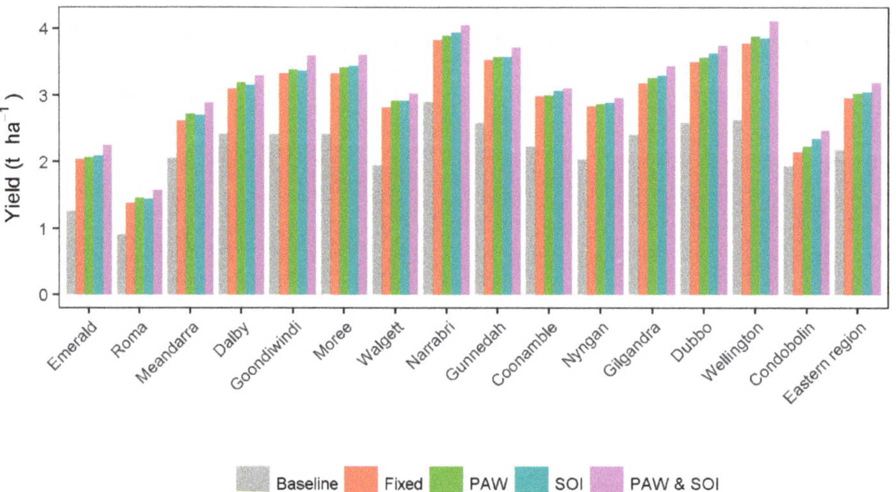

Figure 5. Simulated mean yields for the baseline, fixed adaption, and all the studied tactical adaptation scenarios related to pre-sown soil water and SOI forecast. The baseline corresponds to simulated yield for a medium-season cultivar Janz sown at semi-optimum sowing date (21 May) with no extra nitrogen input. The fixed adaptation scenario corresponds to optimised genotype and management across all years for each site. The three tactical adaptation scenarios include specific adaptation to soil PAW (plant available water) groups, SOI classes, and PAW & SOI groups. These adaptations correspond to optimised genotype and management for each PAW group and/or SOI class. SOI phases were determined in March–April, prior to sowing.

Compared to the baseline, the fixed adaptation scenario increased yields in the majority of years in all sites, although yield losses were also observed for a few years in all sites (Figure 4 and Figure S10). Regional yield (average across all sites) thus increased from 2.06 (baseline) to 2.96 t·ha^{-1} (fixed adaptation) (Figure 5). At the site level, long-term average yield ranged from 1.38 to 3.82 t·ha^{-1} for fixed adaptation compared to 0.91 to 2.90 t·ha^{-1} for the baseline, meaning a yield increase from 0.22 to 1.15 t·ha^{-1} (0.78 t·ha^{-1} on average for the whole region; Figure 3). Compared to the baseline, the fixed adaptation strategy corresponded to an earlier sowing with more nitrogen application of, in general, a shorter-maturing genotype in the northern part of the region, and a longer-maturing genotype in the southern part of the region (data not shown).

In terms of gross margin, site long-term increases from the baseline to fixed adaptation scenario ranged from AU\$ 40.5 to 223.0 ha^{-1}, which corresponds to a regional increase of AU\$ 153.00 ha^{-1} on average for the whole region. Across Eastern Australia, fixed adaptation resulted in an AU\$ 440.00 M increase in gross margin compared to the standard current practice considered here.

3.5. Benefits of Tactical Compared to Fixed Adaptation Vary with the Location, the Soil Pre-Sowing Conditions and the SOI Forecast

To explore the potential of tactical adaptation over fixed adaptation, the genotype and management were optimised for pre-sowing PAW- and/or SOI-specific conditions, and then compared to the fixed adaptation scenario. For most sites (Figure 5), slight increases in long-term average yield were simulated when adapting the genotype and management to either pre-sown PAW or SOI solely. Substantial improvements occurred when optimising average yield for both the genotype and the management for each SOI class and PAW group together (PAW & SOI), rather than sole optimisation of either the PAW group or SOI class. It can nevertheless be noted that fewer of the 117 seasons were classified in each of the nine PAW & SOI groups than in each of three PAW groups or the three SOI classes, meaning that the optimised yield is prone to more uncertainty due to the likely reduction in environmental variations within groups of years considered.

However, tactical adaptation scenarios only allowed yield to increase for some of the years compared to the fixed adaptation (Figure 6 and Figures S11–S13). Actually, when adapting the PAW group only ('PAW adaptation'), losses in yield compared to the fixed adaptation occurred relatively frequently, with losses that were typically small, but which could be as substantial as 2 t·ha^{-1} in some locations (Figure 6 and Figure S11). Similar trends and extents were observed for adaption to SOI classes. Adapting to both PAW and SOI tended to increase the frequency of yield gains, especially in poor seasons (i.e., when yield was medium to low in the fixed adaptation scenario), mainly due to better tuning of the crop phenology (maturity type × sowing time) to the considered environmental conditions. However, yield losses compared to the fixed adaptation still occurred frequently in locations such as Wellington, while they were relatively rare in locations like Coonamble and Moree (Figure S13). Note that the unbalanced number of years among PAW groups, SOI classes, and PAW × SOI groups might cause bias in the adaptation values and risks.

Overall, when considering the optimised genotypes and management for each PAW group ('PAW'), long-term average yield and gross margin increased at each site from 0.012 to 0.13 t·ha^{-1}, and from AU\$ 2.63 to 41.9 ha^{-1}, respectively, compared to the fixed adaptation scenario (Figure 7). When considering genotype and management practices optimised for each SOI class ('SOI'), long-term average yield and gross margin at each site increased from 0.037 to 0.17 t·ha^{-1}, and from AU\$ 5.00 to 41.50 ha^{-1}, respectively. Finally, when considering optimised management and cultivar for each combination of PAW group and SOI class ('PAW & SOI'), average yield and gross margin at each site increased from 0.15 to 0.46 t·ha^{-1}, and from AU\$ 34.2.00 to 108 ha^{-1}, respectively, compared to the fixed adaptation. Hammer et al. (1996) studied the adapted values of SOI phases through changing cultivars and nitrogen fertilisation at Goondiwindi, and also found a substantial, although more limited, increase in gross margin (\$26 ha^{-1}) compared with results in this study (\$48 ha^{-1}, Figure 7B) [1].

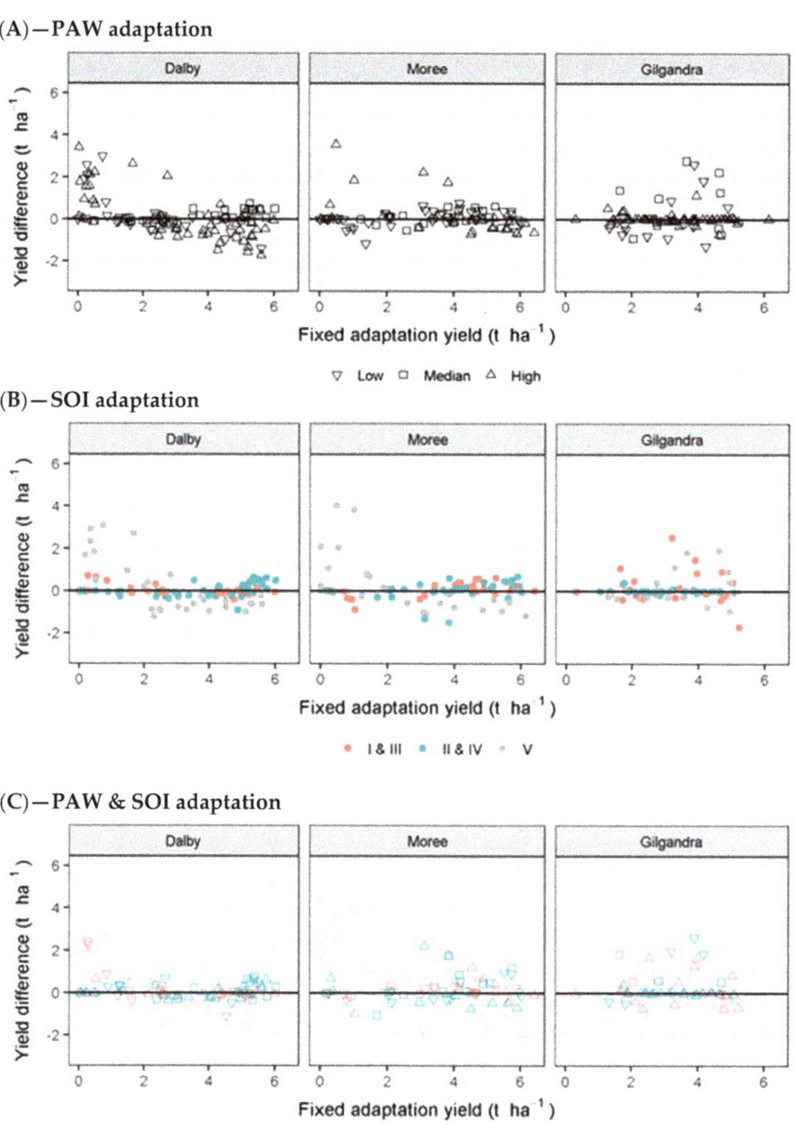

Figure 6. Yield advantage of tactical adaptation scenarios for (**A**) pre-sowing soil PAW (low, median and high), (**B**) SOI classes (I & III, II & IV and V), and (**C**) both PAW & SOI groups over fixed adaptation. The yield difference is calculated for each year. The fixed adaptation scenario corresponds to optimised genotype and management across all years. The three tactical adaptation scenarios correspond to optimised genotype and management for each PAW (plant available water) group and/or SOI class. SOI phases were determined in March–April, prior to sowing. See Figures S11–S13 for other sites.

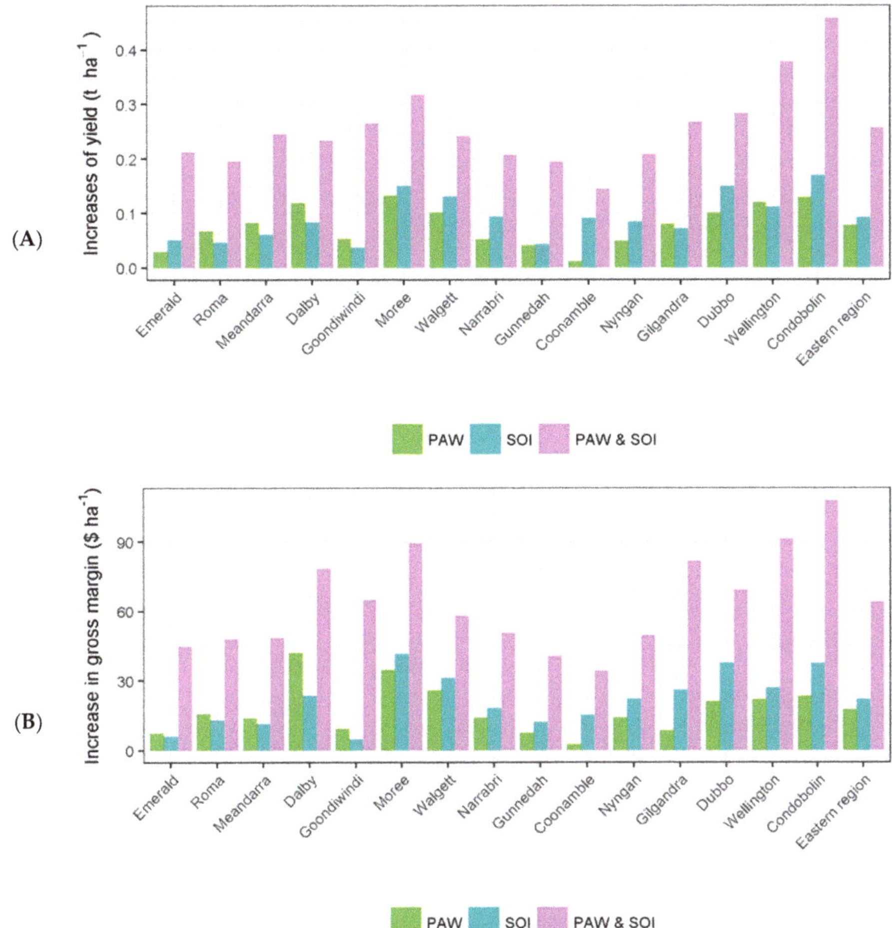

Figure 7. Increase in simulated yield (**A**) and gross margin (**B**) for tactical adaptation options compared to the fixed adaptation in each studied site and the whole Eastern wheatbelt. Increases in yield were averaged for all years from 1900 to 2016. The three tactical adaptation options correspond to optimise long-term yield for either (i) low/medium/high pre-sowing plant available water (PAW), (ii) each class of Southern Oscillation Index (SOI), or (iii) each combination of PAW group and SOI class.

At the regional scale, long-term mean increased in yield and gross margin of PAW & SOI tactical adaptation versus fixed adaptation were 0.26 t·ha^{-1} and AU\$ 63.90 ha^{-1}, respectively. The cumulated regional increase in gross margin was AU\$ 188 M for the Eastern wheatbelt. Note that the increase gross margin in this paper considered fixed nitrogen price, and accounted for neither changing wheat price related to the harvested grain quality, nor to the domestic and/or global market.

3.6. Should Eastern Australian Wheat Producers Adapt Their Decisions Based on SOI Phases?

In Australia, SOI phases impact the climate and crops mostly in the eastern part of the wheatbelt. In this region, wheat management occurs almost exclusively at sowing. While specific adaptation could be more accurate later in the season, i.e., when climate trends in regard to El Niño/La Niña are clearer, early forecast of the SOI phases is required to allow farmers to prepare seeds and plan suitable management. The potential gains of specific adaptation to such pre-sowing forecasts of SOI phases combined with knowledge on initial soil water appeared to be substantial (Figures 5 and 7). While the gains are variable depending on the location, the extra costs for seed companies and farmers to store large stocks of seeds required for specific adaptation may have to be considered, especially at the farm scale. Unpredictable rainfall in the autumn before sowing may also constrain sowing opportunities [1]. That said, the potential gains of specific adaptation appear to be well above what could be expected from most breeding innovations, at least in the short term [40].

Other methods exist to forecast short-term or seasonal climate. For instance, the Australian Bureau of Meteorology produced twice-weekly weather forecasts for a period of 270 days, with a dynamic model called POAMA (http://poama.bom.gov.au/info/poama-2.html) [41–43]. Such seasonal forecasts have been used to look at management strategies of crops [44–46]. To improve these forecasts, which use on a grid of about 250 km, the Bureau of Meteorology is now proposing seasonal forecasts (ACCESS-S) based on ACCESS (Australian Community Climate and Earth System Simulator; http://www.bom.gov.au/australia/charts/about/about_access.shtml) using a 60 km grid. Other indices than SOI or ENSO phases could also be used for specific adaptation, such as the Inter-decadal Pacific Oscillation (IPO) phases [47] or drought environmental types [20,22,48,49] if climate forecasts are sufficiently reliable. For instance, crop models have been used to assess the value of broad and specific adaptations to select sorghum varieties and managements for different types of drought environments [20].

4. Conclusions

In this study, we assessed the value of fixed adaptation (no distinction between the years) and tactical adaptations based on pre-sowing plant available water (PAW) and/or SOI forecasts to increase productivity at given sites. Overall, with our current knowledge, it appears that yield gains can be made from improving cultivar and management strategy both regardless of climate forecast (fixed adaptation), and as a tactical adaptation to pre-sowing soil water conditions and climate forecasts. The benefits of PAW and SOI tactical adaptation could be useful for farmers to adjust farm management practices according to the season, and may be improved with new forecasting climate methods such as the newly developed ACCESS-S model.

Supplementary Materials: The following are available online at http://www.mdpi.com/2225-1154/6/3/77/s1. Figure S1: Cumulative probability distributions of 'summer' average temperature for the three SOI classes, singly and combined ('all years'), Figure S2: Cumulative probability distributions of 'summer' maximum temperature for the three SOI classes, singly and combined ('all years'), Figure S3: Cumulative probability distributions of 'summer' minimum temperature for the three SOI classes, singly and combined ('all years'), Figure S4: Cumulative probability distributions of 'winter' average temperature for the three SOI classes, singly and combined ('all years'), Figure S5: Cumulative probability distributions of 'winter' maximum temperature for the three SOI classes, singly and combined ('all years'), Figure S6: Cumulative probability distributions of 'winter' minimum temperature for the three SOI classes, singly and combined ('all years'), Figure S7: Cumulative probability distributions of 'summer' total rainfall for the three SOI classes, singly and combined ('all years'), Figure S8: Cumulative probability distributions (probability of exceedance) of 'winter' total rainfall for the three SOI classes, singly and combined ('all years'), Figure S9: Probability of last frost day and first heat day for the three SOI classes, singly and combined ('all years'), Figure S10: Simulated yield advantage of fixed adaptation over the baseline scenario, Figure S11: Simulated yield advantage of PAW tactical adaptation scenario over fixed adaptation scenario, Figure S12: Simulated yield advantage of SOI tactical adaptation scenario over fixed adaptation scenario, Figure S13: Simulated yield advantage of PAW & SOI tactical adaptation scenario over fixed adaptation scenario.

Author Contributions: B.Z., S.C. and K.C. conceived and developed the ideas; B.Z. and K.C. performed the data analysis; B.Z., S.C. and K.C. wrote the paper.

Funding: This research received no external funding.

Acknowledgments: This research was funded by the CSIRO and the University of Queensland.

Conflicts of Interest: The authors declare no conflict of interest.

References

1.   Hammer, G.L.; Holzworth, D.P.; Stone, R. The value of skill in seasonal climate forecasting to wheat crop management in a region with high climatic variability. *Aust. J. Agric. Res.* **1996**, *47*, 717–737. [CrossRef]
2.   Chapman, S.C.; Chakraborty, S.; Dreccer, M.F.; Howden, S.M. Plant adaptation to climate change? Opportunities and priorities in breeding. *Crop Pasture Sci.* **2012**, *63*, 251–268. [CrossRef]
3.   Potgieter, A.B.; Hammer, G.L.; Butler, D. Spatial and temporal patterns in Australian wheat yield and their relationship with ENSO. *Aust. J. Agric. Res.* **2002**, *53*, 77–89. [CrossRef]
4.   Meinke, H.; de Voil, P.; Hammer, G.L.; Power, S.; Allan, R.; Stone, R.C.; Folland, C.; Potgieter, A. Rainfall Variability at Decadal and Longer Time Scales: Signal or Noise? *J. Clim.* **2005**, *18*, 89–96. [CrossRef]
5.   Yates, D.; Vervoort, R.W.; Minasny, B.; McBratney, A. The history of using rainfall data to improve production in the grain industry in Australia—From Goyder to ENSO. *Crop Pasture Sci.* **2016**, *67*, 467–479. [CrossRef]
6.   Chenu, K.; Cooper, M.; Hammer, G.L.; Mathews, K.L.; Dreccer, M.F.; Chapman, S.C. Environment characterization as an aid to wheat improvement: Interpreting genotype–environment interactions by modelling water-deficit patterns in North-Eastern Australia. *J. Exp. Bot.* **2011**, *62*, 1743–1755. [CrossRef] [PubMed]
7.   Zheng, B.Y.; Chenu, K.; Dreccer, M.F.; Chapman, S.C. Breeding for the future: What are the potential impacts of future frost and heat events on sowing and flowering time requirements for Australian bread wheat (*Triticum aestivum*) varieties? *Glob. Chang. Biol.* **2012**, *18*, 2899–2914. [CrossRef] [PubMed]
8.   Nicholls, N.; Baek, H.J.; Gosai, A.; Chambers, L.E.; Choi, Y.; Collins, D.; Della-Marta, P.M.; Griffiths, G.M.; Haylock, M.R.; Iga, N.; et al. The El Niño–Southern Oscillation and daily temperature extremes in East Asia and the west Pacific. *Geophys. Res. Lett.* **2005**, *32*, L16714. [CrossRef]
9.   Chambers, L.E.; Griffiths, G.M. The changing nature of temperature extremes in Australia and New Zealand. *Aust. Meteorol. Mag.* **2008**, *57*, 13–35.
10. Lobell, D.B.; Hammer, G.L.; Chenu, K.; Zheng, B.; McLean, G.; Chapman, S.C. The shifting influence of drought and heat stress for crops in Northeast Australia. *Glob. Chang. Biol.* **2015**, *21*, 4115–4127. [CrossRef] [PubMed]
11. Frederiks, T.M.; Christopher, J.T.; Harvey, G.L.; Sutherland, M.W.; Borrell, A.K. Current and emerging screening methods to identify post-head-emergence frost adaptation in wheat and barley. *J. Exp. Bot.* **2012**, *63*, 5405–5416. [CrossRef] [PubMed]
12. Zheng, B.; Chapman, S.C.; Christopher, J.T.; Frederiks, T.M.; Chenu, K. Frost trends and their estimated impact on yield in the Australian wheatbelt. *J. Exp. Bot.* **2015**, *66*, 3611–3623. [CrossRef] [PubMed]
13. Crimp, S.J.; Zheng, B.; Khimashia, N.; Gobbett, D.L.; Chapman, S.; Howden, M.; Nicholls, N. Recent changes in southern Australian frost occurrence: Implications for wheat production risk. *Crop Pasture Sci.* **2016**, *67*, 801–811. [CrossRef]
14. Timmermann, A.; Oberhuber, J.; Bacher, A.; Esch, M.; Latif, M.; Roeckner, E. Increased El Nino frequency in a climate model forced by future greenhouse warming. *Nature* **1999**, *398*, 694–697. [CrossRef]
15. Cai, W.; Wang, G.; Santoso, A.; McPhaden, M.J.; Wu, L.; Jin, F.-F.; Timmermann, A.; Collins, M.; Vecchi, G.; Lengaigne, M.; et al. Increased frequency of extreme La Nina events under greenhouse warming. *Nat. Clim. Chang.* **2015**, *5*, 132–137. [CrossRef]
16. McPhaden, M.J.; Zebiak, S.E.; Glantz, M.H. ENSO as an Integrating Concept in Earth Science. *Science* **2006**, *314*, 1740–1745. [CrossRef] [PubMed]
17. Ashok, K.; Yamagata, T. Climate change: The El Nino with a difference. *Nature* **2009**, *461*, 481–484. [CrossRef] [PubMed]
18. Stone, R.; Hammer, G.L.; Marcussen, T. Prediction of global rainfall probabilities using phases of the Southern Oscillation Index. *Nature* **1996**, *384*, 252–255. [CrossRef]
19. Hammer, G.L.; McLean, G.; Chapman, S.; Zheng, B.; Doherty, A.; Harrison, M.T.; van Oosterom, E.; Jordan, D. Crop design for specific adaptation in variable dryland production environments. *Crop Pasture Sci.* **2014**, *65*, 614–626.
20. Stone, R.; Nicholls, N.; Hammer, G. Frost in Northeast Australia: Trends and influences of phases of the Southern Oscillation. *J. Clim.* **1996**, *9*, 1896–1909. [CrossRef]
21. Woli, P.; Ortiz, B.V.; Johnson, J.; Hoogenboom, G. El Niño–Southern Oscillation Effects on winter wheat in the southeastern United States. *Agron. J.* **2015**, *107*, 2193. [CrossRef]

22. Chenu, K.; Deihimfard, R.; Chapman, S.C. Large-scale characterization of drought pattern: A continent-wide modelling approach applied to the Australian wheatbelt—Spatial and temporal trends. *New Phytol.* **2013**, *198*, 801–820. [CrossRef] [PubMed]

23. Dennett, M.D. Effects of sowing date and the determination of optimum sowing date. In *Wheat: Ecology and Physiology of Yield Determination*; Satorre, E.H., Slafer, G.H., Eds.; Food Products Press: Binghamton, NY, USA, 1999; pp. 123–140.

24. Pook, M.; Lisson, S.; Risbey, J.; Ummenhofer, C.C.; McIntosh, P.; Rebbeck, M. The autumn break for cropping in southeast Australia: Trends, synoptic influences and impacts on wheat yield. *Int. J. Climatol.* **2009**, *29*, 2012–2026. [CrossRef]

25. Flohr, B.M.; Hunt, J.R.; Kirkegaard, J.A.; Evans, J.R. Water and temperature stress define the optimal flowering period for wheat in south-eastern Australia. *Field Crops Res.* **2017**, *209*, 108–119. [CrossRef]

26. Chenu, K.; Porter, J.R.; Martre, P.; Basso, B.; Chapman, S.C.; Ewert, F.; Bindi, M.; Asseng, S. Contribution of Crop Models to Adaptation in Wheat. *Trends Plant Sci.* **2017**, *22*, 472–490. [CrossRef] [PubMed]

27. Holzworth, D.P.; Huth, N.; de Voil, P.G.; Zurcher, E.J.; Herrmann, N.I.; McLean, G.; Chenu, K.; Van Oosterom, E.; Murphy, C.; Moore, A.D.; et al. APSIM—Evolution towards a new generation of agricultural systems simulation. *Environ. Model. Softw.* **2014**, *62*, 327–350. [CrossRef]

28. Zheng, B.Y.; Biddulph, B.; Li, D.; Kuchel, H.; Chapman, S.C. Quantification of the effects of *VRN1* and *Ppd-D1* to predict spring wheat (*Triticum aestivum*) heading time across diverse environments. *J. Exp. Bot.* **2013**, *64*, 3747–3761. [CrossRef] [PubMed]

29. Jeffrey, S.J.; Carter, J.O.; Moodie, K.B.; Beswick, A.R. Using spatial interpolation to construct a comprehensive archive of Australian climate data. *Environ. Model. Softw.* **2001**, *16*, 309–330. [CrossRef]

30. Zheng, B.; Chenu, K.; Doherty, A.; Chapman, S. *The APSIM-Wheat Module (7.5 R3008)*; Agricultural Production Systems Simulator (APSIM) Initiative: Toowoomba, Australian, 2014.

31. Christopher, J.T.; Christopher, M.J.; Borrell, A.K.; Fletcher, S.; Chenu, K. Stay-green traits to improve wheat adaptation in well-watered and water-limited environments. *J. Exp. Bot.* **2016**, *67*, 5159–5172. [CrossRef] [PubMed]

32. Zheng, B.; Holland, E.; Chapman, S.C. A standardized workflow to utilise a grid-computing system through advanced message queuing protocols. *Environ. Model. Softw.* **2016**, *84*, 304–310. [CrossRef]

33. Australian Bureau of Agricultural and Resource Economics (ABARE). *Australian Commodity Statistics 2016*; ABARE: Canberra, Australia, 2016.

34. Stone, R.; Auliciems, A. SOI phase relationships with rainfall in eastern Australia. *Int. J. Climatol.* **1992**, *12*, 625–636. [CrossRef]

35. Alexander, B.; Hayman, P. Can we use forecasts of El Nino and La Nina for frost management in the Eastern and Southern grains belt? In Proceedings of the 14th Agronomy Conference, Adelaide, Australia, 21–25 September 2008.

36. Potgieter, A.B.; Hammer, G.L.; Meinke, H.; Stone, R.C.; Goddard, L. Three putative types of El Niño revealed by spatial variability in impact on Australian wheat yield. *J. Clim.* **2005**, *18*, 1566–1574. [CrossRef]

37. Rimmington, G.M.; Nicholls, N. Forecasting wheat yields in Australia with the Southern Oscillation Index. *Crop Pasture Sci.* **1993**, *44*, 625–632. [CrossRef]

38. Anwar, M.R.; Rodriguez, D.; Liu, D.L.; Power, S.; O'Leary, G.J. Quality and potential utility of ENSO-based forecasts of spring rainfall and wheat yield in south-eastern Australia. *Crop Pasture Sci.* **2008**, *59*, 112–126. [CrossRef]

39. Hayman, P.T.; Whitbread, A.M.; Gobbett, D.L. The impact of El Niño Southern Oscillation on seasonal drought in the southern Australian grainbelt. *Crop Pasture Sci.* **2010**, *61*, 528–539. [CrossRef]

40. Marshall, G.R.; Parton, K.A.; Hammer, G.L. Risk attitude, planting conditions and the value of seasonal forecasts to a dryland wheat grower. *Aust. J. Agric. Econ.* **1996**, *40*, 211–233. [CrossRef]

41. Zhao, M.; Hendon, H.H. Representation and prediction of the Indian Ocean dipole in the POAMA seasonal forecast model. *Q. J. R. Meteorol. Soc.* **2009**, *135*, 337–352. [CrossRef]

42. Marshall, A.G.; Hudson, D.; Hendon, H.H.; Pook, M.J.; Alves, O.; Wheeler, M.C. Simulation and prediction of blocking in the Australian region and its influence on intra-seasonal rainfall in POAMA-2. *Clim. Dyn.* **2014**, *42*, 3271–3288. [CrossRef]

43. Hudson, D.; Marshall, A.G.; Yin, Y.; Alves, O.; Hendon, H.H. Improving intraseasonal prediction with a new ensemble generation strategy. *Mon. Weather Rev.* **2013**, *141*, 4429–4449. [CrossRef]

44. Asseng, S.; McIntosh, P.C.; Wang, G.; Khimashia, N. Optimal N fertiliser management based on a seasonal forecast. *Eur. J. Agron.* **2012**, *38*, 66–73. [CrossRef]

45. Hayman, P.; Cooper, B.; Parton, K.; Alves, O.; Yong, G.; Henry, H.; Scheer, C. Can advances in climate forecasts improve the productive and environmental outcomes from nitrogen fertiliser on wheat? A case study using POAMA for topdressing wheat in South Australia. In Proceedings of the 17th Australian Agronomy Conference, Hobart, Tasmania, 20–24 September 2015; Australian Society of Agronomy: Hobart, Australia, 2015.

46. Rodriguez, D.; de Voil, P.; Hudson, D.; Brown, J.N.; Hayman, P.; Marrou, H.; Meinke, H. Predicting optimum crop designs using crop models and seasonal climate forecasts. *Sci. Rep.* **2018**, *8*, 2231. [CrossRef] [PubMed]

47. Power, S.; Casey, T.; Folland, C.; Colman, A.; Mehta, V. Inter-decadal modulation of the impact of ENSO on Australia. *Clim. Dyn.* **1999**, *15*, 319–324. [CrossRef]

48. Chapman, S.C.; Cooper, M.; Hammer, G.L. Using crop simulation to generate genotype by environment interaction effects for sorghum in water-limited environments. *Aust. J. Agric. Res.* **2002**, *53*, 379–389. [CrossRef]

49. Chenu, K. Characterizing the crop environment—Nature, significance and applications. In *Crop Physiology. Applications for Genetic Improvement and Agronomy*; Sadras, V.O., Calderini, D., Eds.; Academic Press: London, UK, 2015; pp. 321–348. ISBN 978-0-12-417104-6.

Article

Possible Scenarios of Winter Wheat Yield Reduction of Dryland Qazvin Province, Iran, Based on Prediction of Temperature and Precipitation Till the End of the Century

Behnam Mirgol [1] and Meisam Nazari [2,3,*]

[1]   Department of Water Engineering, Faculty of Engineering and Technology, Imam Khomeini International University, 3414896818 Qazvin, Iran; meisam.nazari1991@gmail.com
[2]   Department of Crop Sciences, Faculty of Agricultural Sciences, Georg-August University of Göttingen, Büsgenweg 5, 37077 Göttingen, Germany
[3]   Department of Soil Science, University of Kassel, Nordbahnhofstr. 1a, 37213 Witzenhausen, Germany
*   Correspondence: meisam.nazari@stud.uni-goettingen.de

Received: 31 August 2018; Accepted: 21 September 2018; Published: 23 September 2018

Abstract: The climate of the Earth is changing. The Earth's temperature is projected to maintain its upward trend in the next few decades. Temperature and precipitation are two very important factors affecting crop yields, especially in arid and semi-arid regions. There is a need for future climate predictions to protect vulnerable sectors like agriculture in drylands. In this study, the downscaling of two important climatic variables—temperature and precipitation—was done by the CanESM2 and HadCM3 models under five different scenarios for the semi-arid province of Qazvin, located in Iran. The most efficient scenario was selected to predict the dryland winter wheat yield of the province for the three periods: 2010–2039, 2040–2069, and 2070–2099. The results showed that the models are able to satisfactorily predict the daily mean temperature and annual precipitation for the three mentioned periods. Generally, the daily mean temperature and annual precipitation tended to decrease in these periods when compared to the current reference values. However, the scenarios rcp2.6 and B2, respectively, predicted that the precipitation will fall less or even increase in the period 2070–2099. The scenario rcp2.6 seemed to be the most efficient to predict the dryland winter wheat yield of the province for the next few decades. The grain yield is projected to drop considerably over the three periods, especially in the last period, mainly due to the reduction in precipitation in March. This leads us to devise some adaptive strategies to prevent the detrimental impacts of climate change on the dryland winter wheat yield of the province.

Keywords: CanESM2; HadCM3; precipitation; temperature; winter wheat yield

1. Introduction

The temperature of the Earth is increasing more rapidly than during the previous decades, leading to extensive climate change [1]. The Earth's temperature is projected to maintain its upward trend slightly in the next few decades [1]. A significant rise in the concentration of greenhouse gases such as CO_2, CH_4, N_2O, and water vapor, mainly caused by human activities, has intensified this trend [2]. The concentration of greenhouse gases, volume of ozone, aerosols, and sunspots seem to be the most noticeable reason for temperature variations and climate change in the recent century [3].

More than two billion people live in drylands, constituting nearly 40% of the world's population [4]. Cereals are the major crops cultivated in drylands [5]. Crop production in drylands mainly depends on precipitation during the growing season [6]. Moreover, the rise in temperature has led to exacerbating droughts and a considerable loss in crop yields in arid and semi-arid regions [7].

It is necessary to manage drylands in a sustainable way, by which food security is achieved [8]. To do so, there must be some possible measurements and predictions to protect vulnerable sectors such as agriculture and water resources in drylands [9].

General Circulation Models (GCMs) are the most developed tools for the simulation of general responses to the accumulation of greenhouse gases [10]. Studies have shown that the results of GCMs cannot be exploited directly because they are not accurate enough in describing sub-grid data [10]. Therefore, Statistical Downscaling Models (SDSMs) are one of the tools that have been developed to deal with this problem [11]. SDSMs are the most frequently used models in agricultural research, where some independent variables are measured and collected to predict dependent variables [12]. Tatsumi et al. [13] applied the Hadley Centre Coupled Model (version 3; HadCM3) and Coupled Global Climate Model 3 (CGCM3) to forecast the daily minimum, maximum, and average temperature of Shikoku city in Japan, using downscaling techniques. Their results indicated that the temperature is likely to increase in the Shikoku region, Japan, within the period 2071–2099. In a similar study, Ribalaygua et al. [14] used downscaling techniques to simulate the daily minimum and maximum temperature and daily precipitation in a region located in Spain. Their results showed that maximum and minimum temperatures will rise, while precipitation will decrease in the 21st century. Johns et al. [15], by applying the HadCM3 model, predicted that some regions of Central America and Southern Europe might be moister in the future, whereas Australia may experience a type of drier climate.

In recent years, researchers have studied the potential impacts of climate change on plant growth by using different types of simulation models [16,17]. Russell et al. [18] reported that most of the alterations in wheat yield in the United States are related to climate change. Temperature and precipitation, as two important climatic variables for the evaluation of future grain yield, have been investigated by many researchers. For instance, [16] indicated that the changes in temperature and precipitation within the last 30 years in Mexico had positively impacted on the winter wheat yield. In another study, Landau et al. [19], by applying a multiple-regression model, indicated that the temperature increase led to an improvement in the winter wheat crop characteristics, while the precipitation increase could have negative impacts.

The downscaling of GCMs parameters and studying the possible changes in wheat yield due to climatic effects have been distinctly investigated [14,20]. Lhomme et al. [21], for example, studied the potential effect of climate change on durum wheat yield in Tunisia using the downscaled values of some scenarios. Moreover, the efficiency of the IPCC scenarios has rarely been evaluated and compared [22]. In the present study, the downscaling of two important climatic parameters—temperature and precipitation—was done by the Canadian Earth System Model (CanESM2) and HadCM3 models for the province of Qazvin, located in Iran, where the climate is semi-arid and the dryland farming of winter wheat dominates. Then, the most efficient scenario was chosen to predict the dryland winter wheat yield of the province for the next few decades through a multiple-regression model. The efficiency of the fourth and fifth IPCC scenarios in predicting the temperature and precipitation of the region was also compared.

2. Materials and Methods

2.1. Geography, Climate, and Dryland Farming of the Province

The province of Qazvin has an area of 15,821 km^2, located between 48–45 to 50–50 East of the Greenwich Meridian of longitude and 35–37 to 36–45 North latitude of the Equator. Its average altitude is 1278 m above sea level. It has a semi-arid climate with the annual mean precipitation, daily mean temperature, and relative humidity of 301 mm, 14.2 °C, and 51%, respectively. The province is affected by Siberian and Mediterranean winds, which are considerably important factors in controlling the climate of the province. The geographical situation of the studied area is shown in Figure 1.

The total winter wheat yield of the province is 445 million kg, 364 million kg (82%) of which belongs to irrigated farming and 80.7 million kg (18%) to dryland farming. The total cultivated area for winter wheat is nearly 202,497 ha, 95792 ha and 106,704 ha of which are under irrigated and dryland farming, respectively. The average dryland winter wheat yield of the province is estimated to be 1541 kg ha^{-1}.

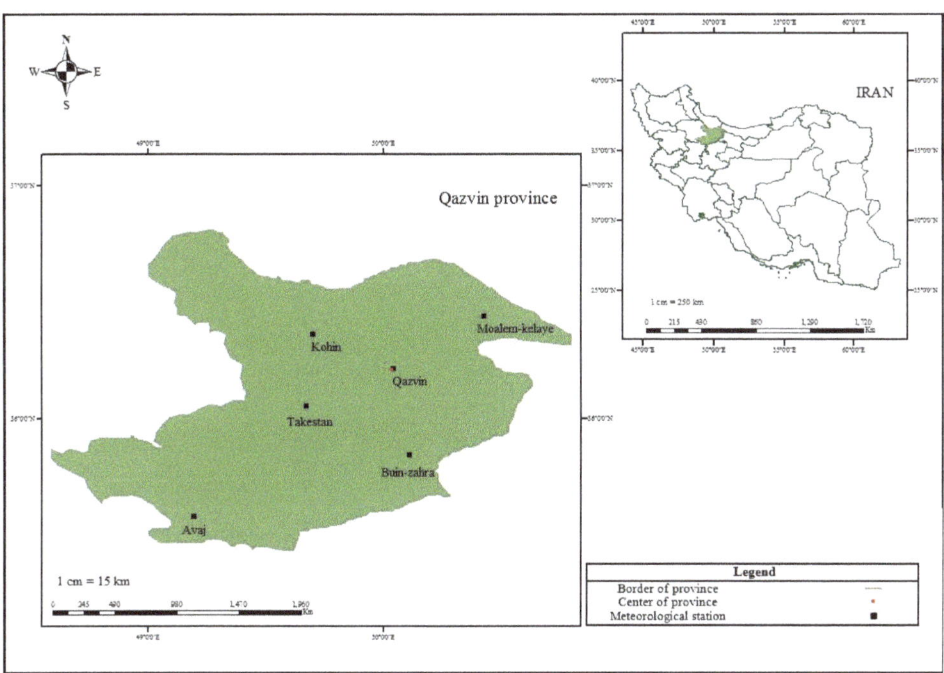

Figure 1. Map of the studied area.

2.2. Methodology

The daily mean temperature and precipitation data for 32 years (1985–2017) were collected from the six meteorological stations in the province (Figure 1). Thereafter, the daily mean temperature and precipitation of all days of all years were calculated separately by the Thiessen polygons method using the software ArcGIS version 10 via Equations (1) and (2):

$$P_a = \frac{\sum p_i A_i}{\sum A_i} \tag{1}$$

$$T_a = \frac{\sum t_i A_i}{\sum A_i} \tag{2}$$

where P_a and T_a are the daily mean precipitation and temperature of the province, respectively; p_i and t_i are the daily mean precipitation and temperature in the station i, respectively; and A_i is the area of the province.

The HadCM3 and CanESM2 models were used to compare the scenarios. HadCM3 has a spatial resolution of 2.5° × 3.75° (latitude by longitude) and the representation produces a grid box resolution of 96 × 73 grid cells. This produces a surface spatial resolution of about 417 km × 278 km, reducing to 295 km × 278 km at 45 degrees North and South. In CanESM2, the long-term time series of standardized daily values are extracted into a one column text file per grid cell. The 128 × 64 grid cells cover global domain according to a T42 Gaussian grid. This grid is uniform along the longitude with a horizontal resolution of 2.81° and is nearly uniform along the latitude of roughly 2.81°. The calibration of the stations (points) against the grid-cells (pixels) was done by the downscaling of the SDSM linear regression model. Data from the years 2006–2015 and 2016–2017 were used for the calibration and validation of both models, respectively. Figures 2 and 3 show the observed versus the simulated values of the temperature and precipitation for the years 2006–2015. Meanwhile, since 26 synoptic variables are considered as predictor variables in these models, having a unique equation was not logically possible because of the accumulated error. To solve this problem, only the predictor variables, being more correlative with the daily mean precipitation and temperature than others, were chosen. Then, the correlation between the variables was detected by Pearson's correlation test ($p < 0.01$) and the most important variables were selected according to the statistical significance between them and the dependent variables ($p < 0.01$). To analyze the climatic data across the study, it was necessary to apply a Statistical Downscaling Model (SDSM). To do so, SDSM version 5.2 was used. SDSM is a decision support tool for assessing local climate change impacts using a powerful statistical downscaling technique. It has the potential to rapidly develop downscaled climatic data [11]. To make statistical connections between the predictor and predicted variables, some regression equations were acquired to predict the climatic variables for the next few periods under the impact of climate change. After acquiring the regression equations and measuring their accuracy, the scenarios were produced through both models for the periods 2010–2039, 2040–2069, and 2070–2099. The properties of these scenarios are indicated in Table 1.

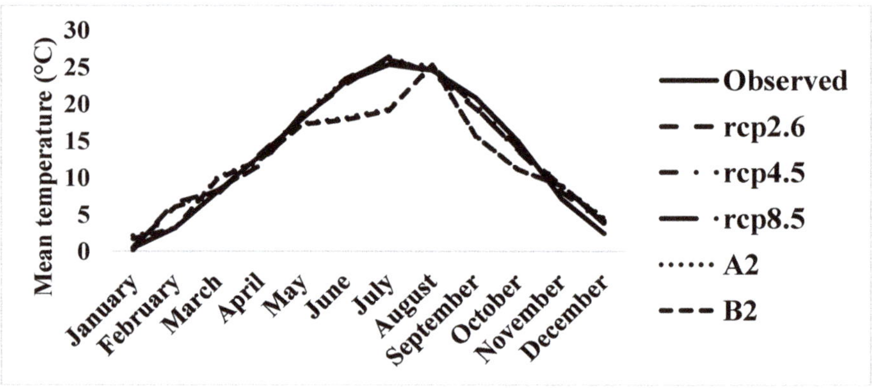

Figure 2. Results of the comparison between the observed and simulated monthly mean temperature values (2006–2015).

Table 1. Properties of the used standard Intergovernmental Panel on Climate Change [10] scenarios.

Models	Scenarios	Properties
CanESM2	rcp2.6	Radiative forcing peaks at 3 W m^{-2} and stabilizes to 2.6 W m^{-2} by the end of 2100; CO_2 concentration is estimated to be 490 ppm by 2100.
	rcp4.5	Radiative forcing is estimated to be 4.5 W m^{-2} by 2100; CO_2 concentration is estimated to be 650 ppm by 2100
	rcp8.5	Radiative forcing is estimated to be 8.5 W m^{-2} by 2100; CO_2 concentration is estimated to be 1370 ppm by 2100
HadCM3	A2	Describes a very heterogeneous world with high population growth, slow economic development, and slow technological change.
	B2	Describes a world with intermediate population and economic growth, emphasizing local solutions to economic, social, and environmental sustainability.

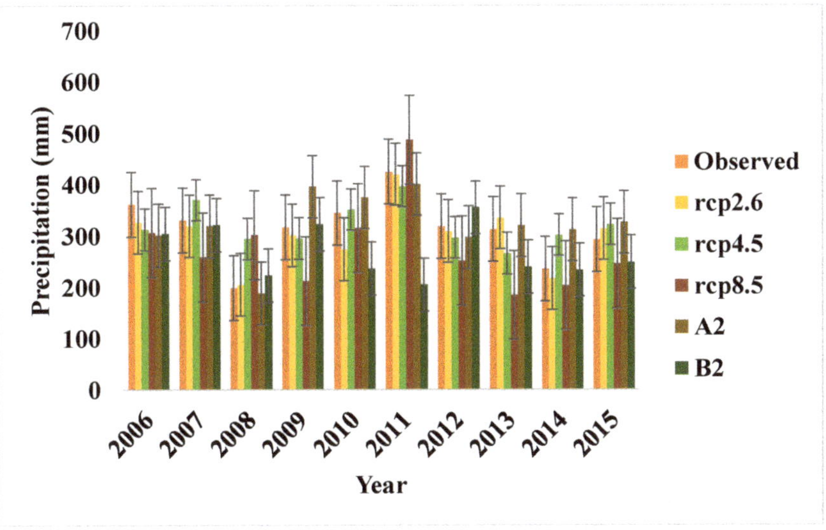

Figure 3. Results of the comparison between the observed precipitation values (2006–2015) and the simulated precipitation values. I = ± SD: standard deviation, the overlapping bars show no significant differences.

The efficiency of the scenarios was compared and the most efficient scenario was recognized through the statistical indicators of Mean Absolute Error (MAE), Root Mean Square Error (RMSE), Nash-Sutcliffe coefficient (NS), Coefficient of Determination (R^2), and Analysis of Variance (at $p < 0.01$) as follows:

$$Z_i = \frac{P_i - \overline{P}}{\sigma_P} \text{ or } Z_i = \frac{O_i - \overline{O}}{\sigma_o} \tag{3}$$

$$MAE = \sum_{i=1}^{n} \left| \frac{P_i - O_i}{n} \right| \tag{4}$$

$$RMSE = \sqrt{\frac{\sum_{i=1}^{n}(P_i - O_i)^2}{n}} \tag{5}$$

$$NS = 1 - \left(\frac{\sum_{i=1}^{n}(O_i - P_i)^2}{\sum_{i=1}^{n}(O_i - \overline{O})^2} \right) \tag{6}$$

$$R^2 = \left[\frac{\frac{1}{n}\sum_{i=1}^{n}(Pi - \overline{P})(Oi - \overline{O})}{\sigma_P \times \sigma_o} \right] \tag{7}$$

where Z_i is the standardized daily mean precipitation or temperature values; O_i and P_i are the observed and simulated daily mean precipitation or temperature values, respectively; $\overline{O}$ is the average of the observed daily mean precipitation or temperature values; $\overline{P}$ is the average of the simulated daily mean precipitation or temperature values; σ_O is the variance of the observed daily mean precipitation or temperature values; σ_P is the variance of the simulated daily mean precipitation and temperature values; and n is the number of data.

Isaaks and Serivastava [23] suggested the MAE and RMSE as statistical indicators able to compare the accuracy of variables. Once the MAE and RMSE values are closer to zero in a scenario, the scenario would be more efficient for predicting climatic variables [24]. When they are exactly 0, it means that there is no error in the predicting task [24]. The Nash-Sutcliffe coefficient (NS) shows to what extent the regression line between the simulated data and measured data can be similar to the regression line

1:1. Its domain is from the negative infinity to 1, and NS = 1 reveals either a complete similarity or a perfect efficiency of a scenario [25]. Meanwhile, R^2 gives information on the correlation between the observed and predicted data and its domain is from 0 to 1 [26]. When R^2 becomes closer to 1, there will be a significant correlation between the data groups [26]. Significant differences between the observed data and values of the predictor scenarios can be distinguished by the analysis of variance [27]. Lack of any significant difference reveals a similarity between the predicted and observed data. In addition, to obtain more appropriate results for the prediction of precipitation, the occurrence of precipitation approach was used. This is a dichotomous method by which the accuracy of whether the occurrence or non-occurrence of precipitation is evaluated. If there is no occurrence of precipitation, then the answer is 'NO', while the answer 'Yes' is a sign of precipitation occurrence [28]. There are four statuses when the observed data are compared with scenario predictions, where a couple of predictions could be true and the remaining predictions could be false. The scenario with a higher percentage of true predictions was selected as the most efficient scenario for predicting the precipitation.

Finally, to predict the dryland winter wheat yield of the province for the next decades and to make a connection between the climatic and yield data for the period 2005–2014, a linear regression model was used. Furthermore, Pearson's correlation test (at $p < 0.01$) between the simulated and observed data, RMSE, and R-square were used to check the regression's validity. All statistical analyses were performed by the software SPSS version 21 (IBM Inc., Chicago, IL, USA).

3. Results

3.1. Temperature Predictions

All three CanESM2 scenarios predicted that the daily mean temperatures would generally increase in the periods 2010–2039, 2040–2069, and 2070–2099 (Table 2). However, the scale of these increases differed by the different scenarios. The scenario rcp2.6 projected that the daily mean temperature of the periods 2010–2039, 2040–2069, and 2070–2099 would be 13.6, 13.9, and 13.9 °C, respectively, which are 0.9, 1.2, and 1.1 °C higher when compared to the observed daily mean temperature. The other scenario rcp4.5 also predicted an increasing trend in the daily mean temperature in the three prospective periods and showed that the mean daily temperature would be 13.4, 14.2, and 14.4 °C in the periods 2010–2039, 2040–2069, and 2070–2099, respectively, each being 0.7, 1.4, and 1.6 °C higher when compared to the observed one. The scenario rcp8.5 predicted the highest temperature trends in comparison with the other two scenarios. It predicted that the mean daily temperature would rise by 13.8, 14.8, and 15.5 °C in the periods 2010–2039, 2040–2069, and 2070–2099, with changes of 1.0, 2.0, and 2.7 °C, respectively, in analogy with the observed value.

Both scenarios (A2 and B2) of HadCM3 generally predicted an increasing daily mean temperature trend for the three future periods in comparison with the observed one, except for scenario B2, which projected a very slightly decreasing trend only for the period 2070–2099 (Table 3). The scenario A2 forecasted that the mean daily temperature would rise to 12.7, 12.8, and 12.8 °C in the periods 2010–2039, 2040–2069, and 2070–2099, being 0.0, 0.1, and 0.2 °C higher, respectively, when compared to the value of the observed period. The mean daily temperatures were projected by the scenario B2 to increase to 12.6 and 12.7 °C in the periods 2010–2039, 2040–2069, respectively. In contrast, it predicted that the mean daily temperature would decrease to 12.6 °C in the period 2070–2099. Accordingly, the predicted temperature changes by scenario B2 are 0.02, 0.05, and −0.04 °C in the periods 2010–2039, 2040–2069, and 2070–2099, respectively, when compared to the observed period.

Table 2. Results of the daily mean temperature predictions of the CanESM2 scenarios for the periods 2010–2039, 2040–2069, and 2070–2099.

Scenarios	Periods	Daily Mean Temperature (°C)
Observed period	1985–2005 (obs)	12.7
rcp2.6	2010–2039 (P1)	13.6
	2040–2069 (P2)	13.9
	2070–2099 (P3)	13.9
	°C change P1 vs. obs	0.9
	°C change P2 vs. obs	1.2
	°C change P3 vs. obs	1.1
rcp4.5	2010–2039 (P1)	13.4
	2040–2069 (P2)	14.2
	2070–2099 (P3)	14.4
	°C change P1 vs. obs	0.7
	°C change P2 vs. obs	1.4
	°C change P3 vs. obs	1.6
rcp8.5	2010–2039 (P1)	13.8
	2040–2069 (P2)	14.8
	2070–2099 (P3)	15.5
	°C change P1 vs. obs	1
	°C change P2 vs. obs	2
	°C change P3 vs. obs	2.7

Table 3. Results of the daily mean temperature predictions of the HadCM3 scenarios for the periods 2010–2039, 2040–2069, and 2070–2099.

Scenarios	Periods	Mean Temperature (°C)
Observed period	1985–2005 (obs)	12.7
A2	2010–2039 (P1)	12.7
	2040–2069 (P2)	12.8
	2070–2099 (P3)	12.8
	°C change P1 vs. obs	0
	°C change P2 vs. obs	0.1
	°C change P3 vs. obs	0.2
B2	2010–2039 (P1)	12.6
	2040–2069 (P2)	12.7
	2070–2099 (P3)	12.6
	°C change P1 vs. obs	0.02
	°C change P2 vs. obs	0.05
	°C change P3 vs. obs	−0.04

3.2. Precipitation Predictions

Overall, the three scenarios of CanESM2 projected a diminishing trend in the annual precipitation for the future periods 2010–2039, 2040–2069, and 2070–2099, when compared to the observed period (Table 4). However, the scenario rcp2.6 projected a less decreasing trend in the annual precipitation for the period 2070–2099. The scenario rcp2.6 predicted that the annual precipitation would drop to 287 and 277 mm in the periods 2010–2039 and 2040–2069, respectively, and decrease to 296 mm in the period 2070–2099. The projected annual precipitation by the scenario rcp4.5 would be 258, 264, and 293 mm in the periods 2010–2039, 2040–2069, and 2070–2099, respectively. The other scenario rcp8.5 forecasted that the annual precipitation would be 283, 278, and 278 mm for the periods 2010–2039, 2040–2069, and 2070–2099, respectively.

Scenario A2 of HadCM3 predicted a decreasing trend in the annual precipitation for the periods 2010–2039, 2040–2069, and 2070–2099, in analogy with the observed period (Table 5). The annual precipitation projected by scenario A2 would be 340, 292, and 276 mm for the periods 2010–2039, 2040–2069, and 2070–2099, respectively. Scenario B2 also forecasted that the annual precipitation for the periods 2010–2039 and 2040–2069 would be 310 and 321 mm, respectively, when compared to the observed period, which conveys a reducing trend. In contrast, it projected an increased annual precipitation of 875 mm for the period 2070–2099, which will be noticeably higher than the observed amount.

Table 4. Results of the annual precipitation predictions of the CanESM2 scenarios for the periods 2010–2039, 2040–2069, and 2070–2099.

Scenarios	Periods	Precipitation (mm)
Observed period	1985–2005 (obs)	346
rcp2.6	2010–2039 (P1)	287
	2040–2069 (P2)	277
	2070–2099 (P3)	296
	% change P1 vs. obs	−18
	% change P2 vs. obs	−21
	% change P3 vs. obs	−15
rcp4.5	2010–2039 (P1)	258
	2040–2069 (P2)	264
	2070–2099 (P3)	293
	% change P1 vs. obs	−29
	% change P2 vs. obs	−26
	% change P3 vs. obs	−16
rcp8.5	2010–2039 (P1)	283
	2040–2069 (P2)	278
	2070–2099 (P3)	278
	% change P1 vs. obs	−20
	% change P2 vs. obs	−21
	% change P3 vs. obs	−21

Table 5. Results of the annual precipitation predictions of the HadCM3 scenarios for the periods 2010–2039, 2040–2069, and 2070–2099.

Scenarios	Periods	Precipitation (mm)
Observed period	1985–2005 (obs)	346
A2	2010–2039 (P1)	340
	2040–2069 (P2)	292
	2070–2099 (P3)	276
	% change P1 vs. obs	−1
	% change P2 vs. obs	−16
	% change P3 vs. obs	−22
B2	2010–2039 (P1)	310
	2040–2069 (P2)	321
	2070–2099 (P3)	875
	% change P1 vs. obs	−10
	% change P2 vs. obs	−7
	% change P3 vs. obs	86

3.3. Comparison of the Scenarios

The variance analysis results showed a higher efficiency for the RCP scenarios than the A and B scenarios in predicting the daily mean temperature of the region (Table 6), because there was no statistically significant difference between the temperature values simulated by the RCPs and the observed values (at $p < 0.01$), while the temperature values simulated by A and B significantly differed from the observed ones (at $p < 0.01$). Among the three scenarios of the model CanESM2, rcp2.6 was selected as the most efficient scenario for predicting the daily mean temperature, as it had the highest Nash-Sutcliffe coefficient and R^2 value and the lowest MAE and RMSE values when compared to scenarios rcp4.5 and rcp8.5.

The results of variance analysis indicated that all scenarios were efficient enough to predict the annual precipitation of the region (Table 7), since no statistically significant difference was found between the simulated and observed values (at $p < 0.01$). The scenario rcp2.6 displayed the lowest values for both MAE and RMSE. Moreover, it showed the highest Nash-Sutcliffe coefficient and R^2 value. Thus, it was selected as the best scenario for predicting the annual precipitation. In addition, the scenarios of CanESM2 simulated closer annual precipitation values to the observed values than the HadCM3 scenarios (Table 8). The CanESM2 scenarios resulted in higher values of true predictions and lower values of false prediction than the scenarios of HadCM3. The indicators provided in Table 8 also, in general, confirmed the excellence of scenario rcp2.6 for predicting the annual precipitation.

Together, these indicators showed a relatively higher efficiency for the CanESM2 scenarios than the HadCM3 scenarios in predicting the daily mean temperature and annual precipitation of the region.

Table 6. Results of the efficiency evaluation of the used scenarios for the daily mean temperature predictions.

Models	Scenarios	MAE	RMSE	Nash-Sutcliffe	R^2	Analysis of Variance
CanESM2	rcp2.6	0.348	0.445	0.808	0.8177	
	rcp4.5	0.355	0.45	0.801	0.8047	0.772 [ns]
	rcp8.5	0.362	0.461	0.795	0.8174	
HadCM3	A2	0.0529	0.0658	0.707	0.7346	0.000 **
	B2	0.0523	0.0654	0.706	0.7380	

ns: no-significant; **: significant at $p < 0.01$.

Table 7. Results of the efficiency evaluation of the used scenarios for the annual precipitation predictions.

Models	Scenarios	MAE	RMSE	Nash-Sutcliffe	Analysis of Variance
CanESM2	rcp2.6	0.434	1.297	−2.139	
	rcp4.5	0.442	1.298	−3.154	0.279 [ns]
	rcp8.5	0.45	1.351	−8.576	
HadCM3	A2	0.444	1.33	−7.243	0.453 [ns]
	B2	0.442	1.299	−3.222	

ns: no-significant.

Table 8. Occurrence of precipitation under the used scenarios.

Occurrences	CanESM2			HadCM3	
	rcp8.5	rcp4.5	rcp2.6	B2	A2
Hit (hit event)	390	395	366	406	425
CN (correct Negative)	1832	1827	1856	1816	1797
Miss (miss event)	1246	1225	1250	1191	1159
FA (false alarm events)	184	205	180	239	271
% true prediction ($\frac{Hit+CN}{n}$)	44.79	44.35	44.25	43.72	43.37
% false prediction ($\frac{Miss+FN}{n}$)	55.2	55.64	55.75	56.27	56.62

3.4. Yield Predictions

The results of the regression analysis and Pearson's correlation test showed that the precipitation in March was the most effective factor for the dryland winter wheat yield of the region (Table 9). The prediction results indicated that the yield would noticeably reduce to 1176, 984, and 890 kg ha^{-1} in the periods 2010–2039, 2040–2069, and 2070–2099, respectively (Table 10). The reduction percentage in the above-mentioned periods is predicted to be −22, −34, and −41%, respectively. These reductions in the yield are consistent with the reductions in the mean precipitation in March during the three prospective periods (Figure 4). The reduction in the yield in the periods 2040–2069 and 2070–2099 will be more severe than that of the period 2010–2039, which is in line with a more severe reduction in the precipitation in March than in the former periods.

Table 9. Regression and correlation results of the yield and precipitation data.

Crop	Regression Model	R	R^2	RMSE (%)	Significance Level	Predictor Model
winter wheat	Forward	0.78	0.62	18.82	0.012 *	Y = 20.883X + 625.846

*: significant at $p < 0.05$ where Y is dryland winter wheat yield; X is the precipitation in March; and the constant numbers are Y-intercepts.

Table 10. Results of the dryland winter wheat yield predictions for the periods 2010–2039, 2040–2069, and 2070–2099.

Crop	Cropping Year	Grain Yield (kg ha^{-1})
	2010–2011 (obs)	1512
	2010–2039 (P1)	1176
	2040–2069 (P2)	984
Winter wheat	2070–2099 (P3)	890
	% change P1 vs. obs	−22
	% change P2 vs. obs	−34
	% change P3 vs. obs	−41

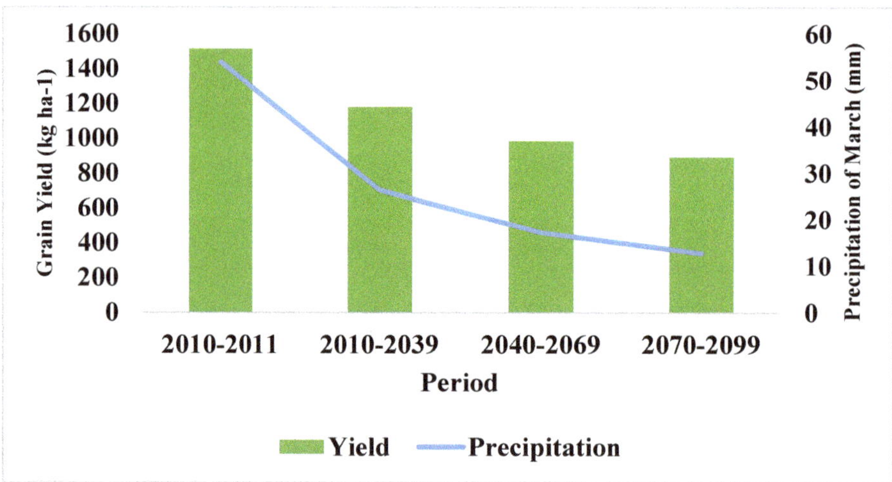

Figure 4. Relationship between the yield reduction and rcp2.6-induced precipitation of March in the three future periods.

4. Discussion

4.1. Temperature Predictions

GCMs have widely been used for predicting future temperature trends. Van Vuuren et al. [29] showed that the mean temperature was likely to increase in the future in many parts of the world. For instance, Basheer et al. [30] claimed that the climate over the Dinder River Basin would be warmer in the upcoming decades. Majhi and Pattnayak [31] also revealed that there would be a gradual temperature increase in Nabarangpur district at the end of the 21st century. Our results also indicated that the temperature would generally increase in the three investigated periods; however, the magnitude of these increases are dependent on the scenarios applied. The CanESM2 scenarios postulated a higher variability in the predicted temperature values than the HadCM3 scenarios. In addition, the temperature changes predicted by CanESM2 were noticeably higher than those predicted by HadCM3. Such different trends have also been observed by [22], who compared some GCMs such as HadCM3 and CanESM2. These diverse trends could have been due to the different scenarios used, as was the case for the study of [32]. Among the CanESM2 scenarios, rcp8.5 and rcp4.5 predicted the highest temperature values, respectively, whilst rcp2.6 projected the lowest ones. These results are in line with the findings of [22]. The greatest temperature values predicted by scenarios rcp8.5 and rcp4.5 seem plausible due to the underlying physical laws to simulate the ongoing increases in the radiative forcing and CO_2 concentrations by the end of the 21st century. In contrast, rcp2.6 simulated a lower radiative forcing towards the end of the 21st century as well as lower CO_2 concentrations.

4.2. Precipitation Predictions

All scenarios, except B2, revealed that there would be a reduction in the annual precipitation in all investigated periods. Scenarios rcp4.5 and rcp8.5 projected the maximum and the minimum reductions in the annual precipitation, respectively, which was a very similar result to what [33] concluded. Scenario B2 projected substantial increases in the annual precipitation for the period 2070–2099. Moreover, scenario rcp2.6 projected a less decreased annual precipitation for the aforementioned period. One study has shown that there is a possibility for a reduction in the rivers' ice thickness in winter and a slight increase in the discharge during the break up from May to June in Siberia [34]. This phenomenon can be caused by extreme warming around Siberia in the period 2070–2099. To confirm this notion, Shiklomanov et al. [35] predicted an increased mean temperature trend for Siberia by the late 21st century. The province of Qazvin is extremely affected by Siberian winds. Therefore, the increased and less decreased annual precipitation projections for the period 2070–2099 by scenarios B2 and rcp2.6 might be logical. Nevertheless, the properties of the scenarios used could be among other reasons for the different precipitation results achieved. Scenarios rcp2.6 and B2 more optimistically simulated the future projections when compared to the other scenarios used. For instance, rcp2.6 predicted a radiative forcing of 3 W m^{-2} and a CO_2 concentration of 490 ppm; and B2 described a world with intermediate population and economic growth, emphasizing local solutions to economic, social, and environmental sustainability. Thus, a more optimistic simulation of the annual precipitation of the region could have been another possible reason for the increased and less decreased precipitation values predicted. Vallam and Qin [22], using a statistical downscaling technique, also showed that scenarios rcp2.6 and B2 could predict either increased or at least lesser decreased rainfall percentage for Frankfurt (Germany), Singapore, and Miami (USA) in the 2080s when compared to the other scenarios used. However, the CanESM2-derived RCP scenarios led to great variabilities in predicting future meteorological variables, especially rainfall in arid regions [22]. This might be another plausible reason for the increase (14%) in the annual precipitation predicted by rcp2.6.

4.3. Yield Predictions

Studies have shown that there is a significant correlation between winter wheat yield and the climatic variables [16]. Thus, the most efficient scenario (rcp2.6) in predicting both temperature and precipitation was applied to predict the dryland winter wheat yield of the province. The results of the Pearson's correlation test indicated that the precipitation in March was the most effective factor on yield ($r = 0.78$, $p < 0.01$). A study on the effects of precipitation on dryland cereals yield in three provinces of Iran was performed, where the climate is semi-arid [36]. The results of the study showed that the yield of dryland winter wheat was significantly correlated to precipitation, especially the precipitation in April. In the province of Qazvin, dryland winter wheat is at the tillering stage in March (personal communication with the farmers). It seems that the lower precipitation in March could lead to a lower number of head-bearing tillers and lack of the opportunity for their survival, finally resulting in lower grain yields. Karimi [37] investigated the effects of precipitation during the tillering of dryland winter wheat in Iran and reported a significant impact on the final grain yield. Even though agricultural factors such as soil, fertilizers, and other climatic variables like radiation could also be effective, Lobell [16] indicated that precipitation had a more considerable influence on dryland farming. Meanwhile, the value of R^2 between the observed and simulated data was 0.62, meaning that the yield was 62% dependent on the annual precipitation and the other 38% was dependent on other unspecified factors. The percentage of RMSE was about 18% between the observed and simulated data, which was an acceptable value that showed the adequate accuracy of the predictions [38]. Moreover, the observed reductions in the precipitation in March during the three future periods could have been due to shifts in the seasons due to warmer temperatures of the areas by which the studied region is affected. As mentioned earlier, the temperature of Siberia has been projected to rise by the late 21st century [35]. Since the province of Qazvin is extremely affected by Siberian winds, it is plausible that these winds will alter the seasons of this province.

5. Conclusions

In this study, the downscaling of two important climatic variables—temperature and precipitation—was done by the CanESM2 and HadCM3 models for the province of Qazvin, located in Iran. The used scenarios were able to predict the daily mean temperature and annual precipitation for the three different future periods 2010–2039, 2040–2069, and 2070–2099. The CanESM2 scenarios seemed to be more efficient than the HadCM3 scenarios in simulating the future temperature and precipitation trends of the region. Generally, the region's daily mean temperature tended to increase and the annual precipitation tended to decrease in the three prospective periods investigated. However, scenarios rcp2.6 and B2, respectively, predicted that the precipitation would decrease less or even increase in the third period (2070–2099). Scenario rcp2.6 was assumed to be the most efficient to predict the dryland winter wheat yield of the province for the upcoming decades. The grain yield was projected to considerably decrease in the three periods, especially in the last period. The yield reductions are assumed to mainly be due to the decrease in precipitation in March during the investigated periods. Some adaptive strategies to prevent the detrimental impacts of climate change on the province dryland wheat yield include the cultivation of resistant winter wheat varieties to drought as well as earlier sowing dates. The authors would like to recommend the comparative use of the applied CanESM2 and HadCM3 scenarios to predict climatic variables of other semi-arid regions.

Author Contributions: Conceptualization, B.M.; Methodology, B.M. and M.N.; Software, B.M. and M.N.; Validation, B.M. and M.N.; Formal Analysis, B.M. and M.N.; Investigation, B.M. and M.N.; Resources, B.M. and M.N.; Data Curation, B.M. and M.N.; Writing-Original Draft Preparation, M.N.; Writing-Review & Editing, B.M. and M.N.; Visualization, B.M. and M.N.; Supervision, B.M. and M.N.; Project Administration, B.M.; Funding Acquisition, B.M.

Funding: This research received no external funding.

Acknowledgments: The authors would like to acknowledge the personnel of the Qazvin Meteorological Organization for providing the meteorological data. Mohammad Eteghadipour is also acknowledged for his useful scientific guides.

Conflicts of Interest: The authors declare no conflicts of interest.

References

1. Thomas, R.K.; Zhang, R.; Horowitz, L.W. Prospects for a prolonged slowdown in global warming in the early 21st century. *Nat. Communities* **2016**. [CrossRef]
2. Nozawa, T.; Nagashima, T.; Shiogama, H.; Crooks, S.A. Detecting natural influence on surface air temperature change in the early twentieth century. *Geophys. Res. Lett.* **2005**, *32*, L20719. [CrossRef]

3. Santer, B.D.; Taylor, K.E.; Wigley, T.M.; Johns, T.C.; Jones, P.D.; Karoly, D.J.; Mitchell, J.F.B.; Oort, A.H.; Penner, J.E.; Ramaswamy, V.; et al. A search for human influences on the thermal structure of the atmosphere. *Nature* **1996**, *382*, 39–46. [CrossRef]

4. White, R.P.; Nackoney, J. Drylands, People, and Ecosystem Goods and Services: A Web-Based Geospatial Analysis. 2003. Available online: http://pdf.wri.org/drylands.pdf (accessed on 17 June 2018).

5. LADA. *Guidelines for Land Use System Mapping*; Technical Report; FAO: Rome, Italy, 2008.

6. Wang, X.; Cai, D.; Wu, H.; Hoogmoed, W.B.; Oenema, O. Effects of variation in rainfall on rainfed crop yields and water use in dryland farming areas in China. *Arid Land Res. Manag.* **2016**, *30*, 1–24. [CrossRef]

7. Andreadis, K.M.; Lettenmaier, D.P. Trends in 20th century drought over the continental United States. *Geophys. Res. Lett.* **2006**, *33*, L10403. [CrossRef]

8. UNEP. Sourcebook of Alternative Technologies for Freshwater Augmentation in West Asia. 2000. Available online: http://www.unep.or.jp (accessed on 15 June 2018).

9. Gan, T.Y. Reducing vulnerability of water resources of Canadian Prairies to potential droughts and possible climate warming. *Water Resour. Manag.* **2000**, *14*, 111–135. [CrossRef]

10. IPCC. *Climate Change: Impacts, Adaptation and Vulnerability*; Cambridge University Press: New York, NY, USA, 2007.

11. Wilby, R.L.; Dawson, C.W.; Barrow, E.M. SDSM- a decision support tool for the assessment of regional climate change impacts. *Environ. Modell. Softw.* **2002**, *17*, 147–159. [CrossRef]

12. Gulden, K.U.; Neşe, G. A Study on Multiple Linear Regression Analysis. *Procedia Soc. Behav. Sci.* **2013**, *106*, 234–240.

13. Tatsumi, K.; Oizumi, T.; Yamashiki, Y. Introduction of daily minimum and maximum temperature change signals in the Shikoku region using the statistical downscaling method by GCMs. *Hydrol. Res. Lett.* **2013**, *7*, 48–53. [CrossRef]

14. Ribalaygua, J.; Pino, M.R.; Pórtoles, J.; Roldán, E.; Gaitán, E.; Chinarro, D.; Torres, L. Climate change scenarios for temperature and precipitation in Aragón (Spain). *Sci. Total Environ.* **2013**, *463–464*, 1015–1030. [CrossRef] [PubMed]

15. Johns, T.C.; Gregory, J.M.; Ingram, W.J.; Johnson, C.E.; Jones, A.; Lowe, J.A.; Mitchell, J.F.B.; Roberts, D.L.; Sexton, D.M.H.; Stevenson, D.S. Anthropogenic climate change for 1860 to 2100 simulated with the HadCM3 model under updated emissions scenarios. *Clim. Dyn.* **2003**, *20*, 583–612. [CrossRef]

16. Lobell, D.B.; Ortiz Monasterio, J.I.; Addams, C.L.; Anser, G.P. Soil, climate and management impacts on regional wheat productivity in Mexico from remote sensing. *Agric. For. Meteorol.* **2002**, *114*, 31–43. [CrossRef]

17. Lobell, D.B.; Asseng, S. Comparing estimates of climate change impacts from process-based and statistical crop models. *Environ. Res. Lett.* **2017**, *12*, 015001. [CrossRef]

18. Russell, K.; Chad, L.; Rebecca, M.L.; David, V.S. Impact of Climate Change on Wheat Production in Kentucky. *Plant Soil Sci. Res. Rep.* **2014**, *2*. [CrossRef]

19. Landau, S.; Mitchell, R.; Barnett, V.; Colls, J.J.; Craigon, J.; Payne, R.W. A parsimonious, multiple- regression model of wheat yield response to environment. *Agric. For. Meteorol.* **2000**, *101*, 151–166. [CrossRef]

20. Bin, W.D.; Liu, L.; O'Leary, G.J.; Asseng, S.; Macadam, I.; Lines-Kelly, R.; Yang, X.; Clark, A.; Crean, J.; Sides, T.; et al. Australian wheat production expected to decrease by the late 21st century. *Glob. Chang. Biol.* **2018**. [CrossRef]

21. Lhomme, J.P.; Mougou, R.; Mansour, M. Potential impact of climate change on durum wheat cropping in Tunisia. *Clim. Chang.* **2009**, *96*, 549–564. [CrossRef]

22. Vallam, P.; Qin, X.S. Projecting future precipitation and temperature at sites with diverse climate through multiple statistical downscaling schemes. *Theor. Appl. Climatol.* **2017**. [CrossRef]

23. Isaaks, E.H.; Serivastava, R.M. *An introduction to applied Geostatistics*; Oxford University Press: New York, NY, USA, 1989.

24. Chai, T.; Draxler, R.R. Root mean square error (RMSE) or mean absolute error (MAE)?—Arguments against avoiding RMSE in the literature. *Geosci. Model Dev.* **2014**, *7*, 1247–1250. [CrossRef]

25. Nash, J.E.; Sutcliffe, J.V. River flow forecasting through conceptual models part I—A discussion of principles. *J. Hydrol.* **1970**, *10*, 282–290. [CrossRef]

26. Gujarati, D.N.; Porter, D.C. *Basic Econometrics*, 5th ed.; Tata McGraw-Hill Education: New York, NY, USA, 2009; pp. 73–78.

27. Armstrong, R.A.; Eperjesi, F.; Gilmartin, B. The application of analysis of variance (ANOVA) to different experimental designs in optometry. *Ophthalmic Physiol. Opt.* **2002**, *22*, 248–256. [CrossRef] [PubMed]

28. Roberts, N.M.; Lean, H.W. Scale-selective verification of rainfall accumulations from high-resolution forecasts of convective events. *Mon. Weather Rev.* **2008**, *136*, 78–97. [CrossRef]

29. Van Vuuren, D.P.; Meinshause, M.; Plattner, G.K.; Joos, F.; Strassmann, K.M.; Smith, S.J.; Reilly, J.M. Temperature increase of 21st century mitigation scenarios. *Proc. Natl. Acad. Sci. USA* **2008**, *105*, 15258–15262. [CrossRef] [PubMed]

30. Basheer, A.K.; Lu, H.; Omer, A.; Ali, A.B.; Abdelghader, A.M.S. Impacts of climate change under CMIP5 RCP scenarios on the streamflow in the Dinder River and ecosystem habitats in Dinder National Park, Sudan. *Hydrol. Earth Syst. Sci.* **2016**, *20*, 1331–1353. [CrossRef]

31. Majhi, S.; Pattnayak, K.C.; Pattnayak, R. Projections of rainfall and surface temperature over Nabarangpur district using multiple CMIP5 models in RCP 4.5 and 8.5 scenarios. *Int. J. Appl. Res.* **2016**, *2*, 399–405.

32. Mekonnen, D.F.; Disse, M. Analyzing the future climate change of Upper Blue Nile River Basin (UBNRB) using statistical down scaling techniques. *Hydrol. Earth Syst. Sci.* **2016**, *22*, 2391–2408. [CrossRef]

33. Aung, M.T.; Shrestha, S.; Weesakul, S.; Shrestha, P.K. Multi-model climate change projections for Belu River Basin, Myanmar under representative concentration pathways. *J. Earth Sci. Clim. Chang.* **2016**, *7*, L323.

34. Costard, F.; Gautier, E.; Brunstein, D.; Hammadi, J.; Fedorov, A.; Yang, D.; Dupeyrat, L. Impact of the global warming on the fluvial thermal erosion over the Lena River in Central Siberia. *Geophys. Res. Lett.* **2007**, *34*, L14501. [CrossRef]

35. Shiklomanov, N.I.; Streletskiy, D.A.; Swales, T.B.; Kokorev, V.A. Climate change and stability of urban infrastructure in Russian permafrost regions: Prognostic assessment based on GCM climate projections. *Geogr. Rev.* **2017**, *107*, 125–142. [CrossRef]

36. Bannayan, M.; Lotfabadi, S.S.; Sanjani, S.; Mohamadian, A.; Aghaalikhani, M. Effects of precipitation and temperature on crop production variability in northeast Iran. *Int. J. Biometeorol.* **2011**, *55*, 387–401. [CrossRef] [PubMed]

37. Karimi, M. Drought during growing season of 1997–8 and its effects on wheat production in Iran. *Sonbloe J.* **1999**, *30*, 1–7.

38. Rinaldy, M.; Losavio, N.; Flagella, Z. Evaluation of OILCROP-SUN model for sunflower in southern Italy. *Agric. Syst.* **2003**, *78*, 17–30. [CrossRef]

Perspective

New Breeding Techniques for Greenhouse Gas (GHG) Mitigation: Plants May Express Nitrous Oxide Reductase

Jordan J. Demone [1,†], Shen Wan [1,†], Maryam Nourimand [1], Asbjörn Erik Hansen [1], Qing-yao Shu [2] and Illimar Altosaar [1,*]

[1] Department of Biochemistry, Microbiology, and Immunology, University of Ottawa, Ottawa, ON K1H 8M5, Canada; jdemone@uottawa.ca (J.J.D.); swa2@uottawa.ca (S.W.); mnourima@uottawa.ca (M.N.); aerik.hansen@gmail.com (A.E.H.)

[2] National Key Laboratory of Rice Biology, Institute of Crop Sciences, Zhejiang University, Hangzhou 310058, China; qyshu@zju.edu.cn

* Correspondence: altosaar@uottawa.ca; Tel.: +1-613-562-5800 (ext. 6371)

† These authors contributed equally to this work.

Received: 4 August 2018; Accepted: 25 September 2018; Published: 27 September 2018

Abstract: Nitrous oxide (N_2O) is a potent greenhouse gas (GHG). Although it comprises only 0.03% of total GHGs produced, N_2O makes a marked contribution to global warming. Much of the N_2O in the atmosphere issues from incomplete bacterial denitrification processes acting on high levels of nitrogen (N) in the soil due to fertilizer usage. Using less fertilizer is the obvious solution for denitrification mitigation, but there is a significant drawback (especially where not enough N is available for the crop via N deposition, irrigation water, mineral soil N, or mineralization of organic matter): some crops require high-N fertilizer to produce the yields necessary to help feed the world's increasing population. Alternatives for denitrification have considerable caveats. The long-standing promise of genetic modification for N fixation may be expanded now to enhance dissimilatory denitrification via genetic engineering. Biotechnology may solve what is thought to be a pivotal environmental challenge of the 21st century, reducing GHGs. Current approaches towards N_2O mitigation are examined here, revealing an innovative solution for producing staple crops that can 'crack' N_2O. The transfer of the bacterial nitrous oxide reductase gene (*nosZ*) into plants may herald the development of plants that express the nitrous oxide reductase enzyme (N_2OR). This tactic would parallel the precedents of using the molecular toolkit innately offered by the soil microflora to reduce the environmental footprint of agriculture.

Keywords: radiative warming; atmospheric phytoremediation; N_2O; nitrous oxide reductase; N_2OR; *nosZ*; fertilizer; crop breeding; transgenic; GHG

---

1. Introduction—Nitrous Oxide Continues to Bloom Unabated

Atmospheric nitrogen (N) deposition is a pressing matter for climate change scientists concerned with the increasing danger that nitrous oxide (N_2O), a noxious greenhouse gas (GHG), poses. Reactive nitrogen (Nr)—ammonia (NH_3), nitrogen oxides (NO_x), nitrates (NO_3^-), and N_2O—enters the biosphere from its original form of atmospheric N as at least three derivatives: gas, dry deposit, and precipitation (wet deposition) [1,2]. The sources of N_2O are largely anthropogenic [3]. Many crops must receive N-based fertilizer to reach yield targets, which is supplied by inorganic fertilizers and animal manure [4]. In an effort to boost the yield in crop staples like wheat, corn, and soybeans, farmers apply N fertilizers at rates and times that are not always properly synchronized with crop demand [5]. While crops thrive when fertilized, experimental analysis has demonstrated that up to

40% of fertilizer N can be lost via leaching [6,7]. Other routes of N loss include soil erosion, NH_3 volatilization and oxidation, and bacterial/fungal denitrification [8], although N losses through NH_3 volatilization are higher than those via N leaching [9]. Around 62% of total global N_2O issues from natural and agricultural soils, and the bulk of this production, mainly results from the processes of bacterial nitrification and denitrification [10].

Nr compounds enter the atmosphere through biological processes, but the invention of the Haber-Bosch process in 1908 was a critical moment for the sudden increase in Nr and GHG production globally [11]. This process of artificial N-fixation allowed for the large-scale reduction of N_2 to NH_3, producing massive amounts of synthetic N-based fertilizers that supported dramatic increases in high-yield farming [12]. This process now accounts for 80% of anthropogenic N-fixation (the remaining 20% resulting from combustion [13], with anthropogenic N-fixation in turn accounting for 60% of global N-fixation [14]). Haber-Bosch remains the industry standard synthetic N fertilizer today and as a result, has contributed to the ~2% increase in atmospheric levels of N_2O [15,16]. This effect is also magnified by the global emissions of N_2O produced by fossil fuel combustion [17] and the natural ability of legumes to fix N through symbiotic relationships with soil bacteria [18].

N_2O is the third most prevalent GHG, behind carbon dioxide (CO_2) and methane (CH_4) [19]. The concentration of this gas in the atmosphere has been steadily increasing since the early 1900s (Figure 1), and it is 265 times more radiative than CO_2 [19]. N_2O also has an atmospheric lifetime of 121 years; by comparison, CH_4 has an atmospheric lifetime of only 12 years, but CO_2 also has a long half-life and can take anywhere from 20–200 years to be absorbed by the ocean [19], compounding the 'greenhouse gas' effect. Since chlorofluorocarbons (CFCs) were banned in 1989, N_2O has become the leading cause of ozone layer depletion [20].

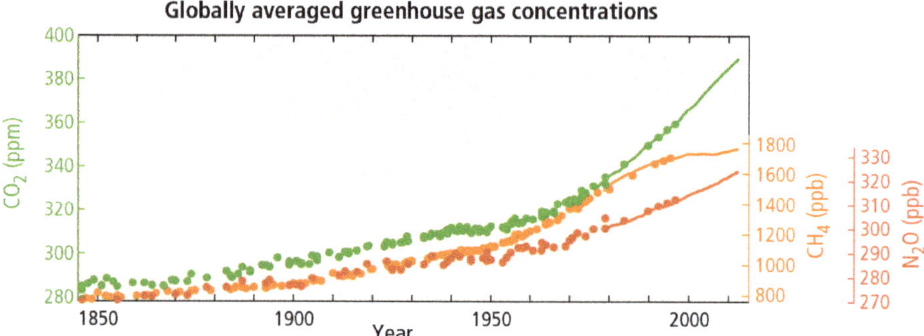

Figure 1. GHG levels since 1850. The green line represents the increase in CO_2 concentration since 1850; the orange line represents the increase in CH_4 concentration since 1850; lastly, the red line represents the increase in N_2O since 1850 [19].

N_2O emission results from the coupled oxidation and reduction of N performed by heterotrophic [21] (and some autotrophic) soil proteobacteria: (1) the nitrification pathway is catalyzed by autotrophs (*Nitrosomonas* spp. and other genera [22]) and also heterotrophs, and involves the oxidation of NH_3/ammonium(NH_4^+) to nitrite (NO_2^-) [23] and nitric oxide (NO) [24]), which is followed by the oxidation of NO_2^- to NO_3^- by *Nitrobacter* spp. [25]; and (2) the denitrification pathway, whereby NO_3^- is reduced to N_2O and ultimately inert N_2 gas [26]. As many as a third of soil bacterial species [27] lack the *nosZ* gene that reduces N_2O to inert N_2 [28], which leads to a sizeable amount of incomplete denitrification reactions and the subsequent buildup of N_2O since it is an obligate intermediate [29]. This N_2O diffuses out of the soil and into the atmosphere, contributing to the greenhouse effect, contaminating water, and leading to serious human health implications [30,31].

2. Combating GHGs: Current N_2O Mitigation Strategies and Limitations

Demands for crop-borne food must be met, and so researchers must address the hazards of N-based fertilizers [32]. There are multiple N_2O mitigation strategies either currently in commercial use or in development (summarized in Table 1).

Table 1. Summary of current N_2O mitigation strategies.

Strategy	Mechanism of Action	Pros	Cons
(1) Conservation tillage and crop rotation [33]	Tillage, rotation of N-fixing crops, cover cropping [33]	Prevent NH_3 volatilization and eventual N_2O emissions [34,35]	Unreliable N_2O mitigation [36,37]. Yield reduction [38]. Not effective at scrubbing N_2O from the air
(2) Best management practices (BMPs) [39]	Correct source, placement, time, and rate of fertilization [40]. Proper irrigation (fertigation) [41]	Proven to reduce N_2O emissions [41] and other N losses [42]	Technical constraints [43]
(3) EENFs [44]	Multiple types: stable, short-release (SRFs), and constant-release (CRFs); rely on enrichment of chemical inhibitors or coated N-compounds that are released into the soil over a period of time [45]; urease inhibitors (UIs) [46]	Proven to reduce N_2O emissions [47,48]	Inconsistent yields from year to year [48]. More expensive than standard N fertilizers [49]. Long lifetime of N-compounds in soil can lead to NH_3 volatilization [50,51]. Not effective at scrubbing N_2O from the air
(4) Synthetic N_2O mitigators	SNIs suppress activity of nitrifying bacteria in the soil [52]. SDIs operate by unknown mechanism [44,53,54]	SNIs and SDIs reduce N_2O emissions [52,54]	Effectiveness depends on environmental conditions, prefer low temperature and sandy soils [55]. Not effective at scrubbing N_2O from the air
(5) Biological N_2O mitigators	BNIs suppress activity of nitrifying bacteria in the soil by releasing compounds that inhibit NH_3-oxidizing pathways [56]. BDIs inhibit nitrate reductase to inhibit N_2O production [57]	BNIs demonstrated to reduce N_2O emission [56]; BDIs inhibit denitrification and can conceivably mitigate N_2O emissions [57]	BNI-exuding plants must be grown in rotation with other crops [58]. Little work done on BDI-exuding plants [57]. Not effective at scrubbing N_2O from the air
(6) Microbial bioremediation	Proper water table management to facilitate growth of rhizobia [59]; inoculation of plant roots with genetically modified N_2O-cracking rhizobia [60,61]	Enables plants to degrade contaminants in the soil; N_2O-cracking rhizobia demonstrated to reduce N_2O emissions [60,61]	Most effective on crops that naturally cultivate a rhizosphere of N_2O-reducing [62] microorganisms, i.e., soybean [63]. Not effective at scrubbing N_2O from the air
(7) Rhizosecretion	Transformation of amenable crops to express recombinant bacterial proteins that reduce N_2O [64]	Plants that secrete N_2O-cracking enzyme could target N_2O in soil [64]	Plant transformation is a time-consuming process [65]. Bacterial proteins may not function efficiently in heterologous hosts [66]. Not effective at scrubbing N_2O from the air
(8) Atmospheric phytoremediation	Transformation of amenable crops with genes expressing recombinant bacterial proteins that reduce N_2O [67]	Arm crops and other plant species to mop up N_2O in the atmosphere [67], including N_2O emitted by other non-agricultural sources	Plant transformation is a time-consuming process [65]. Bacterial genes may not function in a heterologous system [66]. Not yet experimentally validated via gas analysis

BDI, biological denitrification inhibitor; BNI, biological nitrification inhibitor; EENFs, enhanced efficiency nitrogen fertilizers; SDI, synthetic denitrification inhibitor; SNI, synthetic nitrification inhibitor.

(1) **Conservation tillage and crop rotation**. Mechanical incorporation (tillage) of N-based fertilizer into the soil may also be effective [68], but this is affected by many other parameters, such as the method of N application (i.e., broadcast vs surface urea ammonium nitrate). These techniques also result in a reduced yield [38]. Conservation tillage increases N_2O emissions compared with no-till and conventional tillage techniques using broadcast application, while tillage in general does not reduce N_2O emissions produced from surface urea ammonium nitrate-treated fields [69]. Other studies have shown that conservation tillage *reduces* N_2O emissions [70], underscoring the lack of reliability of this N management technique [36,37]. Crop rotation with N-acquisitive plant species can also reduce N_2O emissions following the application of high N-fertilizer treatment [33]; cover cropping can also control N_2O emissions, but the results are often variable and in some cases can increase N_2O emissions [71];

(2) **Best management practices (BMP)** [39]. Such nitrogen use efficiency techniques are myriad and involve simple steps such as proper fertilizer placement, timing of fertilizer application, the right type of N-compound, and so on. Others involve the proper incorporation of N-compounds into the soil so that they may be taken up by the plant more effectively and will be less likely to volatilize [72]. Fertigation, a technique involving careful irrigation of fields following the application of N fertilizer, is effective at mitigating N_2O emissions [41]. Such knowledge-based N management practices have been shown to be effective at both increasing crop yield and reducing immediate N_2O emissions [73], but some approaches may also increase N_2O production in the long term [55]. Their effectiveness also depends heavily on proper practices put in place by the farmers themselves, which requires proper training [43];

(3) **Fertilizer management using enhanced efficiency nitrogen fertilizers (EENFs)**. These fertilizer cocktails are concocted in such a way that they prevent the volatilization of NH_3 and inhibit nitrification/denitrification [46]. EENFs generally fall into one of three categories: (a) stabilized fertilizers, which contain nitrification and/or urease inhibitors; (b) slow-release fertilizers (SRFs), whereby the N source in the fertilizer is released over time from encapsulated granules, although the release rates can be variable; and (c) controlled-release fertilizers (CRFs), where the release rate is constant [45]. Urease inhibitors (UIs) are also a common EENF component. N-(*n*-butyl) thiophosphoric triamide (NBPT), phenylphosphorodiamidate (PPD), and hydroquinone are used worldwide and act by inhibiting the bacterial hydrolysis of urea into NH_3 in fertilizer [46,74,75]. UIs are typically used in conjunction with nitrification inhibitor (NIs) for maximum effectiveness [76,77], but NBPT alone can reduce N_2O emissions from N-treated soil [78]. There is controversy regarding the effectiveness of EENFs; while reductions in N_2O emissions from the soil have been recorded [47,48], recent studies have shown that crop yields are only marginally higher when EENFs are used in place of standard N fertilizers [79]. Those studies that demonstrated reduced N_2O emissions also reported inconsistent results from year to year [50]. Questionable effectiveness notwithstanding, EENFs are more expensive than conventional N-containing fertilizers and require special handling and storage [49,80], which are all features that make these fertilizers less attractive to farmers;

(4) **Synthetic N_2O mitigators**. Synthetic nitrification inhibitors (SNIs) and UIs are both used in EENFs and can be applied to crops in conjunction with standard N fertilizer. NIs inhibit the activity of *Nitrosomonas* to block the nitrification of N in fertilizer (the oxidation of NH_3 to hydroxylamine via ammonia monooxygenase (AMO)) [23,52]. The efficacy of the inhibitors is also dependent on environmental conditions, as they are unstable; 3,4-dimethylpyrazole phosphate (DMPP), for example, exhibited reduced activity in hot, dry conditions [81]. The use of these inhibitors can also lead to less than desirable results: DMPP and 3-methylpyrazole 1,2,4-triazole (3MP + TZ) have been shown to increase N_2O emissions in vegetable crop systems, as the inhibitors promote the buildup of N in the fraction of the soil most available to bacteria during the breakdown of vegetative matter. Synthetic denitrification inhibitors (SDIs) suppress denitrification via unknown mechanisms [82], although some are known to inhibit the activity of fungal copper reductase [83].

SDIs nitrapyrin [84], toluidine [54], and acetylene [44] all effectively mitigate N_2O emission, albeit with toxic side-effects [55], and they do not technically inhibit nitric oxide reductase;

(5) **Biological N_2O mitigators**. This category is comprised of compounds produced by plants that inhibit enzymes in either the bacterial nitrification or denitrification pathway. The exploitation of such inhibiting root exudates is another intriguing approach towards N_2O mitigation [82]. Biological nitrification inhibitors (BNIs) are compounds that block the activity of NO_2^- producing enzymes. The roots of the tropical grass *Brachiaria humidicola* exude brachialactone, a compound that can mitigate N_2O emission from soil [85]. Attempts at developing BNI-producing cultivated wheat by crossing *Triticum aestivum* with BNI-producer *Leymus racemosus*, a wild wheat, have imparted some BNI activity, but also made the lines susceptible to rust infection [86]. The use of BNIs as an effective N_2O mitigator is also severely limited by the fact that the enactor of nitrification is a plant itself and cannot be applied to growing crops, although growing *B. humidicola* in rotation with maize saw a four-fold increase in yield [87]. Biological denitrification inhibitors (BDIs) are a relatively new discovery. Currently, the only example of such an inhibitor is the procyanidin produced by the invasive Fallopia spp. (Asian knotweed). This compound has been demonstrated to be an allosteric inhibitor of *Pseudomonas brassicacearum* nitrate reductase and while it does reduce denitrification in the soil, it has not yet been proven to mitigate N_2O levels [57];

(6) **Microbial bioremediation** [88]. The success of N fertilizer management techniques and proper irrigation is largely due to the creation of a microsphere conducive to denitrifying bacteria flourishing [89]. Proper water table management techniques can promote the growth of N_2O-cracking bacteria in the soil and reduce N_2O emissions from the managed soil regions [59]. Another type of microbial bioremediation takes advantage of the ability of certain bacterial species to inhabit the root nodules of leguminous crops. Field peas [62], broad beans [90], and soybean [63] house bacteria (or rhizobia) that fix N and, unfortunately, also produce N_2O gas. While maintaining the rhizosphere, N_2O emissions can be mitigated by inoculating the roots of leguminous plants with rhizobia modified to express higher levels of a bacterial N_2O-cracking enzyme [60]. Genetically engineered strains of *Bradyrhizobium japonicum* have been used to inoculate the roots of soybean and reduced N_2O emissions [61]. Needless to say, this method is far more effective on crops that naturally cultivate a rhizosphere of N_2O-reducing microorganisms. It is also another technique that cannot target atmospheric N_2O;

(7) **Rhizosecretion**. This is a biotechnology-based approach, involving the transformation of amenable crop plants with genes expressing recombinant bacterial proteins that reduce N_2O by secreting N_2O-cracking enzymes [64,91]. Plants can be engineered to express proteins under the control of promoters that induce hairy root formation in plants. This rooting response results from the presence of the *rolABCD* genes from *Agrobacterium rhizogenes*, the bacterium that induces hairy root disease [92]. The rhizosecretion expression system harnesses the ability of *A. rhizogenes* to both target gene expression to the roots and to increase root biomass, subsequently increasing the amount of recombinant protein secreted into the soil [91]. Tobacco plants expressing a bacterial N_2O-cracking enzyme tagged for secretion under the control of the *A. rhizogenes rolD* promoter have been successful in demonstrating reducing activity [64,93]. Gas analysis was not performed to confirm that these plants mitigated N_2O emission. Ultimately, this approach arrives at a similar problem as other 'rhizoremediative' techniques: the N_2O-reducing ability of such a transgenic plant would be limited to the rhizosphere. This system would not have access to the bulk of N_2O gas, much of which comes from other sources;

(8) **Atmospheric phytoremediation using genetically engineered plants**. The potential of transgenic plants for environmental phytoremediation is well-documented: several fungal and bacterial oxidoreductases have been functionally expressed in plants as phytoremediation strategies including pentaerythritol tetranitrate reductase [94], mercuric reductase [95], and arsenate reductase [96]. This type of plant-based decontamination strategy provides advantages,

such as stable cultivation and control of the remediant organism and atmospheric exposure of the gas-cracking enzyme [97].

Atmospheric phytoremediation may ameliorate problems created by the other N_2O mitigation strategies described. The concept here is to develop crops with the ability to "crack" N_2O in both the soil and the atmosphere by incorporating the bacterial *nosZ* gene into their genomes. This gene encodes the nitrous oxide reductase enzyme (N_2OR), an oxidoreductase that catalyzes the removal of N_2O from the atmosphere, a process performed naturally by both denitrifying and non-denitrifying bacteria in the soil [98]. While conventional N_2O mitigation strategies aim to control N_2O production at earlier stages in the nitrification/denitrification pathway, this approach will target the atmospheric sum of N_2O emitted by all sources (Figure 2).

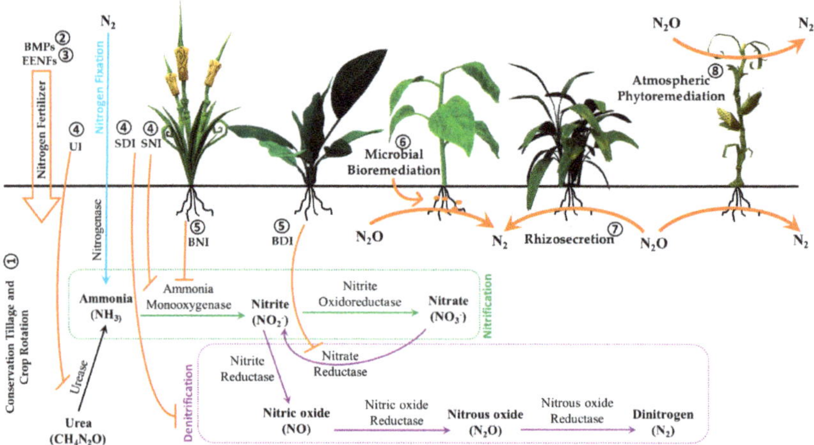

Figure 2. Nitrification-denitrification pathway and overview of current N_2O mitigation strategies. Orange arrows and lines show eight N_2O mitigation strategies described in Table 1. Green arrows show nitrification and purple arrows represent denitrification reactions. BDI, biological denitrification inhibitor; BMPs, best management practices; BNI, biological nitrification inhibitor; EENFs, enhanced efficiency nitrogen fertilizers; SDI, synthetic denitrification inhibitor; SNI, synthetic nitrification inhibitor; UI, urease inhibitor. **O** Encircled numbers refer to Table 1 strategies.

3. Nitrous Oxide Reductase—An Orphaned Soil Protein?

The *nosZ* gene can be categorized as either 'clade I' or 'clade II' based on sequence and *nos* operon organization, including the lack of an accessory *nosR* gene in the clade II members [99]. Clade II *nosZ* genes are also known as 'atypical' *nos* genes since they are found in non-denitrifying bacterial species. The N_2OR enzyme that the clade II gene encodes catalyzes the same reaction performed by the clade I-encoded enzyme, but has a higher affinity for N_2O [100], an important factor to consider when conceptualizing the development of an *nosZ*-expressing plant.

N_2OR is a multi-copper protein encoded by the *nosZ* gene (which is accompanied by an operon cluster of additional genes (*nosRDFYL*) [101]) and is the only enzyme that can catalyze the conversion of N_2O into N_2. The first active N_2OR was characterized from the soil bacterium *Pseudomonas stutzeri* and similar enzyme structures were resolved in bacterial species *Marinobacter hydrocarbonoclasticus* (formerly *Pseudomonas nautica*) (Figure 3), *Achromobacter cyclocastes*, and *Paracoccus denitrificans*. N_2OR is a head-to-tail homodimer and each monomer contains two domains: an electron transferring domain (binuclear Cu_A centre) and a catalytic domain (tetranuclear Cu_Z centre) [102]. There is some variability between the species regarding Cu_Z bridging and cupric coordination in the catalytic centre, suggesting that N_2OR substrate binding is species-specific. Regardless, the catalytic mechanism of N_2O reduction in N_2OR is still unclear [103].

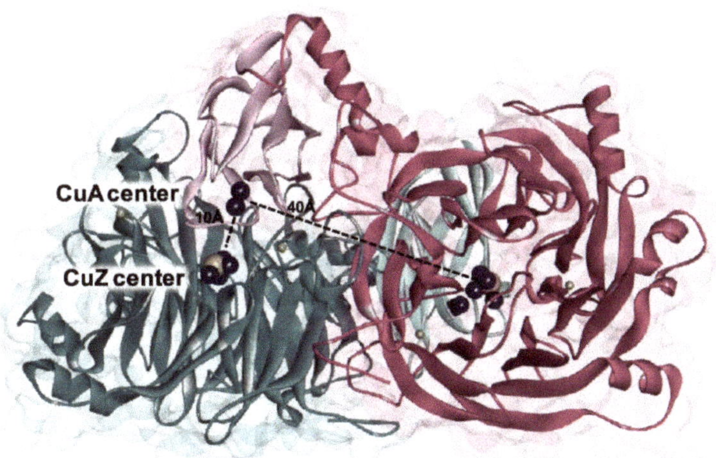

Figure 3. Structure of *Marinobacter hydrocarbonoclasticus* nitrous oxide reductase (N_2OR) homodimer. N_2OR is organized as a head-to-tail homodimer. Monomers are coloured differently so that they can be distinguished. In both monomers, the N-terminal domain is dark-coloured. The N-terminal domain forms a seven-bladed β-propeller fold that coordinates the catalytic tetranuclear active site Cu_Z through seven histidine residues at its hub. The C-terminal domain forms a cupredoxin fold and binds the dinuclear mixed-valent Cu_A centre [104].

The proven ability of N_2OR to "crack" the N_2O molecule raises the question of why the protein has not yet been incorporated into a commercially available transgenic cropping choice for environmentally motivated producers and small-plot farmers. Work has been done on this gene and its potential role in plant biotechnology since it was originally isolated in 1998 from the anaerobic soil bacterium *A. cyclocastes* [105,106], but it has yet to be converted into a commercially valuable tool. In this sense, N_2OR may be considered an "orphaned" protein, neglected among a veritable molecular toolkit of genes in the soil microflora [107,108]. Such forays into integrating soil and air sciences are demonstrative of the possibilities of what the soil microbiome offers biotechnologists [27]; it has already been discussed regarding the N-management possibilities offered by the microbiome and the current practice of 'bioprospecting' is also revealing a plethora of beneficial bacterial products, which is only accelerating thanks to whole-system approaches involving computational analyses [109].

Web of Science reports that between 1900 and 1991, there are no records binned under the combined topics "nitrous oxide reductase" and "microb*". The scientific literature blossomed from its first occurrence of 1992 to the present day, witnessing at least 175 publications dealing with the science of this important enzyme in our total environment. The scientific community waited until 1996 to start discussing denitrification in a plant context, according to these same search terms. With the search terms "nitrous oxide reductase" and "plant", the scientific record shows that soil microbiologists have taken a growing interest in the movement of N into the atmosphere (Figure 4). It is encouraging to note that in the same time period, the linkage between N_2OR and climate began its nascent phase.

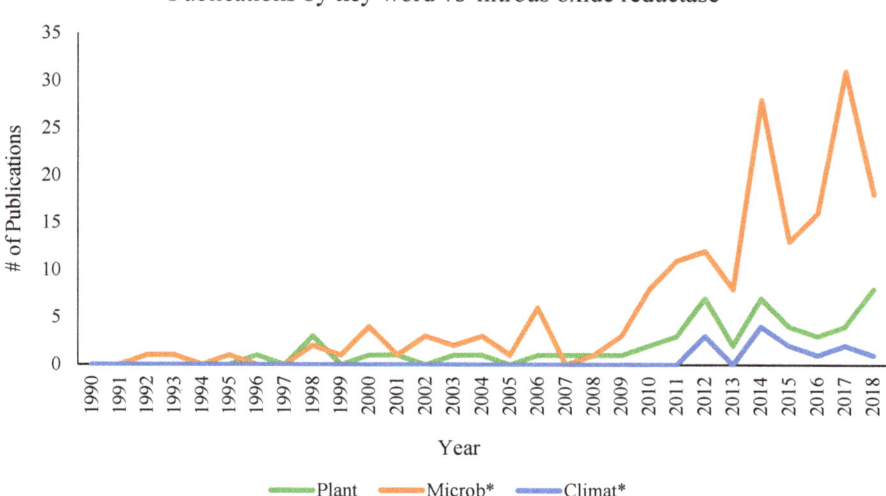

Figure 4. Nitrous oxide reductase-related publications released since 1990 on Web of Science (Clarivate Analytics). Publications by key word vs "nitrous oxide reductase" from 1990 to 2018. The orange line indicates "nitrous oxide reductase" + "microb*"; green: "nitrous oxide reductase" + "plant"; blue: "nitrous oxide reductase" + "climat*".

4. Catch Me If You Can: Can Plants Catalytically Convert N_2O *in planta*?

Rather than a 'cat and robin redbreast' conundrum, we are confronted with an opportunity to deploy protein engineering to ensure that more N_2OR molecules are attracted to the substrate binding site of the copper enzyme. Protein engineering offers ways to sidestep the challenges of expressing a complex bacterial protein in a plant [110]. There are potential issues with a recombinant metalloprotein like N_2OR, such as whether the ABC transporter can assemble within a plant cell, or the plant can incorporate copper into the electron transferring and catalytic domains [111,112]. It is possible to re-engineer N_2OR and produce a functional product [66], so there is precedent for designing an artificial metalloenzyme through rational protein design. This approach may be key to engineering a plant-compatible N_2OR protein.

A principle challenge associated with imparting N_2OR functionality to plants is that transforming the *nosZ* sequence alone may not be effective [113]; in *P. stutzeri*, the transcription of *nosZ* was dependent on the *nosDFY* genes being expressed, as they encode components of a putative ABC transporter system for the biogenesis of the Cu_Z centre [114]. Therefore, catalytically active N_2OR may not be produced when only *nosZ* is expressed in a heterologous host [28]. Nevertheless, a model N_2O-expressing plant has been engineered [64,93]. The clade I *nosZ* gene from soil bacterium *Pseudomonas stutzeri* was successfully expressed in a heterologous system—in this case, the tobacco plant (*Nicotiana tabacum*). In those proof-of-concept experiments the *nosZ*-expressing tobacco plants reduced 826 µg N_2O/min/gram of leaf tissue [115]. Assuming the tobacco yield to be 0.50 tonne/ha [116], the calculated N_2O-cracking ability of the *nosZ*-expressing tobacco could be as high as 600 kg of N_2O/ha/day [115], or 60 tonnes/ha/year (100 day growing season). This value surpasses the calculated N_2O flux of 0.05–1.98 kg N_2O/ha/year [117]. In other words, if every tobacco plant in the world produced N_2OR, this industrial crop (6.6 million tonnes were produced worldwide in 2016 [118]) could conceivably crack 785 Tg of N_2O (1 Tg = 1 million metric tonnes) during an average growing season of 100 days, far surpassing the estimated ~30 Tg of N_2O emitted per year [119]. Such catalytic capacity would give the 'Stop Smoking' campaigns a whole new flavour.

Although these transgenic plants produced a functional N_2OR enzyme, no gas analysis was performed to quantifiably ensure that these plants could reduce N_2O to N_2 using a recombinant N_2OR. In the future, it is imperative that such analyses be performed to properly judge the efficacy of such a gene-stacking trait system for atmospheric phytoremediation.

An associated issue rests with *P. stutzeri* being an anaerobic species that produces enzymes that function optimally in a low-oxygen environment. While expressing *nosZ* in plants to reduce N_2O appears to be an elegant solution, the N_2OR enzyme was not evolutionarily engineered to be functional in the presence of oxygen. Most soil bacteria that produce N_2OR do so in an anaerobic environment [102].

In the past five years, studies have identified several prokaryotic species that may express an oxygen-compatible N_2OR. Aerobic N_2O reducers may be undertaking an important role in mitigating the amounts of N_2O emitted to the atmosphere in events of oxic-to-anoxic transitions, but these systems have not yet been validated in plants. Here, we discuss two candidates for an oxygen-compatible *nosZ* expression system: clade II-*nosZ* member *Gemmatimonas aurantiaca gen nov.*, spp. *nov.* strain T-27, a polyphosphate-accumulating soil aerobe that is strongly represented in many oxygen-rich soil samples [120]; and *Azospira oryzae*, another clade II N-fixing bacterium originally isolated from the roots of rice (*Oryza sativa*) [121]. N_2O reduction by the *G. aurantiaca* strain T-27 was observed in both the absence and presence of oxygen [120]. The inability of this organism to consume N_2O in the complete absence of oxygen and the observed oxygen-induced activation of *nosZ* expression compels one to consider *in planta* overexpression, whereby the diurnal fluctuation of photosynthetic oxygen production may offer an egress for N_2O accumulation. The *A. oryzae* strains I09 and I13 also show more rapid N_2OR recovery rates and tolerance against oxygen inhibition than *P. stutzeri* [121] and so may be appropriate candidates for crop plant transformation and N_2OR expression.

If the ideal *nosZ* sequence were to be identified and transformed into commercially important crop plants, the benefits would be numerous and profound: seed-borne GHG technology foresees the transgenic cassette passed on from generation to generation, meaning that constant application of the beneficial catalyst would not be required (as with NI application and rhizoremediation); the expression of *nosZ* in the aerial tissues of the plants allows the reducing enzyme to confront N_2O much more easily than when the enzyme is expressed in the soil.

5. Novel Breeding Task: "Gas Cracking" Plants

The challenge of expressing heterologous bacterial proteins in plants necessitates codon optimization due to differences in GC content and codon bias with eukaryotes [122]. Altering the codon bias (or applying 'directed evolution' [123]) of a bacterial gene to be expressed in plants has been highly successful: *P. stutzeri nosZ* in tobacco [115], 5-enolpyruvylshikimate-3-phosphate (EPSP) synthase from *Agrobacterium tumefaciens* in Roundup Ready crops [124], and *Bacillus thuringiensis Cry* genes in maize [125] and rice [126]. Indeed, the global advance promulgating engineered crops is pillared on today's artificial intelligence-guided plant codon optimization rules offered by both large and small boutique DNA houses. However, there has been success expressing native bacterial sequences in plants, i.e., in the case of cotton expressing the native sequence of the *P. stutzeri* gene *ptxd* (*PHOSPHONATE DEHYDROGENASE*) [127,128]. One can dare to fathom how a universally-functional *nosZ* expression system could conceivably redirect some aspects of GHG mitigation research. Such a plant transformation cassette could theoretically be applied to any plant—wheat, rice, soybean, peat moss [129]—recruiting these species for the purpose of denitrification mitigation.

Even with an effective *nosZ* expression system, there are additional challenges in developing *nosZ*-expressing plant lines. There are relatively few powerful monocot-optimized expression systems available [130] (although *Bt* corn, LibertyLink wheat, and Roundup Ready wheat can attest to the effectiveness of the 35S promoter system in monocots), and there is difficulty in transforming monocots [65]. With the advent of new plant transformation technologies like the soil bacterium

Ochrobactrum haywardense [131] and the *BABYBOOM/WUSCHEL2* system [132], the production of genetically modified crops with stacked or pyramided GHG genes may be expedited in the near future.

6. Conclusions—Challenges to the Future Success of *nosZ*

We must address what may be the greatest challenge of all for the modern molecular plant breeder: convincing the general public that transgenic crops may be beneficial for all the plant-planet's denizens, as modified crops that enter the food stream may appear unpopular in some boroughs. Regardless, there is a clear, urgent need to control soil N_2O losses due to the detrimental effects of this potent GHG in the atmosphere. Climate-smart crops should be given a crack at directly addressing this issue and tackling climate change. Such GHG-reducing plant lines, endowed with the ability to catalytically "crack" N_2O in the air, could be vital in the battle to shift public perception towards the acceptance of "GMOs" in agricultural research.

Involvement of N_2O in climate change and global warming has been the subject of increasing investigations due to its potential heat-trapping properties [3]. N_2O emission from soil is primarily the result of an incomplete enzymatic reaction which is mediated by the bacterial enzyme, N_2OR [98]. Therefore, in the late 1990s [105,106], the development of N_2OR-positive transgenic plants was proposed as an environmental phytoremediation strategy with promise to remove N_2O from soil and the atmosphere (Figure 2). However, producing a foreign protein in a plant cell is often a serious challenge. For example, different codon usage [133] and cellular properties between eukaryotic and prokaryotic cells are considered as unknown aspects of this strategy. At least two key questions need to be addressed in future studies to probe the probability for success of this green gene de-toxic tactic for accelerating the destruction of nitrous oxide via canopy catalysis: (1) Which candidate is the best source-organism to donate *nosZ* sequence for plant transformation? Activity of bacterial N_2OR is associated with the anaerobic conditions in soil [101], whereas the plant cell is mostly an aerobic environment. Photosynthesis and respiration cause different levels of oxygen content in plant cells in a diurnal cycle which is not consistent with the enzymatic activity of N_2OR in anaerobic soil bacteria. Therefore, selecting obligate or facultative aerobic bacteria containing active N_2OR enzymes as 'the source code' would be pivotal; (2) Which plant cell compartment is the best destination for targeting N_2OR accumulation? The native enzyme N_2OR in bacteria is directed to the periplasm, where Cu chaperones provide enough Cu for the assembly of metal centres [134]. The absence of periplasmic space in plant cells reinforces the notion that subcellular localization of N_2OR may influence its enzymatic activity *in planta*. Moreover, the important role of Cu in the functional assembly of N_2OR posits whether the transformation of bacterial *nosDFY*, along with *nosZ*, is essential for a functional enzyme. Urgent exploration of how the cellular pool of metal nutrients and proteins (pseudo chaperones) in eukaryotic cells may suffice to activate N_2OR *in planta* may compel the use of such climate-smart plants.

Author Contributions: J.J.D. constructed the initial paper draft; Q.-y.S., J.J.D., S.W., A.E.H., M.N., and I.A. contributed to study design and data analysis; J.J.D., S.W., and I.A. wrote the paper with input from A.E.H., and M.N.; M.N. constructed Figure 2; Q.-y.S. conceptualized the original approach to amplifying soil denitrification *in planta*; I.A. coordinated and supervised the project; all authors contributed to and approved the final draft.

Funding: This research was funded by a grant from the Government of Ontario in 2018. Q.-Y.S. is a grateful recipient of The Rockefeller Foundation's Career Biotechnology Fellowship.

Acknowledgments: We thank three anonymous reviewers for taking the time to critique and strengthen this manuscript. We are grateful to Quinn Ingram for technical expertise with plant growth.

Conflicts of Interest: The authors declare no competing interests.

References

1.   Dentener, F.; Drevet, J.; Lamarque, J.F. Nitrogen and sulfur deposition on regional and global scales: A multimodel evaluation. *Glob. Biogeochem. Cycles* **2006**, *20*, 1–21. [CrossRef]

2. Vet, R.; Artz, R.S.; Carou, S. A global assessment of precipitation chemistry and deposition of sulfur, nitrogen, sea salt, base cations, organic acids, acidity and pH, and phosphorus. *Atmos. Environ.* **2014**, *93*, 3–100. [CrossRef]

3. Forster, P.; Ramaswamy, V. *Changes in Atmospheric Constituents and in Radiative Forcing*; Solomon, S.Q.D., Manning, M., Chen, Z., Marquis, M., Averyt, K.B., Tignor, M., Miller, H.L., Eds.; Cambridge University Press: Cambridge, UK, 2007; pp. 129–234.

4. Reay, D.S. Fertilizer 'solution' could turn local problem global. *Nature* **2004**, *427*, 485. [CrossRef] [PubMed]

5. Shapiro, C.A.; Ferguson, R.B.; Hergert, G.W.; Wortmann, C.S.; Walters, D.T. *Fertilizer Suggestions for Corn*; EC117; Nebraska Extension: Lincoln, NE, USA, 2008; pp. 1–6.

6. Froment, M.A.; Chalmers, A.G.; Collins, C.; Grylls, J.P. Rotational set-aside; influence of vegetation and management for one-year plant covers on soil mineral nitrogen during and after set-aside at five sites in England. *J. Agric. Sci.* **1999**. [CrossRef]

7. Quinton, J.N.; Govers, G.; Van Oost, K.; Bardgett, R.D. The impact of agricultural soil erosion on biogeochemical cycling. *Nat. Geosci.* **2010**, *3*, 311–314. [CrossRef]

8. Zhu, Z.L.; Cai, G.X.; Simpson, J.R.; Zhang, S.L.; Chen, D.L.; Jackson, A.V.; Freney, J.R. Processes of nitrogen loss from fertilizers applied to flooded rice fields on a calcareous soil in north-central China. *Fertil. Res.* **1988**, *18*, 101–115. [CrossRef]

9. Zhou, M.; Zhu, B.; Brüggemann, N.; Dannenmann, M.; Wang, Y.; Butterbach-Bahl, K. Sustaining crop productivity while reducing environmental nitrogen losses in the subtropical wheat-maize cropping systems: A comprehensive case study of nitrogen cycling and balance. *Agric. Ecosyst. Environ.* **2016**, *231*, 1–14. [CrossRef]

10. Zumft, W.G. Cell biology and molecular basis of denitrification. *Microbiol. Mol. Biol. Rev.* **1997**, *61*, 533–616. [PubMed]

11. Gu, B.; Chang, J.; Min, Y.; Ge, Y.; Zhu, Q.; Galloway, J.N.; Peng, C. The role of industrial nitrogen in the global nitrogen biogeochemical cycle. *Sci. Rep.* **2013**, *3*, 2579. [CrossRef] [PubMed]

12. Erisman, J.W.; Sutton, M.A.; Galloway, J.; Klimont, Z.; Winiwarter, W. How a century of ammonia synthesis changed the world. *Nat. Geosci.* **2008**, *1*, 636–639. [CrossRef]

13. Fowler, D.; Pyle, J.A.; Raven, J.A.; Sutton, M.A. The global nitrogen cycle in the twenty-first century: Introduction. *Philos. Trans. R. Soc. Lond. B Biol. Sci.* **2013**, *368*, 20130164. [CrossRef] [PubMed]

14. Kanakidou, M.; Myriokefalitakis, S.; Daskalakis, N.; Fanourgakis, G.; Nenes, A.; Baker, A.R.; Tsigaridis, K.; Mihalopoulos, N. Past, present, and future atmospheric nitrogen deposition. *J. Atmos. Sci.* **2016**, *73*, 2039–2047. [CrossRef]

15. Richardson, D.; Felgate, H.; Watmough, N.; Thomson, A.; Baggs, E. Mitigating release of the potent greenhouse gas N_2O from the nitrogen cycle—Could enzymic regulation hold the key? *Trends Biotechnol.* **2009**, *27*, 388–397. [CrossRef] [PubMed]

16. Burney, J.A.; Davis, S.J.; Lobell, D.B. Greenhouse gas mitigation by agricultural intensification. *Proc. Natl. Acad. Sci. USA* **2010**, *107*, 12052–12057. [CrossRef] [PubMed]

17. Dignon, J.; Hameed, S. Global emissions of nitrogen and sulfur-oxides from 1860 to 1980. *Japca* **1989**, *39*, 180–186. [CrossRef]

18. Larsson, L.; Ferm, M.; Kasimir-Klemedtsson, A.; Klemedtsson, L. Ammonia and nitrous oxide emissions from grass and alfalfa mulches. *Nutr. Cycl. Agroecosyst.* **1998**, *51*, 41–46. [CrossRef]

19. Intergovernmental Panel on Climate Change. *Climate Change 2014: Mitigation of Climate Change*; Working Group III Contribution to the IPCC Fifth Assessment Report. 9781107654815; Cambridge University Press: Cambridge, UK, 2015.

20. Ravishankara, A.R.; Daniel, J.S.; Portmann, R.W. Nitrous oxide N_2O: The dominant ozone-depleting substance emitted in the 21st century. *Science* **2009**, *326*, 123–125. [CrossRef] [PubMed]

21. Chen, Z.; Ding, W.; Xu, Y.; Müller, C.; Rütting, T.; Yu, H.; Fan, J.; Zhang, J.; Zhu, T. Importance of heterotrophic nitrification and dissimilatory nitrate reduction to ammonium in a cropland soil: Evidences from a 15N tracing study to literature synthesis. *Soil Biol. Biochem.* **2015**, *91*, 65–75. [CrossRef]

22. Ruser, R.; Schulz, R. The effect of nitrification inhibitors on the nitrous oxide (N_2O) release from agricultural soils—A review. *J. Plant Nutr. Soil Sci.* **2015**, *178*, 171–188. [CrossRef]

23. Lees, H. Hydroxylamine as an intermediate in nitrification. *Nature* **1952**, *169*, 156–157. [CrossRef]

24. Caranto, J.D.; Lancaster, K.M. Nitric oxide is an obligate bacterial nitrification intermediate produced by hydroxylamine oxidoreductase. *Proc. Natl. Acad. Sci. USA* **2017**, *114*, 8217–8222. [CrossRef] [PubMed]

25. Cébron, A.; Garnier, J. *Nitrobacter* and *Nitrospira* genera as representatives of nitrite-oxidizing bacteria: Detection, quantification and growth along the lower Seine River (France). *Water Res.* **2005**, *39*, 4979–4992. [CrossRef] [PubMed]

26. Shaw, L.J.; Nicol, G.W.; Smith, Z.; Fear, J.; Prosser, J.I.; Baggs, E.M. *Nitrosospira* spp. can produce nitrous oxide via a nitrifier denitrification pathway. *Environ. Microbiol.* **2006**, *8*, 214–222. [CrossRef] [PubMed]

27. Kole, M.M.; Page, W.J.; Altosaar, I. Distribution of *Azotobacter* in Eastern Canadian soils and in association with plant rhizospheres. *Can. J. Microbiol.* **1988**, *34*, 815–817. [CrossRef]

28. Philippot, L.; Andert, J.; Jones, C.M.; Bru, D.; Hallin, S. Importance of denitrifiers lacking the genes encoding the nitrous oxide reductase for N_2O emissions from soil. *Glob. Chang. Biol.* **2011**, *17*, 1497–1504. [CrossRef]

29. Zumft, W.G.; Kroneck, P.M.H. Respiratory transformation of nitrous oxide (N_2O) to dinitrogen by *Bacteria* and *Archaea*. *Adv. Microb. Physiol.* **2007**, *52*, 107–227. [PubMed]

30. Brodsky, J.B.; Cohen, E.N. Adverse effects of nitrous oxide. *Med. Toxicol.* **1986**, *1*, 362–374. [CrossRef] [PubMed]

31. Fluegge, K. Does environmental exposure to the greenhouse gas, N_2O, contribute to etiological factors in neurodevelopmental disorders? A mini-review of the evidence. *Environ. Toxicol. Pharmacol.* **2016**, *47*, 6–18. [CrossRef] [PubMed]

32. Edgerton, M.D. Increasing crop productivity to meet global needs for feed, food, and fuel. *Plant Physiol.* **2009**, *149*, 7–13. [CrossRef] [PubMed]

33. Abalos, D.; van Groenigen, J.W.; De Deyn, G.B. What plant functional traits can reduce nitrous oxide emissions from intensively managed grasslands? *Glob. Chang. Biol.* **2018**, *24*, e248–e258. [CrossRef] [PubMed]

34. Terman, G.L. Volatilization losses of nitrogen as ammonia from surface-applied fertilizers, organic amendments, and crop residues. *Adv. Agron.* **1980**, *31*, 189–223.

35. Holcomb, J.C.; Sullivan, D.M.; Horneck, D.A.; Clough, G.H. Effect of irrigation rate on ammonia volatilization. *Soil Sci. Soc. Am. J.* **2011**, *75*, 2341–2347. [CrossRef]

36. Van Kessel, C.; Venterea, R.; Six, J.; Adviento-Borbe, M.A.; Linquist, B.; van Groenigen, K.J. Climate, duration, and N placement determine N_2O emissions in reduced tillage systems: A meta-analysis. *Glob. Chang. Biol.* **2013**, *19*, 33–44. [CrossRef] [PubMed]

37. Mei, K.; Wang, Z.; Huang, H.; Zhang, C.; Shang, X.; Dahlgren, R.A.; Zhang, M.; Xia, F. Stimulation of N_2O emission by conservation tillage management in agricultural lands: A meta-analysis. *Soil Till. Res.* **2018**, *182*, 86–93. [CrossRef]

38. Pittelkow, C.M.; Liang, X.; Linquist, B.A.; van Groenigen, K.J.; Lee, J.; Lundy, M.E.; van Gestel, N.; Six, J.; Venterea, R.T.; van Kessel, C. Productivity limits and potentials of the principles of conservation agriculture. *Nature* **2015**, *517*, 365–368. [CrossRef] [PubMed]

39. Bruulsema, T.; Lemunyon, J.; Herz, B. Know your fertilizer rights. *Crops Sci.* **2009**, *42*, 13–18.

40. Bruulsema, T.W.; Ketterings, Q. *Best Management Practices for Fertilizers on Northeastern Dairy Farms*; Fertilize BMP Item 30-3220; Ref. 08052; International Plant Nutrition Institute: Norcross, GA, USA, 2008.

41. Guardia, G.; Cangani, M.T.; Andreu, G.; Sanz-Cobena, A.; García-Marco, S.; Álvarez, J.M.; Recio-Huetos, J.; Vallejo, A. Effect of inhibitors and fertigation strategies on GHG emissions, NO fluxes and yield in irrigated maize. *Field Crops Res.* **2017**, *204*, 135–145. [CrossRef]

42. Cayuela, M.L.; Aguilera, E.; Sanz-Cobena, A.; Adams, D.C.; Abalos, D.; Barton, L.; Ryals, R.; Silver, W.L.; Alfaro, M.A.; Pappa, V.A.; et al. Direct nitrous oxide emissions in Mediterranean climate cropping systems: Emission factors based on a meta-analysis of available measurement data. *Agric. Ecosyst. Environ.* **2017**, *238*, 25–35. [CrossRef]

43. Sanz-Cobena, A.; Lassaletta, L.; Aguilera, E.; Del Prado, A.; Garnier, J.; Billen, G.; Iglesias, A.; Sanchez, B.; Guardia, G.; Abalos, D.; et al. Strategies for greenhouse gas emissions mitigation in Mediterranean agriculture: A review. *Agric. Ecosyst. Environ.* **2017**, *238*, 5–24. [CrossRef]

44. Hynes, R.; Knowles, R. Production of nitrous oxide by *Nitrosomonas europaea*: Effects of acetylene, pH, and oxygen. *Can. J. Microbiol.* **2011**, *30*, 1397–1404. [CrossRef]

45. Halvorson, A.D.; Snyder, C.S.; Blaylock, A.D.; Del Grosso, S.J. Enhanced-efficiency nitrogen fertilizers: Potential role in nitrous oxide emission mitigation. *Agron. J.* **2014**, *106*, 715–722. [CrossRef]

46. Thapa, R.; Chatterjee, A.; Awale, R.; McGranahan, D.A.; Daigh, A. Effect of enhanced efficiency fertilizers on nitrous oxide emissions and crop yields: A meta-analysis. *Soil Sci. Soc. Am. J.* **2016**, *80*, 1121–1134. [CrossRef]

47. Shoji, S.; Delgado, J.; Mosier, A.; Miura, Y. Use of controlled release fertilizers and nitrification inhibitors to increase nitrogen use efficiency and to conserve air and water quality. *Commun. Soil Sci. Plant Anal.* **2001**, *32*, 1051–1070. [CrossRef]

48. McTaggart, I.P.; Tsuruta, H. The influence of controlled release fertilisers and the form of applied fertiliser nitrogen on nitrous oxide emissions from an andosol. *Nutr. Cycl. Agroecosyst.* **2003**, *67*, 47–54. [CrossRef]

49. Sanz-Cobena, A.; Abalos, D.; Meijide, A.; Sanchez-Martin, L.; Vallejo, A. Soil moisture determines the effectiveness of two urease inhibitors to decrease N_2O emission. *Mitig. Adapt. Strateg. Glob. Chang.* **2016**, *21*, 1131–1144. [CrossRef]

50. Kim, M.; Zhang, Z.; Okano, H.; Yan, D.; Groisman, A.; Hwa, T. Need-based activation of ammonium uptake in *Escherichia coli*. *Mol. Syst. Biol.* **2012**, *8*, 616. [CrossRef] [PubMed]

51. Soares, J.; Cantarella, H.; Menegale, M. Ammonia volatilization losses from surface-applied urea with urease and nitrification inhibitors. *Soil Biol. Biochem.* **2012**, *52*, 82–89. [CrossRef]

52. McCarty, G.W. Modes of action of nitrification inhibitors. *Biol. Fertil. Soils* **1999**, *29*, 1–9. [CrossRef]

53. Liu, C.; Wang, K.; Zheng, X. Effects of nitrification inhibitors (DCD and DMPP) on nitrous oxide emission, crop yield and nitrogen uptake in a wheat-maize cropping system. *Biogeosciences* **2013**, *10*, 2427–2437. [CrossRef]

54. Bollag, J.-M.; Kurek, E.J. Nitrite and nitrous oxide accumulation during denitrification in the presence of pesticide derivatives. *Appl. Environ. Microbiol.* **1980**, *39*, 845–849. [PubMed]

55. Lam, S.K.; Suter, H.; Mosier, A.R.; Chen, D. Using nitrification inhibitors to mitigate agricultural N_2O emission: A double-edged sword? *Glob. Chang. Biol.* **2016**, *23*, 485–489. [CrossRef] [PubMed]

56. Subbarao, G.V.; Nakahara, K.; Hurtado, M.P.; Ono, H.; Moreta, D.E.; Salcedo, A.F.; Yoshihashi, A.T.; Ishikawa, T.; Ishitani, M.; Ohnishi-Kameyama, M.; et al. Evidence for biological nitrification inhibition in *Brachiaria* pastures. *Proc. Natl. Acad. Sci. USA* **2009**, *106*, 17302. [CrossRef] [PubMed]

57. Bardon, C.; Poly, F.; Piola, F.; Pancton, M.; Comte, G.; Meiffren, G.; Haichar Fel, Z. Mechanism of biological denitrification inhibition: Procyanidins induce an allosteric transition of the membrane-bound nitrate reductase through membrane alteration. *FEMS Microbiol. Ecol.* **2016**, *92*, fiw034. [CrossRef] [PubMed]

58. Subbarao, G.V.; Sahrawat, K.L.; Nakahara, K.; Rao, I.M.; Ishitani, M.; Hash, C.T.; Kishii, M.; Bonnett, D.G.; Berry, W.L.; Lata, J.C. A paradigm shift towards low-nitrifying production systems: The role of biological nitrification inhibition (BNI). *Ann. Bot.* **2013**, *112*, 297–316. [CrossRef] [PubMed]

59. Jacinthe, P.-A.; Dick, W.A.; Brown, L.C. Bioremediation of nitrate-contaminated shallow soils and waters via water table management techniques: Evolution and release of nitrous oxide. *Soil Biol. Biochem.* **2000**, *32*, 371–382. [CrossRef]

60. Itakura, M.; Tabata, K.; Eda, S.; Mitsui, H.; Murakami, K.; Yasuda, J.; Minamisawa, K. Generation of *Bradyrhizobium japonicum* mutants with Increased N_2O reductase activity by selection after introduction of a mutated *dnaQ* gene. *Appl. Environ. Microbiol.* **2008**, *74*, 7258–7264. [CrossRef] [PubMed]

61. Itakura, M.; Uchida, Y.; Akiyama, H.; Hoshino, Y.T.; Shimomura, Y.; Morimoto, S.; Tago, K.; Wang, Y.; Hayakawa, C.; Uetake, Y.; et al. Mitigation of nitrous oxide emissions from soils by *Bradyrhizobium japonicum* inoculation. *Nat. Clim. Chang.* **2012**, *3*, 208–212. [CrossRef]

62. Houlden, A.; Timms-Wilson, T.M.; Day, M.J.; Bailey, M.J. Influence of plant developmental stage on microbial community structure and activity in the rhizosphere of three field crops. *FEMS Microbiol. Ecol.* **2008**, *65*, 193–201. [CrossRef] [PubMed]

63. Xu, Y.; Wang, G.; Jin, J.; Liu, J.; Zhang, Q.; Liu, X. Bacterial communities in soybean rhizosphere in response to soil type, soybean genotype, and their growth stage. *Soil Biol. Biochem.* **2009**, *41*, 919–925. [CrossRef]

64. Wan, S.; Johnson, A.M.; Altosaar, I. Expression of nitrous oxide reductase from *Pseudomonas stutzeri* in transgenic tobacco roots using the root-specific *rolD* promoter from *Agrobacterium rhizogenes*. *Ecol. Evol.* **2012**, *2*, 286–297. [CrossRef] [PubMed]

65. Sood, P.; Bhattacharya, A.; Sood, A. Problems and possibilities of monocot transformation. *Biol. Plant* **2011**, *55*, 1–15. [CrossRef]

66. Matsubara, T.; Frunzke, K.; Zumft, W.G. Modulation by copper of the products of nitrite respiration in *Pseudomonas perfectomarinus*. *Biol. Fertil. Soils* **1982**, *149*, 816–823.

67. Wan, S.; Ward, T.L.; Altosaar, I. Strategy and tactics of disarming GHG at the source: N$_2$O reductase crops. *Trends Biotechnol.* **2012**, *30*, 410–415. [CrossRef] [PubMed]
68. Rochette, P.; Angers, D.A.; Chantigny, M.H.; Gasser, M.-O.; MacDonald, J.D.; Pelster, D.E.; Bertrand, N. NH$_3$ volatilization, soil concentration and soil pH following subsurface banding of urea at increasing rates. *Can. J. Soil Sci.* **2013**, *93*, 261–268. [CrossRef]
69. Venterea, R.T.; Burger, M.; Spokas, K.A. Nitrogen oxide and methane emissions under varying tillage and fertilizer management. *J. Environ. Qual.* **2005**, *34*, 1467–1477. [CrossRef] [PubMed]
70. Chen, G.; Kolb, L.; Cavigelli, M.A.; Weil, R.R.; Hooks, C.R.R. Can conservation tillage reduce N$_2$O emissions on cropland transitioning to organic vegetable production? *Sci. Total Environ.* **2018**, *618*, 927–940. [CrossRef] [PubMed]
71. Mitchell, D.C.; Castellano, M.J.; Sawyer, J.E.; Pantoja, J. Cover crop effects on nitrous oxide emissions: Role of mineralizable carbon. *Soil Sci. Soc. Am. J.* **2013**, *77*, 1765. [CrossRef]
72. Snyder, C.S.; Bruulsema, T.W.; Jensen, T.L.; Fixen, P.E. Review of greenhouse gas emissions from crop production systems and fertilizer management effects. *Agric. Ecosyst. Environ.* **2009**, *133*, 247–266. [CrossRef]
73. Abalos, D.; Jeffery, S.; Sanz-Cobena, A.; Guardia, G.; Vallejo, A. Meta-analysis of the effect of urease and nitrification inhibitors on crop productivity and nitrogen use efficiency. *Agric. Ecosyst. Environ.* **2014**, *189*, 136–144. [CrossRef]
74. Cantarella, H.; Otto, R.; Soares, J.R.; Silva, A.G.d.B. Agronomic efficiency of NBPT as a urease inhibitor: A review. *J. Adv. Res.* **2018**, *13*, 19–27. [CrossRef] [PubMed]
75. Bremner, J.M. Recent research on problems in the use of urea as a nitrogen fertilizer. *Fertil. Res.* **1995**, *42*, 321–329. [CrossRef]
76. Zaman, M.; Zaman, S.; Nguyen, M.L.; Smith, T.J.; Nawaz, S. The effect of urease and nitrification inhibitors on ammonia and nitrous oxide emissions from simulated urine patches in pastoral system: A two-year study. *Sci. Total Environ.* **2013**, *465*, 97–106. [CrossRef] [PubMed]
77. Ding, W.X.; Yu, H.Y.; Cai, Z.C. Impact of urease and nitrification inhibitors on nitrous oxide emissions from fluvo-aquic soil in the North China Plain. *Biol. Fertil. Soils* **2011**, *47*, 91–99. [CrossRef]
78. Singh, S.; Bakshi, B.R. Accounting for the biogeochemical cycle of nitrogen in input-output life cycle assessment. *Environ. Sci. Technol.* **2013**, *47*, 9388–9396. [CrossRef] [PubMed]
79. Rose, T.J.; Quin, P.; Morris, S.G.; Kearney, L.J.; Kimber, S.; Rose, M.T.; Van Zwieten, L. No evidence for higher agronomic N use efficiency or lower nitrous oxide emissions from enhanced efficiency fertilisers in aerobic subtropical rice. *Field Crops Res.* **2018**, *225*, 47–54. [CrossRef]
80. Chaves, B.; Opoku, A.; De Neve, S.; Boeckx, P.; Van Cleemput, O.; Hofman, G. Influence of DCD and DMPP on soil N dynamics after incorporation of vegetable crop residues. *Biol. Fertil. Soils* **2006**, *43*, 62–68. [CrossRef]
81. Yang, M.; Fang, Y.; Sun, D.; Shi, Y. Efficiency of two nitrification inhibitors (dicyandiamide and 3,4-dimethypyrazole phosphate) on soil nitrogen transformations and plant productivity: A meta-analysis. *Sci. Rep.* **2016**, *6*, 1–10.
82. Coskun, D.; Britto, D.T.; Shi, W.; Kronzucker, H.J. How plant root exudates shape the nitrogen cycle. *Trends Plant Sci.* **2017**, *22*, 661–673. [CrossRef] [PubMed]
83. Matsuoka, M.; Kumar, A.; Muddassar, M.; Matsuyama, A.; Yoshida, M.; Zhang, K.Y. Discovery of fungal denitrification inhibitors by targeting copper nitrite reductase from *Fusarium oxysporum*. *J. Chem. Inf. Model.* **2017**, *57*, 203–213. [CrossRef] [PubMed]
84. Liu, T.; Liang, Y.; Chu, G. Nitrapyrin addition mitigates nitrous oxide emissions and raises nitrogen use efficiency in plastic-film-mulched drip-fertigated cotton field. *PLoS ONE* **2017**, *12*, e0176305. [CrossRef] [PubMed]
85. Byrnes, R.C.; Nùñez, J.; Arenas, L.; Rao, I.; Trujillo, C.; Alvarez, C.; Arango, J.; Rasche, F.; Chirinda, N. Biological nitrification inhibition by *Brachiaria* grasses mitigates soil nitrous oxide emissions from bovine urine patches. *Soil Biol. Biochem.* **2017**, *107*, 156–163. [CrossRef]
86. Subbarao, G.V.; Arango, J.; Masahiro, K.; Hooper, A.M.; Yoshihashi, T.; Ando, Y.; Nakahara, K.; Deshpande, S.; Ortiz-Monasterio, I.; Ishitani, M.; et al. Genetic mitigation strategies to tackle agricultural GHG emissions: The case for biological nitrification inhibition technology. *Plant Sci. Int. J. Exp. Plant Biol.* **2017**, *262*, 165–168. [CrossRef] [PubMed]

87. Karwat, H.; Moreta, D.; Arango, J.; Núñez, J.; Rao, I.; Rincón, Á.; Rasche, F.; Cadisch, G. Residual effect of BNI by *Brachiaria humidicola* pasture on nitrogen recovery and grain yield of subsequent maize. *Plant Soil.* **2017**, *420*, 389–406. [CrossRef]

88. Kuiper, I.; Lagendijk, E.L.; Bloemberg, G.V.; Lugtenberg, B.J.J. Rhizoremediation: A beneficial plant-microbe interaction. *Mol. Plant Microbe Interact.* **2004**, *17*, 6–15. [CrossRef] [PubMed]

89. Qian, J.H.; Doran, J.W.; Weier, K.L.; Mosier, A.R.; Peterson, T.A.; Power, J.F. Soil denitrification and nitrous oxide losses under corn irrigated with high-nitrate groundwater. *J. Environ. Qual.* **1997**, *26*, 348–360. [CrossRef]

90. Köpke, U.; Nemecek, T. Ecological services of faba bean. *Field Crops Res.* **2010**, *115*, 217–233. [CrossRef]

91. Gaume, A.; Komarnytsky, S.; Borisjuk, N.; Raskin, I. Rhizosecretion of recombinant proteins from plant hairy roots. *Plant Cell Rep.* **2003**, *21*, 1188–1193. [CrossRef] [PubMed]

92. Gelvin, S.B. Crown gall disease and hairy root disease. *Plant Physiol.* **1990**, *92*, 281–285. [CrossRef] [PubMed]

93. Wan, S.; Mottiar, Y.; Johnson, A.M.; Goto, K.; Altosaar, I. Expression of the *nos* operon proteins from *Pseudomonas stutzeri* in transgenic plants to assemble nitrous oxide reductase. *Transgenic Res.* **2012**, *21*, 593–603. [CrossRef] [PubMed]

94. French, C.E.; Rosser, S.J.; Davies, G.J.; Nicklin, S.; Bruce, N.C. Biodegradation of explosives by transgenic plants expressing pentaerythritol tetranitrate reductase. *Nat. Biotechnol.* **1999**, *17*, 491–494. [CrossRef] [PubMed]

95. Rugh, C.L.; Wilde, H.D.; Stack, N.M.; Thompson, D.M.; Summers, A.O.; Meagher, R.B. Mercuric ion reduction and resistance in transgenic *Arabidopsis thaliana* plants expressing a modified bacterial *merA* gene. *Proc. Natl. Acad. Sci. USA* **1996**, *93*, 3182–3187. [CrossRef] [PubMed]

96. Nahar, N.; Rahman, A.; Nawani, N.N.; Ghosh, S.; Mandal, A. Phytoremediation of arsenic from the contaminated soil using transgenic tobacco plants expressing *ACR2* gene of *Arabidopsis thaliana*. *J. Plant Physiol.* **2017**, *218*, 121–126. [CrossRef] [PubMed]

97. Heaton, A.C.P.; Rugh, C.L.; Wang, N.J.; Meagher, R.B. Phytoremediation of mercury- and methylmercury-polluted soils using genetically engineered plants. *Soil Sediment Contam.* **1998**, *7*, 497–509. [CrossRef]

98. Maeda, K.; Hanajima, D.; Toyoda, S.; Yoshida, N.; Morioka, R.; Osada, T. Microbiology of nitrogen cycle in animal manure compost. *Microb. Biotechnol.* **2011**. [CrossRef] [PubMed]

99. Sanford, R.A.; Wagner, D.D.; Wu, Q.; Chee-Sanford, J.C.; Thomas, S.H.; Cruz-Garcia, C.; Rodriguez, G.; Massol-Deya, A.; Krishnani, K.K.; Ritalahti, K.M.; et al. Unexpected nondenitrifier nitrous oxide reductase gene diversity and abundance in soils. *Proc. Natl. Acad. Sci. USA* **2012**, *109*, 19709–19714. [CrossRef] [PubMed]

100. Yoon, S.; Nissen, S.; Park, D.; Sanford, R.A.; Löffler, F.E. Nitrous oxide reduction kinetics distinguish bacteria harboring clade I *NosZ* from those harboring clade II *NosZ*. *Appl. Environ. Microbiol.* **2016**, *82*, 3793–3800. [CrossRef] [PubMed]

101. Cuypers, H.; Berghofer, J.; Zumft, W.G. Multiple *nosZ* promoters and anaerobic expression of *nos* genes necessary for *Pseudomonas stutzeri* nitrous oxide reductase and assembly of its copper centers. *Biochim. Biophys. Acta* **1995**, *1264*, 183–190. [CrossRef]

102. Rosenzweig, A.C. Nitrous oxide reductase from Cu_A to Cu_Z. *Nat. Struct. Biol.* **2000**, *7*, 169–171. [CrossRef] [PubMed]

103. Wang, Y.; Wang, Z.; Duo, Y.; Wang, X.; Chen, J.; Chen, J. Gene cloning, expression, and reducing property enhancement of nitrous oxide reductase from *Alcaligenes denitrificans* strain TB. *Environ. Pollut.* **2018**, *239*, 43–52. [CrossRef] [PubMed]

104. Carreira, C.; Pauleta, S.R.; Moura, I. The catalytic cycle of nitrous oxide reductase—The enzyme that catalyzes the last step of denitrification. *J. Inorg. Biochem.* **2017**, *177*, 423–434. [CrossRef] [PubMed]

105. Inatomi, K.I. Analysis of the nitrous oxide reduction genes, *nosZDFYL*, of *Achromobacter cycloclastes*. *DNA Res.* **1998**, *5*, 365–371. [CrossRef] [PubMed]

106. Inatomi, K.-I. Gene Encoding Nitrous Oxide Reductase and the Nitrous Oxide Reductase. Patent Application JP H1175842A, 29 August 1997.

107. Cheng, J.; Romantsov, T.; Engel, K.; Doxey, A.C.; Rose, D.R.; Neufeld, J.D.; Charles, T.C. Functional metagenomics reveals novel β-galactosidases not predictable from gene sequences. *PLoS ONE* **2017**, *12*, e0172545. [CrossRef] [PubMed]

108. Hill, P.; Heberlig, G.W.; Boddy, C.N. Sampling terrestrial environments for bacterial polyketides. *Molecules* **2017**, *22*, E707. [CrossRef] [PubMed]

109. Kolter, R.; Chimileski, S. The end of microbiology. *Environ. Microbiol.* **2018**. [CrossRef] [PubMed]

110. Lin, Y.W. Rational design of metalloenzymes: From single to multiple active sites. *Coord. Chem. Rev.* **2017**, *336*, 1–27. [CrossRef]

111. Zumft, W.G. Biogenesis of the bacterial respiratory Cu_A, Cu-S enzyme nitrous oxide reductase. *J. Mol. Microbiol. Biotechnol.* **2005**, *10*, 154–166. [CrossRef] [PubMed]

112. Dell'Acqua, S.; Pauleta, S.R.; Moura, I.; Moura, J.J.G. The tetranuclear copper active site of nitrous oxide reductase: The Cu_Z center. *J. Biol. Inorg. Chem.* **2011**, *16*, 183–194. [CrossRef] [PubMed]

113. Kusnadi, A.R.; Nikolov, Z.L.; Howard, J.A. Production of recombinant proteins in transgenic plants: Practical considerations. *Biotechnol. Bioeng.* **1997**, *56*, 473–484. [CrossRef]

114. Honisch, U.; Zumft, W.G. Operon structure and regulation of the *nos* gene region of *Pseudomonas stutzeri*, encoding an ABC-Type ATPase for maturation of nitrous oxide reductase. *J. Bacteriol.* **2003**, *185*, 1895–1902. [CrossRef] [PubMed]

115. Wan, S.; Greenham, T.; Goto, K.; Mottiar, Y.; Johnson, A.M.; Staebler, J.M.; Zaidi, M.A.; Shu, Q.; Altosaar, I. A novel nitrous oxide mitigation strategy: Expressing nitrous oxide reductase from *Pseudomonas stutzeri* in transgenic plants. *Can. J. Plant Sci.* **2014**, *94*, 1013–1023. [CrossRef]

116. Andri, K.B.S.; Santosa, P.; Arifin, Z. An empirical study of supply chain and intensification program on Madura tobacco industry in East Java. *Int. J. Agric. Res.* **2011**, *6*, 58–66. [CrossRef]

117. Sgouridis, F.; Ullah, S. Soil greenhouse gas fluxes, environmental controls, and the partitioning of N_2O sources in UK natural and seminatural land use types. *J. Geophys. Res. Biogeosci.* **2017**, *122*, 2617–2633. [CrossRef]

118. FAO. FAOSTAT. 2016; Available online: fao.org.

119. Intergovernmental Panel on Climate Change. *Climate Change 2014: Synthesis Report*; Contribution of Working Groups I, II and III to the Fifth Assessment Report of the Intergovernmental Panel on Climate Change; Pachauri, R.K., Meyer, L.A., Eds.; Cambridge University Press: Cambridge, UK, 2014.

120. Park, D.; Kim, H.; Yoon, S. Nitrous oxide reduction by an obligate aerobic bacterium, *Gemmatimonas aurantiaca* strain T-27. *Appl. Environ. Microbiol.* **2017**, *80*, e00502–e00517. [CrossRef] [PubMed]

121. Suenaga, T.; Aoyagi, T.; Hosomi, M.; Hori, T.; Terada, A. Draft genome sequence of *Azospira* sp. strain I13, a nitrous oxide-reducing bacterium harboring clade II type *nosZ*. *Genome Announc.* **2018**, *6*, e00414–e00418. [CrossRef] [PubMed]

122. Plotkin, J.B.; Kudla, G. Synonymous but not the same: The causes and consequences of codon bias. *Nat. Rev. Genet.* **2011**, *12*, 32–42. [CrossRef] [PubMed]

123. Hartley, B.S. Evolution of enzyme structure. *Proc. R. Soc. Lond. B Biol. Sci.* **1979**, *205*, 443–452. [CrossRef] [PubMed]

124. Padgette, S.R.; Kolacz, K.H.; Delannay, X.; Re, D.B.; LaVallee, B.J.; Tinius, C.N.; Rhodes, W.K.; Otero, Y.I.; Barry, G.F.; Eichholtz, D.A.; et al. Development, identification, and characterization of a glyphosate-tolerant soybean line. *Crop Sci.* **1995**, *35*, 1451–1461. [CrossRef]

125. Sardana, R.; Dukiandjiev, S.; Giband, M.; Cheng, X.; Cowan, K.; Sauder, C.; Altosaar, I. Construction and rapid testing of synthetic and modified toxin gene sequences *CryIA* (*b&c*) by expression in maize endosperm culture. *Plant Cell Rep.* **1996**, *15*, 677–681. [PubMed]

126. Cheng, X.; Sardana, R.; Kaplan, H.; Altosaar, I. *Agrobacterium*-transformed rice plants expressing synthetic *cryIA(b)* and *cryIA(c)* genes are highly toxic to striped stem borer and yellow stem borer. *Proc. Natl. Acad. Sci. USA* **1998**, *95*, 2767–2772. [CrossRef] [PubMed]

127. Pandeya, D.; Campbell, L.A.M.; Nunes, E.; Lopez-Arredondo, D.L.; Janga, M.R.; Herrera-Estrella, L.; Rathore, K.S. *ptxD* gene in combination with phosphite serves as a highly effective selection system to generate transgenic cotton (*Gossypium hirsutum* L.). *Plant Mol. Biol.* **2017**, 566–567. [CrossRef] [PubMed]

128. Pandeya, D.; López-Arredondo, D.L.; Janga, M.R.; Campbell, L.M.; Estrella-Hernández, P.; Bagavathiannan, M.V.; Herrera-Estrella, L.; Rathore, K.S. Selective fertilization with phosphite allows unhindered growth of cotton plants expressing the *ptxD* gene while suppressing weeds. *Proc. Natl. Acad. Sci. USA* **2018**, *115*, E6946–E6955. [CrossRef] [PubMed]

129. Liimatainen, M.; Voigt, C.; Martikainen, P.J.; Hytönen, J.; Regina, K.; Óskarsson, H.; Maljanen, M. Factors controlling nitrous oxide emissions from managed northern peat soils with low carbon to nitrogen ratio. *Soil Biol. Biochem.* **2018**, *122*, 186–195. [CrossRef]
130. Park, S.-H.; Yi, N.; Kim, Y.S.; Jeong, M.-H.; Bang, S.-W.; Choi, Y.D.; Kim, J.-K. Analysis of five novel putative constitutive gene promoters in transgenic rice plants. *J. Exp. Bot.* **2010**, *61*, 2459–2467. [CrossRef] [PubMed]
131. Anand, A.; Bass, S.H.; Bertain, S.M.; Cho, H.-J.; Kinney, A.J. Ochrobactrum-Mediated Transformation of Plants. US Patent App. WO/2017/040343, 26 August 2016.
132. Hofmann, N.R. A breakthrough in monocot transformation methods. *Plant Cell* **2016**, *28*, 1989. [CrossRef] [PubMed]
133. Webster, G.R.; Teh, A.Y.H.; Ma, J.K.C. Synthetic gene design—The rationale for codon optimization and implications for molecular pharming in plants. *Biotechnol. Bioeng.* **2017**, *114*, 492–502. [CrossRef] [PubMed]
134. Wunsch, P.; Herb, M.; Wieland, H.; Schiek, U.M.; Zumft, W.G. Requirements for Cu-A and Cu-S center assembly of nitrous oxide reductase deduced from complete periplasmic enzyme maturation in the nondenitrifier *Pseudomonas putida*. *J. Bacteriol.* **2003**, *185*, 887–896. [CrossRef] [PubMed]

Article

The Nexus of Weather Extremes to Agriculture Production Indexes and the Future Risk in Ghana

Abdul-Aziz Ibn Musah *, Jianguo Du, Thomas Bilaliib Udimal and Mohammed Abubakari Sadick

School of Management, Jiangsu University, 301 Xuefu Road, Zhenjiang 212013, Jiangsu, China; jgdu2005@163.com (J.D.); tbudimal2007@yahoo.com (T.B.U.); oluzerpublications@yahoo.com (M.A.S.)
* Correspondence: dooziizu@gmail.com; Tel.: +86-18605242822

Received: 18 September 2018; Accepted: 25 October 2018; Published: 31 October 2018

Abstract: The agricultural industry employs a large workforce in Ghana and remains the primary source of food security and income. The consequences of extreme weather in this sector can be catastrophic. A consistent picture of meteorological risk and adaptation patterns can lead to useful information, which can help local farmers make informed decisions to advance their livelihoods. We modelled historical data using extreme value theory and structural equation modelling. Subsequently, we studied extreme weather variability and its relationship to composite indicators of agricultural production and the long-term trend of weather risk. Minimum and maximum annual temperatures have negligible heterogeneity in their trends, while the annual maximum rainfall is homogenous in trend. Severe rainfall affects cereals and cocoa production, resulting in reduced yields. Cereals and cocoa grow well when there is even distribution of rainfall. The return levels for the next 20–100 years are gradually increasing with the long-term prediction of extreme weather. Also, heavy rains affect cereals and cocoa production negatively. All indicators of agriculture had a positive relationship with maximum extreme weather.

Keywords: extreme weather; agriculture production; return level; extreme value theory; weather; risk

---

1. Introduction

Many developing countries particularly those in the tropical regions are sensitive to changing the climate, especially where temperatures are already threatening agricultural production [1–3]. They have restricted access to a human and physical asset that can mitigate its effects [4]. These difficulties are often manifold by the lack of connection to new technologies and established markets [2,4]. Ghana is an example of a country facing these challenges. The irrigated land for agricultural use covers only 1% of farmland, and the majority of the farmers are entirely dependent upon seasonal rainfall [5–7].

This concern about the changing climate is due to its negative impact on the living conditions of humankind. Developing nations, particularly Ghana, is increasingly concerned about the changing climate because they are more vulnerable compared to developed nations. Climate change is a significant issue of risk to sustainable growth in Africa. As such, the efforts of African countries to realise the Millennium Development Goals can be considered as an offer if the adverse effects of climate change are taken seriously by Africa nations. Generally, African states contribute very little to climate change yet they bear the major brunt of it. Also, the Africa continent is more vulnerable to the effects of this changing climate as a result of its excessive reliance on rainfed agriculture, and extreme poverty [8]. The critical long-term effects of climate variation include: change in precipitation leading to reduced agricultural production, reduced food security, deterioration of water security, and reduction of fish stocks due to high temperature and displacement. Also, sea-level upsurge due to climate variation affects coastal areas greatly. The adverse effects of climate change in the form of a reduction in agricultural output ultimately lead to a delay in the development of African countries where a more substantial part of national income comes from agriculture. Also, the agricultural sector functions as a basis of livelihood for most people in Africa [8].

To tackle climate change, Ghana signed the United Nations Framework Convention on Climate Change (UNFCCC) at the Earth Summit in Rio de Janeiro in June 1992, following the adoption of the Convention on 9 May 1992 [9]. In Ghana, three critical physical effects of climate change identified include temperature change, precipitation change, and sea level rise [7]. According to a report [10], there is a shift in the rainfall regime in Ghana towards a longer dry season and vanishing wet season. Despite the signing of the Convention by Ghana, the country continues to face the adverse effects of climate change in the area of health, agricultural, already depletion of coastal areas, and low water levels. For example the country's only hydroelectric dam (which produces 80% of the national electricity supply) due to lower rainfall [11]. The consequence of climate change on the Ghanaian economy is due to the lack of environmental adaptation strategies and the socio-economic costs of adapting those strategies to mitigate the effects of climate change.

Climate change affects the transport system in the areas that are heavily dependent on weather conditions [12,13]. According to Reference [14], climate change adversely affects the critical elements of food production such as soil, water, and biodiversity. As a result, Ghana's economic dependence on areas (as energy, agriculture and forestry) which are particularly susceptible to the changing climate makes it more prone to the adverse effects of weather. In this vein, it is essential to carry out studies on the changing climate and its volatility in Ghana.

Specifically, this article examines the following.

- ➢ Examining the trends in extreme maximum rainfall and extreme high/low temperature
- ➢ Assessing the variability and weather risk of extreme maximum/minimum
- ➢ Analysis of the relationship of extreme weather to agriculture production indexes

  - Effect of exceptionally high rainfall on agriculture production indexes
  - Effect of extremely high temperature on agriculture production indexes
  - Impact of low temperature on agriculture production indexes

Rare weather conditions like severe rainfall, extreme temperature (and heat waves), or strong winds, may have significant effects on sectors such as agriculture and health, which may result in severe risk to human life [15]. Further, risks of extreme heat and drought depend not only on the severity of the event but also on the sensitivity and vulnerability of the exposure system [16].

The existing studies only show regional climate parameters and how the joints of their scales occur. We contend that the environmental parameters if could serve as a tool for eliminating human disasters if their extreme conditions are well understood and managed correctly [17]. Focusing on the regional research, particularly climate system, the influence of climate change and uncertainty in weather conditions could alter and transform societal and institutional behaviours [18,19].

Substantial studies concede extreme value theory as a method that estimates rare event whiles generalised extreme value distribution (GEVD) is capable in determining the probability of events occurrence that fall outside of an observed data range. Given this, GEVD has attracted attention in diverse areas of research such as climatology data analysis [20–23]. Issues relating to Extreme Value Theory gradually implemented in practical covariate approach of non-stationary conditions [15,20,24–28]. An investigation by [29–31] on daily rainfall at various observation sites in West Africa revealed an increasing trend of yearly maximum rainfall. Research has shown variations in extreme rainfall [30]. Thus extreme rain is related to a decline in annual precipitation intensity. In weather forecasting, efforts are made to predict the impact of weather conditions on food security [32]. Such reviews can help planners provide adequate protection and adaptation solutions that contribute to the resilience of the population and the reduction of socio-economic disasters. In the world over, 33% of observed crop production modifiability emanates from a change in climate thereby, a cause of variations of crop yield in Africa [33–35]. The intra-inter yearly rainfall and temperature show considerable effects on crops production and therefore ensures food safety [36].

Similar studies demonstrate that rainfall and temperature adversely affect crop yield. It calls for authorities in Africa to enforce sustainable food security policies [37,38]. In a period of severe soil moisture, flowering development stagnates [39]. Research has shown that drought is inimical to the growth of cocoa. Therefore, there is a causality between rainfall and cocoa yield [40]. Analogously, the sustenance of a bumper harvest is positively related to rainfall distribution than the total amount of rainfall received annually [41]. However, Reference [42] argues on the positive and negative causality of crops production in Ghana.

The yearly rainfall in cocoa growing areas in Ghana is more than 2000 mm. Also, two rainfall seasons are recorded from April to July and September to November, where July to August faces relative dry weather with high humidity condition. There is a dry weather condition between a second month and the eleventh month of the annual calendar [40,43]. Variations in climate pose a threat to the health of animals, and unfavourable heat affects them reproductively [44,45].

The 21st century saw a decline in yields ranging from 2.5% and 10% as temperature rises in some agronomic species [46]. The results of the evaluation of the effect temperature on crop yield at various levels indicate a decrease in yield. For example, the decline in barley production is due to the low temperatures during the vegetative stages and represents about 42% of low yield. The different seasons with low temperatures and high rainfall are unusable conditions for the potato, resulting in reduced yields [47].

Ascertained by [48–50], climate change due to the uncertainty of precipitation has a significant impact on agriculture production. On this account, this study introduces a different dimension into the analysis of weather effects on agriculture by looking at the extremes conditions of temperature and rainfall hence; we aim to fill this gap in the literature.

Given the increasing occurrences of climate change, there is a need for researchers to consider extreme conditions that often occur due to climate variability and its related events. Relying on climate variation in a whole without considering the specifics thus, minimum and maximum extremes have resulted in a situation where policies are formulated but not directed at specific extreme effects. This study looks at weather variability concerning maximum and minimum extreme conditions to enhance the formulation of targeted policies to help curb their impact on agriculture production. Further, we have investigated the relationship between extreme weather events and agriculture production indexes and assessed agricultural risk using extreme value theory (EVT) and structural equation modelling, which are different from previous studies.

2. Materials and Methods

2.1. Climate Change and Variability in Ghana

The regional scenarios of seasonal precipitation and temperature changes in 32 regions globally analysed by (IPCC, 2014) show the current variations in climate and the range of variations in 30-year period predicted by GCM, focused in 2025, 2055, and 2085. This background information is critical in explaining the probable effects of climate variation on livestock and crop production.

The IPCC approximate that the past period saw temperatures increased by an average of 0.6 °C. The preceding 25 years, there was no observation of atmospheric temperatures from 1995–2006, 11 out of 12 was the warmest years [51]. Countries are beginning to experience consequences related to global warmings, such as the long-term drought within the Sahel zone in Africa and the expansion of the malaria transmission belt of tropical Africa [52]. Universally, the figure noted for weather-associated natural adversities is fast increasing. From the 1960s, accounts of natural risks have tripled. During 2007, fifteen (14) out of fourteen (15), "emergency appeals" for emergency public-spirited assistance were in the areas of storms, droughts and floods, five times more than in the prior year [53]. Ghana's, climate variation is experiencing increasing unpredictable rainfall and temperatures in all regions [54].

Also, global warming is predicted to show variations in rainfall patterns, acidification, and moisture [55]. In this context, the global effect of climate variation on global life-assistance systems remains uncertain. Some parts experience extreme precipitation resulting in flooding; for example, the Mediterranean areas are experiencing a decline that could result in drought conditions [55]. By some reports [55], the anticipation of global average temperatures will rise between 1.4–5.8 °C by close of the century, as sea levels, increase as melting glaciers melt. Observations recently, however, indicate that many predictions concerning climate change are near the higher limit of the IPCC estimates. Sea levels, for example, have exceeded the IPCC estimates of up to 30 cm [56].

Based on a study by Reference [57], is establish that an estimated 35% of the entire land in Ghana is affected by increasing desertification. The unexpected variability of precipitation patterns is observed for years in Ghana as affirmed by Reference [58]. With the historical data, precipitation was mostly high in the 1960s, but fell to low levels by the end of 1970s and then rose again in 1980s. This fall in precipitation patterns is still prevalent currently, as Reference [59], with 20 years of data, observed this; temperatures are rising throughout Ghana and is precipitation decreasing and becoming gradually unpredictable. The effects of changing climate are anticipated to be severe in Ghana, even though there are rises and fall in both yearly temperatures and precipitation. Conceding to the World Bank's projection, the temperature trend from 2010–2050 shows warming in almost the highest-temperature parts of Ghana, including the North and the Upper Regions.

Nevertheless, the region with the lowest temperature is the Brong Ahafo region. These are base on different climate scenarios [58]. For example, looking at the scenario, it was recognised that the temperatures of the three northern regions would increase by 2.1–2.4 °C by 2050. On the contrary, the predicted increase in Ashanti, West, East, Volta, and Central regions ranges from 1.7–2.0 °C and those of Brong Ahafo 1.3–1.6 °C.

We also reviewed the latest temperature and precipitation forecasts from the Intergovernmental Panel on Climate Change (IPCC) [60] to simulate the impact of climate change on agricultural production in Ghana. These projections are on Phase Five of the Coupled Model Inter-comparison Project (CMIP5), which brings together the results of 39 different global models. We used projections for West Africa until 2035. According to the first scenario, the most optimistic, the temperature should increase by 0.7 degrees and precipitation by 8%. These increases represent the expected minimum increase in temperature and the maximum expected increase in precipitation. The second scenario concerns the median increase in temperature (0.9 degrees) and precipitation (1%). The third scenario, the least optimistic, concerns the maximum expected increase in temperature (1.5 degrees) and the maximum decrease in precipitation (4%). A meta-analysis of crop yield response to climate change, using local average temperature as an indicator of change, concluded that global warming at 2 °C could lead to an increase in wheat, rice, and maize yields, with yields subsequently decreasing with increased warming. The AR4 also showed that crop-level adaptations had a markedly positive effect on all crops, regions, and warming levels [61].

According to Reference [62], Tables 1 and 2 show some of the climate changes in Ghana and the corresponding time periods.

Table 1. The projections of precipitation in Ghana.

Location	Climate Type	Forecast Changes
Accra	Coastal Savanna Zone	From 52% decreases to 44% increases in wet season rainfall by the year 2080.
Kumasi	Deciduous Forest Zone	From 48% decreases to 45% increases in wet season rainfall by the year 2080. Based on their A2 scenario, which generally shows the largest greenhouse gas (GHG) impact, predicts the weakest increase in wet season rainfall, 1.13%.
Tarkwa	Rain Forest Zone	From 45% decreases to 31% increases in wet season rainfall.
Techiman	Forest-Savanna Transition Zone	From 46% decreases to 36% increases in wet season rainfall. The A2 scenario, which generally shows the largest GHG impact, predicts the largest decrease in wet season rainfall, −2.94%.
Tamale	Guinea Savanna Zone	From 36% decreases to 32% increases in wet season rainfall consistent trend toward decreased rainfall.
Walembelle	Northern Guinea Savanna Zone	From 25% decreases to 24% increases in wet season rainfall
Bawku	Sudan Savanna Zone	Range from 28% decreases to 30% increases in wet season rainfall.

Source: Extracted from [8,43].

Table 2. Temperature projections in various climate stations in Ghana.

Location	Climate Type	Temperature Projections	
		Wet Season	Dry Season
Accra	Coastal Savanna Zone	1.68 ± 0.38 °C by 2050 2.54 ± 0.75 °C by 2080	1.74 ± 0.60 °C by 2050 2.71 ± 0.91 °C by 2080
Kumasi	Deciduous Forest Zone	1.71 ± 0.39 °C by 2050 2.60 ± 0.77 °C by 2080	1.81 ± 0.68 °C by 2050 2.83 ± 1.04 °C by 2080.
Tarkwa	Rain Forest Zone	1.69 ± 0.37 °C by 2050 2.56 ± 0.75 °C by 2080	1.76 ± 0.67 °C by 2050 2.76 ± 1.01 °C by 2080.
Techiman	Forest-Savanna Transition Zone	1.77 ± 0.43 °C by 2050 2.71 ± 0.85 °C by 2080	1.95 ± 0.79 °C by 2050 3.05 ± 1.20 °C by 2080.
Tamale	Guinea Savanna Zone	1.84 ± 0.46 °C by 2050 2.83 ± 0.91 °C by 2080	2.05 ± 0.75 °C by 2050 3.18 ± 1.18 °C by 2080.
Walembelle	Northern Guinea Savanna Zone	1.92 ± 0.52 °C by 2050 2.96 ± 0.98 °C by 2080	2.10 ± 0.71 °C by 2050 3.27 ± 1.11 °C by 2080.
Bawku	Sudan Savanna Zone	1.92 ± 0.53 °C by 2050 2.97 ± 0.98 °C by 2080	2.11 ± 0.68 °C by 2050 3.25 ± 1.08 °C by 2080

Source: Extracted from [8,43].

2.2. Seasonal Changes of Precipitation and Temperature

The climate of Ghana is tropical, with a dry season in winter and a rainy season during the summer due to an African monsoon. The duration of the rains varies according to the ecological zones. As shown in Figure 1, the rainy season is usually from May to September to the north, from April to October in the centre, and from April to November to the south. However, on the east coast, the rainy season is shorter than the rains from April to June, with no rainfall in July and August, and it picks up slightly in September and October. The south is the coolest part of Ghana, where it has more than

1500 mm (per year), and even more the small west coast, where it reaches 2000 mm (80 inches) per year. The north is the driest in Ghana, where rainfall is about 1000 mm (40 inches) per year and the east coast, including the city of Accra, where it falls below 800 mm (31.5 in).

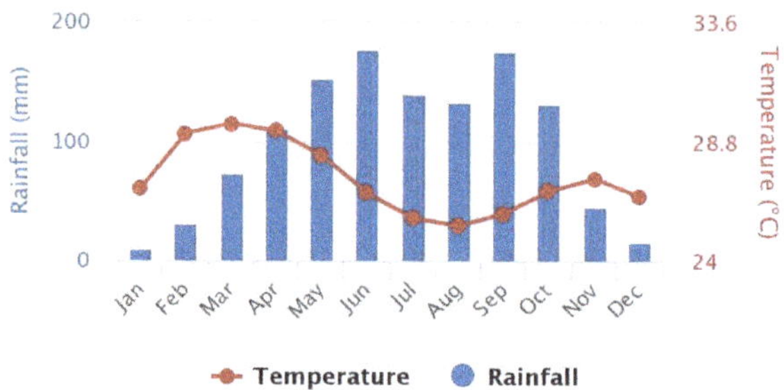

Figure 1. The Monthly trend of temperature and rainfall in Ghana.

2.3. The trend of Climate Change in Ghana

Ghana is located in West Africa, bordered to the north by Burkina Faso, east to Togo, west to Ivory Coast, and south to the Gulf of Guinea. It is located between 4.50 degrees north and 11.50 degrees north and longitude 3.50° west and 1.30° east. The country has an area of 239,460 Km^2 and a surface area of 8520 Km^2 as seen in Figure 2. The country has a population of around 24 million since 2010, with an annual growth rate of about 2.5% [63]. Young people dominate this population. The main exports are cocoa, gold, wood, diamonds, bauxite, manganese, and hydroelectricity. Until recently, the country also began to export crude oil. In 1991/92, the poverty level in Ghana reached 51.7 per cent, and this figure has steadily declined in recent years to 39.5 per cent in 1998/99, 28.5 per cent in 2005/06, and 24.2 per cent in 2012/2013. The country enjoys a high temperature while the average annual temperature is between 24 °C and 30 °C. Despite the average annual temperature, temperatures may be 18 °C and 40 °C in the southern and northern parts of Ghana. Rainfall in Ghana is generally declining from south to north. A more prosperous region in Ghana is the far southwest, with an annual rainfall of about 2000 mm. However, the annual rainfall in northern Ghana is less than 1100 mm. The country has two major systems of rain: the double-twin system and the single maximum regime. For the maximum binary system, the maximum periods are from April to July and from September to November in southern Ghana. While the only maximum system is from May to October in northern Ghana, the prolonged drought lasts from November to May. Over the years, temperatures have risen in all ecological regions of Ghana, while rainfall levels have generally declined and standards have steadily increased [9].

Despite dramatic improvements in technology and crop yields, food production continues to depend heavily on the climate because solar radiation, temperature and rainfall are the critical factors of increase in crop production. The climate is affected by the diseases of plant and the spread of pests, including the supply and demand for irrigation water. For instance, in recent decades, the ongoing drought in the Sahel has caused a continued deterioration in food production [64] in Ghana. The effect of the changing climate on crops was in 1990, where the crop has suffered or decreased. Also, due to drought, climate indicators such as rainfall and average mean temperature are associated with crop change [57]. Table 3 below presented climate change variations experienced.

Figure 2. Location Map of Ghana.

Table 3. Climatic variations experienced in Ghana.

Time Period	Climatic Variations
January–July 1976	Scorching weather conditions
1983–1984	Drought: A yearlong of bushfires
October–December 1989	Scorching weather conditions
1991	Lots of rains throughout the year
1995	About 40 days of intensive rains
2004	Noticeable are frigid winds during March–April (Easter) and November–January was very cold weather
2005	Cold periods resulting in animal deaths
August 2006	One week of intensive rains, and
2007	Lots of rains in August and September.

Source: Extracted from [62].

2.4. The Generalized Extreme Value Distribution (GEVD)

The GEVD is part of the family of continuous distribution functions that allows a continuous range of shapes and consists of classes of distribution functions such as Gumbel, Fréchet, and Weibull. Considering the Fisher-Tippett Gnedenko theorem, the GEVD is a limit-form distribution function, which maximises the maxima of the sequence of random variable considered as independent and identical distributed (i.i.d.). It, therefore, models the maximum of a finite sequence of random variables. The combined model of maxima is by Equation (1):

$$G_{\gamma,\mu,\sigma} = \exp\left\{-\left(1 + \gamma\left(\frac{x-\mu}{\sigma}\right)^{-\frac{1}{\gamma}}\right)\right\} \text{ with, } \gamma \neq 0,\ \sigma > 0 \text{ and } \gamma\left(\frac{x-\mu}{\sigma}\right) > 0$$
$$\sigma > 0 \tag{1}$$

The derivative of Equation (1), give a probability density function in Equation (2) as:

$$g_{\gamma,\mu,\sigma} = \frac{1}{\sigma}\left(1+\gamma\left(\frac{x-\mu}{\sigma}\right)\right)^{-1-\frac{1}{\gamma}} \exp\left\{-\left(1+\gamma\left(\frac{x-\mu}{\sigma}\right)\right)^{-\frac{1}{\gamma}}\right\}, \; \gamma \neq 0 \tag{2}$$

where μ and σ are the location and scale parameters, respectively [20].

The GEVD shape parameter γ also termed as the extreme value index. The decay rate of GEVD seen as γ^{-1}. If $\gamma > 0$ for a class of distributions, G fits distributions as; the heavy-tailed Fréchet distribution, Cauchy, Student's t, Pareto class, and mixture other distributions. G fit into the short-tailed Weibel distribution, uniform, and beta distribution if $\gamma < 0$. G fits the right-tailed Gumbel distributions (normal, exponential, gamma, and lognormal) if $\gamma = 0$ [65–67].

2.5. Maximum Likelihood Estimation for GEVD

The assumption that $X_1, \ldots, X_m$ follows an (i.i.d) and also from generalised extreme value distribution with parameter when $\gamma \neq 0$ the log-likelihood function given as:

$$\text{Provided that } 1+\gamma\left(\frac{x_{(i)}-\mu}{\sigma}\right) > 0 \text{ for } i = 1, 2, \ldots, m \tag{3}$$

$$l(\mu,\sigma,\gamma) = -m\ln\sigma - (1+1/\gamma)\sum_{i=1}^{m}\ln\left[1+\gamma\left(\frac{x_{(i)}-\mu}{\sigma}\right)\right] - \sum_{i=1}^{m}\left[1+\gamma\left(\frac{x_{(i)}-\mu}{\sigma}\right)\right]^{-1/\gamma} \tag{4}$$

Parameters combination that deviates from the above conditions (Equation (3)), i.e., in a configuration where at least one of the observed data exceeds the endpoint of the distribution (Equation (4)), the likelihood is zero, and the log-likelihood is equal to $-\infty$. This case $\gamma = 0$ requires separate treatment with GEVD's Gumbel restriction leading to logarithmic log-likelihood as in Equation (5);

$$l(\mu,\sigma) = -m\ln\sigma - \sum_{i=1}^{m}\left(\frac{x_{(i)}-\mu}{\sigma}\right) - \sum_{i=1}^{m}\exp\left\{-\left(\frac{x_{(i)}-\mu}{\sigma}\right)\right\} \tag{5}$$

Equations (2) and (3) are differentiated and maximised concerning the parameter vector (μ,σ,γ), Solving for (μ,σ,γ), results to the maximum likelihood estimates for the whole GEVD model [20,28,68,69]. Maximum likelihood estimation offers the advantage of estimation of the three parameters together and applicable to the series of maxima per block [70].

Model Checking for GEVD

The model fit of GEVD measure after estimating the parameters by utilising residual plots function as defined by Equation (6),

$$res = \begin{cases} \left(1+\frac{\gamma}{\sigma}(x-\mu)\right)^{-1/\gamma} & if\ \gamma = 0 \\ \exp\left[-\exp\left(-\frac{x-\mu}{\sigma}\right)\right] & if\ \gamma \neq 0 \end{cases} \tag{6}$$

Ascertain by Reference [20] conversion of data to unit exponential distributed residuals is on the null assumption that GEVD fits the data.

2.6. Return Period or Level Estimates

The frequency of extreme quantiles incidence estimated with a fixed value of return level. The return level is the mean number of events taking place within a unit period, e.g., one year [71]. Return levels are essential for prediction purposes and estimated from stationary models. The expected return time is the number of time (years) one is expected to wait on average before the observation of another extreme event of at least the same intensity. If a threshold exceedance of a given probability of an observed extreme incidence in any given time (year) is p, then the mean return period T is such that $T = 1/p$.

2.7. Test for Stationarity and Seasonality

The stationarity of the data conducted by the augmented Dickey-Fuller (ADF) stationarity test on the assumption that there is no trend [72]. The quality of convergence of the weather extremes is access using the Kolmogorov–Smirnov (K-S) and Anderson–Darling goodness-of-fit tests. The K-S test, relying on the empirical study of the cumulative distribution function, is used to determine whether the sample is from the hypothesised continuous distribution. The K-S approach is less sensitive for normal distribution [72]. The Anderson-Darling test, an enhancement of the K-S test, compares the fit to the expected cumulative distribution function of the observed cumulative distribution function. This test gives more substantial weight to the tail of the distribution than the K-S test [72].

The assumption is that the data is from a population which is independent identically distribution (i.i.d). The alternative hypothesis is a two-tail test on the assumption that the data follow a monotonic trend. Thus, the following test statistics by Mann-Kendall determine by Equation (7):

$$S = \sum_{k}^{n-1} \sum_{j=k+1}^{n} sgn\left(x_j - x_k\right) \tag{7}$$

with sgn the signum function.

3. Methodology

This paper analyses past Composite Indexes in Agriculture ranges from 1961–2016: crops production, cocoa production, livestock production, cereal production, and food production in Ghana. The data also consider records of maximum rainfall, maximum temperature, and minimum temperature value as weather indicators from January 1965 till July 2016. We sourced the data from the Ghana Meteorological Agency for climate data, and agriculture production indexes from the Food and Agriculture Organization also in Ghana. Rainfall and temperature are assumed to be the primary determinants of weather in Ghana as seen in Figure 1. The first task was to check for stationarity of the weather variables using Augmented Dickey-Fuller (ADF) unit root test and then the Mann-Kendall Trend Test of seasonality. It was necessary to apply methods that explicitly allow for testing non-stationarity in the distribution parameters of climate variables [20].

Next step was to model from the dataset of the weather indicators employing the Block Maximum Method for the weather extremes under Generalized Extreme Value Distribution (GEVD). There were two approaches to the modelling of Block minima data for the minimum temperature. Either the GEVD for minima fitted to this data or the data negated and the GEVD for maxima fitted [20]. The latter approach was adopted since the Extremes Toolkit does not include a routine to estimate the GEVD for minima directly. The block maxima method is a parametric approach to Extreme Value Theory. It entails fitting the GEVD to a specific group of maximum values chosen in a given sample of data. It focuses on the statistical behaviour of the largest or smallest value in a sequence of independent random variables. Assume that the sequence is grouped into blocks of size N (with a reasonably large number) and that only the maximum score M_i ($i = 1, 2, 3, \ldots, n$) of each block extracted. Each M_i ($i = 1, 2, 3, \ldots, n$) of the weather indicators is then used to estimate the relationship between the composite indexes of agriculture production.

The mean return period defines the amount of time (e.g., years) that is expected to pass on average before a new extreme with the same or increased intensity. Given the likelihood that events past a certain threshold will follow an extreme of a particular security at any given time (year) is defined as p, then the mean return period T can be calculated as $T = 1/p$.

Food production index includes food crops that are considered edible, and that contain nutrients with the exclusion of coffee and tea because they have no nutritive value although edible (FAO). Figure 3 shows the primary crop food calendar.

Finally, we investigated the relationship between extreme weather events and agriculture production using SEM software to evaluate the potential impacts of weather extremes on Agriculture production. We used SEM regression for the paths equation modelling analysis with the partial least squares (SEM) estimation technique [73]. SEM is a modelling approach with a flexible procedure, which can handle data with missing values, strongly correlated variables, and small samples. SEM-regression works with both continuous and discrete observed variables as indicators. The SEM estimates loading and path parameters between variables and maximises the variance explained for the outcome variables [73].

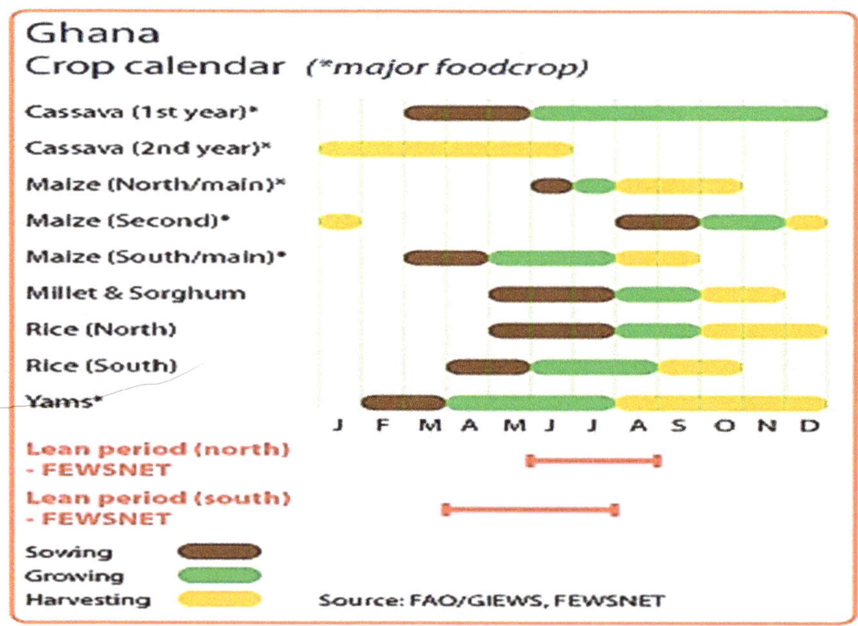

Figure 3. Major food crops calendar in Ghana.

4. Results and Discussion

4.1. Stationarity Test for the Weather Indicators

The ADF test is captured in Table 4 indicating the significance of the p-value statistics. The premise of non-stationary at 1%, 5%, and 10% rejected, and therefore we conclude the stationarity of the weather indicators.

It is reported by scholars that, Mann-Kendall Trend Test of stationarity is reliable and efficient. In line with this, analysing environmental data demands the exposure of movements of events on separate points [74]. Based on this, the test outcome illustrates high or low trends in weather conditions of a particular jurisdiction.

Table 4. Stationarity and Seasonality test.

Augmented Dickey-Fuller Stationarity Test

Test Variable	Test's Critical Values			Test Statistics	*p*-Value
	1%	**5%**	**10%**		
Annual maxi. Rainfall	−3.958	−3.410	−3.127	−16.350	0.0000
Annual maxi. Temperature	−10.007	−3.431	−2.862	−2.567	0.0000
Annual mini. Temperature	−12.482	−3.431	−2.862	−2.567	0.0000

Seasonal Mann-Kendall Trend Test

Series	Statistics		*p*-value		tau	Slope 95% CI
	z (trend)	**z (Het)**	**p (trend)**	**p (Het)**		
Maxi. Rainfall	0.434	22.376	0.664	0.0216	0.0019	0.0044 [−0.0194,0.0308]
Maxi. Temperature	21.842	4.779	<0.001	0.9410	0.1320	0.0318 [0.0286,0.0346]
Mini. Temperature	25.123	23.894	<0.001	0.1320	0.1520	0.0231 [0.0212,0.0250]

95% confidence interval in parenthesis.

In Table 4, the estimated annual trend is 0.0044 mm/year, a yearly increase in the maximum annual rainfall. The *p*-value based on the Kendall seasonal trend test is $p = 0.6640$, which shows no importance. The 95% confidence interval on both sides for the trend (−0.014,0.0307), the chi-square test for heterogeneity (Het) gave a *p*-value of 0.0216. Therefore, there is a difference in the level of a trend in the different seasons of the maximum annual rainfall. As shown in Table 4, the estimated annual trend is 0.0318 degrees Celsius (°C)/year, which is a yearly increase in the yearly maximum temperature. The *p*-value corresponds to the Kendall seasonal test for the $p < 0.001$ trends, indicating that it is statistically significant. The 5% level of significance on both sides for the trend is (0.0286, 0.0346). The chi-square heterogeneity test (Het) provides a *p*-value of 0.9410, so there is no evidence for different sets of stresses at different times of the maximum annual temperature. The estimated annual trend is 0.0231 degrees Celsius (°C)/year, a yearly increase in the maximum annual temperature. The *p*-value of the Kendall seasonal trend test, $p < 0.001$, indicating that it is statistically significant. The 5% level of significance on both sides for the trend is (0.0212,0.0250). The chi-square test for heterogeneity (Het) gives a *p*-value of 0.1318, i.e., no indication of the different trend in different seasons of the minimum annual temperature.

4.2. GEVD Model for Extreme Maximum Rainfall

In Table 5, the estimated return periods of maximum rainfall likely to occur over the next 5, 10, 20, 50 or even 100 years fitted by GEVD. The estimated results are (μ, σ, γ) (149.03,23.98,0.0024), with standard errors (3.758, 2.718, 0.1002). The approximate 95% confidence intervals for the parameters are thus (141.67, 156.39) for μ, (18.65, 29.31) for σ, and (−0.193,0.198) for γ.

Table 5. Generalised extreme value estimates of maximum rainfall.

GEV	Maximum Rainfall		
	Location	**Scale**	**Shape**
Estimates	$\mu = 149.03$	$\sigma = 23.98$	$\gamma = 0.0024$
Std error	3.758	2.718	0.1002
95% CI (normal app)	(141.67,156.39)	(18.65,29.31)	(−0.193,0.198)
Estimated Return Levels	**95% Lower**	**Estimate**	**95% Upper**
5-year return level	173.14	185.06	196.98
10-year return level	186.59	203.13	219.68
20-year return level	196.99	220.50	244.03
50-year return level	206.49	243.04	279.57
100-year return level	210.72	259.95	309.05

The validity and reliability of the extrapolation of GEVD fit is assessed base on the observed data. Four graphical analyses assist with model checking [20,75]. Figure 4 shows diagnostic plots assessing the accuracy of the GEVD model fitted. Neither the quantile plot nor the density plot has any reason to doubt the validity of the fitted model: each drawn set of points is almost linear. The return level plots asymptotically converge to a determinate value due to the positive estimates, with the curve approaches a straight line. The sample variable under consideration provides an adequate representation graphically of the empirical estimates. Finally, the corresponding density estimate appears to be consistent with the density curve. As a result, all four diagnostic diagrams support the GEVD model as in Figure 4 (Top-left: empirical plot; Top-right: empirical quantile plot; Bottom-left: density plot; Bottom-right: return level plot).

The determination of the limiting distribution by maximising the GEV negative log-likelihood for annual maximum rainfall leads to the following function in Equation (8):

$$G(z) = \exp\left\{-\left[1 + 0.00243\left(\frac{z - 23.98}{149.03}\right)\right]^{\frac{-1}{0.00243}}\right\} \tag{8}$$

Equation (8) gave estimates of return levels for 5, 10, 20, 50, and 100-years and their 5% significant level as shown in Table 3. Thus, based on the data from 1965 to 2016, once in 50 years we should expect to see an extreme annual maximum rainfall hit between 206.5 and 279.6 mm. The upper bound of the model prediction for the 50-years Return level of 279.6, but 510 mm extreme rainfall recorded in 1968. Of course, this is undoubtedly extreme beyond regular extreme events, which is not expected based on the model's predictions. Results from Table 2, indicates that extreme maximum rainfall is steadily increasing significantly over the 100 years.

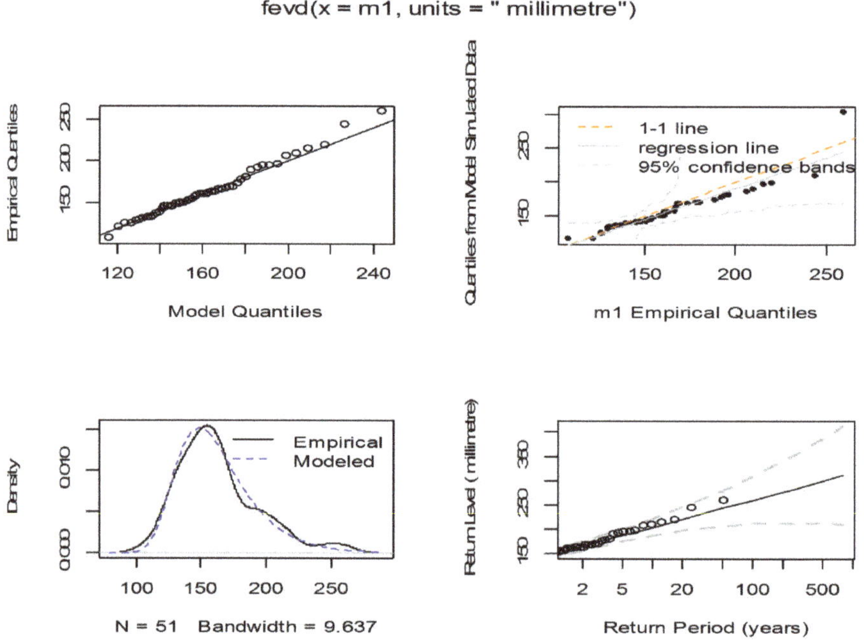

Figure 4. Diagnostic annual maximum rainfall plots.

4.3. GEVD Model for Extreme Maximum Temperature

As shown in Table 6, the estimated return level of maximum temperature likely to occur over the next 5, 10, 20, 50, or even 100 years by fitting these data to the GEVD. The maximum rainfall data yield estimates for (μ, σ, γ) of (41.933,0.892, 0.203), with standard errors (0.137,0.105,0.079). The approximate 95% confidence intervals for the parameters are thus (41.66,42.20) for μ, (0.686,1.098) for σ, and (0.0463,0.359) for γ.

Table 6. GEVD estimates of maximum temperature.

GEVD	Maximum Temperature		
	Location	Scale	Shape
Estimates	$\mu = 42.08$	$\sigma = 0.826$	$\gamma = -0.292$
Std error	0.128	0.0912	0.0942
95% CI (normal app)	(41.664,42.202)	(0.686,1.098)	(0.046,0.359)
Estimated Return Levels	**95% lower**	**Estimate**	**95% upper**
5-year return level	42.82	43.08	43.35
10-year return level	43.16	43.44	43.72
20-year return level	43.39	43.72	44.04
50-year return level	43.59	44.00	44.41
100-year return level	43.67	44.17	44.67

Analytic plots used in estimating the accuracy of the GEVD model fitted to the annual maximum temperature data shown in Figure 5 (Top-left: empirical plot; Top-right: empirical quartile plot; Bottom-left: density plot; Bottom-right: return level plot). All four diagnostic schemes provide support for fitting the GEVD to the maximum annual temperature.

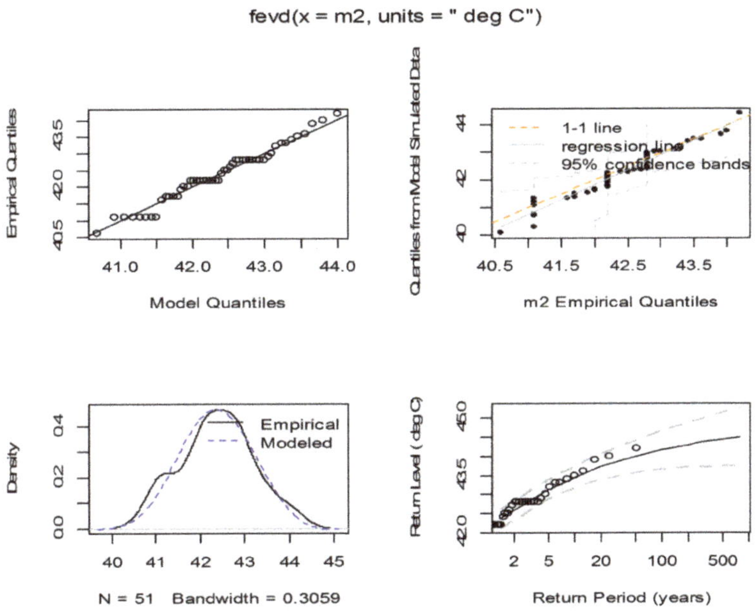

Figure 5. Diagnostic annual maximum temperature plots.

The determination of the limiting distribution by maximising the GEV negative log-likelihood for annual maximum temperature leads to the following function, Equation (2): (μ, σ, γ) of (42.081, 0.826, −0.292)

$$G(z) = \exp\left\{-\left[1 + 0.826\left(\frac{z - 0.892}{42.08}\right)\right]^{\frac{-1}{-0.292}}\right\} \tag{9}$$

From Equation (9), estimates of return periods for 5, 10, 20, 50, and 100-years and their confidence intervals at 95% as shown in Table 6. Thus, based on the data from 1965 to 2016, once in 100 years we should expect to see an extreme annual maximum temperature hit between 43.6 °C and 44.4 °C maximum temperature. The upper bound of the model prediction for the 100-years return is 44.4 °C, but 65 °C extreme annual temperature recorded in 1989. Of course, this is also undoubtedly extreme beyond regular extreme events, which is not expected based on the model's predictions. It is revealed by Table 6, that extreme maximum temperature consistently increasing marginally over the 100 years.

4.4. GEVD Model for Extreme Minimum Temperature

In Table 7 below, the estimated return periods of minimum rainfall likely to occur over the next 5, 10, 20, 50 or even 100 years fitted to the GEVD. The maximum rainfall variable yields estimates for (μ, σ, γ) of $(6.408, 5.261, -0.632)$, with standard errors $(0.817, 0.758, 0.148)$ respectively. Approximate 95% confidence intervals for the parameters are thus $(4.806, 8.011)$ for μ, $(3.774, 6.747)$ for σ, and $(-0.922, -0.342)$ for γ.

Table 7. GEV estimates of Minimum Temperature.

GEV	Minimum Temperature		
	Location	**Scale**	**Shape**
Estimates	$\mu = 6.408$	$\sigma = 5.261$	$\gamma = -0.632$
Std error	0.817	0.758	0.148
95% CI(normal app)	(4.806,8.011)	(3.774,6.747)	(−0.922,−0.342)
Estimated Return Levels	**95% lower**	**Estimate**	**95% upper**
5-year return level	10.355	11.506	12.657
10-year return level	11.882	12.723	13.564
20-year return level	12.761	13.456	14.151
50-year return level	13.233	14.022	14.812
100-year return level	13.333	14.274	15.216

Equation (10) is the determination of the limiting distribution by maximising the GEV negative log-likelihood for annual minimum temperature leads to the following function:

$$G(z) = \exp\left\{-\left[1 - 0.632\left(\frac{z - 5.261}{6.408}\right)\right]^{\frac{-1}{(-0.632)}}\right\} \tag{10}$$

Supposing the relative stability of the GEVD process producing estimates for annual minimum temperature in degree Celsius (°C), the model estimates that the 5-year return level is 11.5 °C with 95% confidence interval (10.4, 12.7). For ten years it is a 12.7 °C extreme minimum temperature with 95% confidence interval (11.9, 13.6), and for 50 years it is 14.0 °C extreme minimum temperature with 95% confidence interval (13.2, 14.8). Thus, based on the data from 1968 to 2016, once in 100 years we should expect to see an extreme annual minimum temperature hit between 13.3 °C and 15.2 °C. For the period under annual extreme minimum temperature, there was no extreme beyond normal extreme events. In Table 7, the extreme minimum temperature is consistently increasing over the 100 years' duration. In Figure 6 (Top-left: empirical plot; Top-right: empirical quartile plot; Bottom-left: density plot; Bottom-right: return level plot), all four diagnostic schemes provide support for fitting the GEVD to the minimum annual temperature.

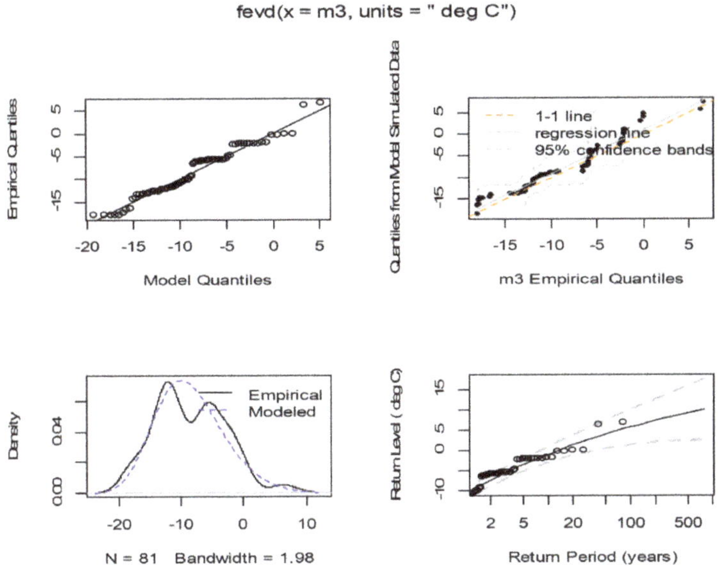

Figure 6. Diagnostic annual minimum temperature plots.

4.5. Return Level

Given 50-year return level for each of the indicators of extreme weather (for the year 2076), the return levels of extreme maximum rainfall in Ghana is higher than 150 mm reaching a warning line of extremely torrential rain, as defined by the Meteorological Service of Ghana. Similarly, the 50 years return level for maximum temperature exceeds 40 °C reaching a warning line of unusual temperature as defined by the Meteorological Service of Ghana. Also, the 50 years return level for extreme minimum temperature is lower than 20 °C reaching a warning line of frigid cold, as defined by the Meteorological Service of Ghana.

4.6. Structural Equation Modeling (SEM)-Regression Analysis

The term "structural equation modelling" (SEM) conveys two significant phases of the process: (a) causal effects under the research epitomised by a lot of structural equations (i.e., regression), and (b) these structural relationships can be presented to enable more specific concepts of theory studying. The assumed model (Figure 7) can then be statistically tested in a simultaneous analysis of the entire variables system to determine its compatibility with the data. If the suitability is appropriate, the model argues for the acceptance of assumed interactions between the variables; if inappropriate, the likelihood of such relationships fails to accept [76]. We chose PLS-SEM in present work for the following reasons: It is suitable for studies of theory construction [77,78]. It is appropriate to assess the sophisticated models of the cause-effect interaction [79,80]. The PLS-SEM assume a non-boundary approach, with fewer restrictions regarding sample size and data distribution [77].

SEM-regression estimation procedure was used to examine the hypothesised relationships as shown in Figure 4 between weather indicators and agriculture production. The results of SEM analysis showed a significant correlation between extreme weather and Agriculture production.

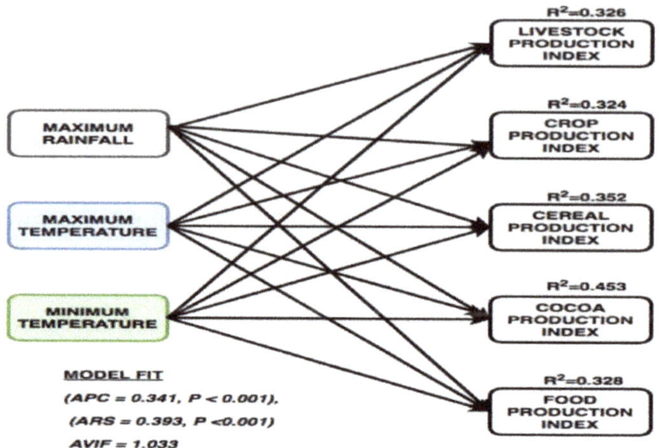

Figure 7. The Conceptual frame of the relationship of extreme weather on agriculture production indexes.

4.6.1. The relationship between Maximum Rainfall and Composite Agriculture Indexes

The analysis as showed in Table 8 is that, Livestock production index ($\beta = -0.1840$, $p = 0.144$), crop production ($\beta = -0.189$, $p < 0.133$), Cereal production ($\beta = -0.266$, $p < 0.031$), Cocoa ($\beta = -0.461$, $p < 0.001$), and food production index ($\beta = -0.190$, $p < 0.131$). Each is influenced by the effect of extreme maximum rainfall negatively on all composite agriculture indexes with no significant effect on crop production, food production, and livestock indexes. There has been a significant effect on cereal production and cocoa production indexes.

Table 8. Standardised Regression Weights and significance of correlations.

Predictor	Outcome	Path Coefficient	*p*-Values
Maximum Rainfall	Livestock Production Index	−0.184	0.144
	Crop production index	−0.189	0.133
	Cereal Production index	−0.266 *	0.031
	Cocoa production	−0.461 ***	<0.001
	Food Production Index	−0.190	0.131
Maximum Temperature	Livestock Production Index	0.305 *	0.015
	Crop production index	0.263 *	0.037
	Cereal Production index	0.276 *	0.025
	Cocoa production	0.424 *	0.023
	Food Production Index	0.268 *	0.033
Minimum Temperature	Livestock Production Index	0.457 ***	<0.001
	Crop production index	0.482 ***	<0.001
	Cereal Production index	0.415 ***	<0.001
	Cocoa production	−0.211 *	0.038
	Food Production Index	0.484 ***	<0.001

Significance of coefficient: *** $p < 0.001$ and * $p < 0.050$.

The results as shown in Table 8, shows each index is influenced by the effect of extreme maximum rainfall negatively, with no significant impact on crop production, food production, and livestock indexes. There has been a considerable effect on cereal production and cocoa production indexes. Maximum extreme rainfall hurts the performance of cereals.

Consequently, a unit increase in maximum extreme rainfall leads to a decrease in cereal production by 0.266 units. Maximum extreme rain leads to filtration of essential nutrients necessary for grain growth. Under such condition, any nutrient whether organic or inorganic leached beyond the reach of the roots, will result in reduced yields.

For cereals to bear maximum yields, rainfall, especially during tasseling for maize, is needed in moderation, inter-sparse with sunlight for maximum yields. Torrential rains do not favour most crops production and most especially cereals. Several studies have shown the importance of rainfall variability in crop production in various spatial scales [33,38].

Excessive rain has an adverse impact on agriculture. These effects run via different mechanisms. Heavy rains and floods have resulted in crop damage and the creation of poor conditions for harvesting, storage and transport of agricultural products. It is not astonishing that maximum rainfall has a negative association with all the variables under consideration, but only cereal and cocoa production indexes are statistically significant. Rainfall affects more variations in cocoa yields from year to year than with any other climatic factor. Trees are prone to a soil water shortage. The rain should be abundant and well distributed throughout the year. The annual precipitation between 1500 mm to 2000 mm is generally preferred. Droughts with rainfall below 100 mm per month should not exceed three months. The flooding of farmland leads to the leaching of nutrients needed for the growth of cocoa trees. If the phenomenon occurs over a period, this often leads to the death of cocoa trees or poor yields are observed [81]. It affects the flowering of cocoa trees and leads to flower aborting in some instances.

4.6.2. The Relationship between Maximum Temperature and Composite Agriculture Indexes

As shown in Table 8, Livestock production index ($\beta = 0.305$, $p = 0.015$), crop production ($\beta = -0.263$, $p = 0.037$), Cereal production ($\beta = 0.276$, $p = 0.025$), Cocoa ($\beta = 0.424$, $p = 0.023$), and food production index ($\beta = 0.268$, $p < 0.033$). Each is influenced significantly by the effect of extreme maximum temperature positively on all agriculture production indexes.

As shown in Table 8, each outcome is influenced significantly by the effect of extreme maximum temperature positively on all agriculture production indexes. The result indicates that a unit change in the maximum temperature will result in about 0.305 change in livestock production index. The nature of Ghana's livestock production immune it from the effects of extreme temperature conditions. Most animals are subject to a free or semi-intensive management system where animals are about to move freely.

Also, most cattle raised in Ghana are more adaptable to the state of the coast. As a result, maximum temperatures in Ghana does not affect them negatively since most of the animal rearing areas are almost in the coastal savannah region where the temperatures are not as high as the actual Sahel regions.

Breeding animals are sensitive to climate change and are severely affected by heat stress with an adverse effect on reproductive function [44,82]. According to Reference [83], high temperature and radiant heat load affect the reproductive rhythm through the hypothalamohypophyseal-ovarian axis. The primary factor in regulating ovarian activity is GnRH of thalamus and gonadotropin, i.e., FSH and LH of the anterior pituitary wall.

Research by [84,85] showed that the LH pulse amplitude and frequency of heat stressed cattle decreased. However, this is not the case in Ghana as shown in the results. Extreme temperatures that result in detrimental conditions not recorded in Ghana. High extreme temperatures hurt the crop production index, cereals production index, cocoa production, and food production index.

The maize pollen viability declines at temperatures above 35 °C [86–88]. Temperature increases in the 21st century may lead to yield losses of between 2.5% and 10% in some agronomic species [46]. Other assessments of crop yield due to temperature have produced different outcomes. Studies conducted by [89,90] showed estimates of yield between 3.8% and 5% decreases According to [90], crop growth for maize, soybeans and cotton will increase gradually with temperatures ranging from 29 °C to 32 °C and then sharply decrease as temperatures rise above this limit. It is however not surprising that maximum temperature in Ghana does not have adverse effects on yields. Maximum temperatures in Ghana is from 29 °C to 32 °C recorded in a dry season where no cultivation is taking place.

The period for production is the rainy season where temperatures hardly get close to 29 °C to 32 °C. Cocoa especially requires much heat, but direct sunshine damages it. As a result, some level of protection is necessary, especially when trees are young. Cocoa trees respond well to moderately high temperatures with a maximum yearly mean of 30 °C to 32 °C [91]. It is however not surprising that maximum temperature associate positively with cocoa production in Ghana where the maximum temperature falls within the acceptable range for cocoa.

4.6.3. The Relationship between Minimum Temperature and Composite Agriculture Indexes

As shown in Table 8, livestock ($\beta = 0.457$, $p < 0.001$), crop ($\beta = 0.482$, $p < 0.001$), Cereal ($\beta = 0.415$, $p < 0.001$), Cocoa ($\beta = -0.211$, $p = 0.038$), and food ($\beta = 0.439$, $p < 0.001$). Each is influenced significantly by the effect of extreme minimum temperature adversely on cocoa and positively on food, livestock, cereal, and crop production indexes.

As shown in Table 8, each outcome is influenced significantly by the effect of extreme minimum temperature adversely on cocoa production index and positively on (food, livestock, cereal, and crop) production index. Except for the cocoa sector, which is associated negatively with minimum extreme temperature the remaining areas are associated positively with low temperature. Average monthly temperatures below 23 °C are considered to suppress flowering.

The range in the average monthly temperature of the mainstream of cocoa-growing regions is found to be from 15 °C to 32 °C and considered to be the optimum for cocoa growth. The absolute minimum for any reasonable period is taken to be 10 °C, below which frost injury is likely [82]. Temperatures below the absolute minimum have a devastating impact on cocoa yields, as the results show.

Low arable yields caused by unfavourable weather conditions during certain stages of the growing season. The effects of unfavourable weather situations have shown reduced arable yields in recent decades. During the vegetative stage, low temperatures cause a reduction in barley yields. Low temperatures account for about 42% of the decrease in yield. Estimates show low temperatures in April, high rainfall in May and a heat wave in July followed by a cold and rainy August created unfavourable growth conditions for potatoes resulting in a decrease in yields [47].

Low yields of corn associated with a combination of low amounts of irradiation during the growing season (64% of low yields) and cold and wet spring (79% of low yields) cause delayed planting and slow biomass growth. Delayed frost has often worsened this situation (36% of low returns). Also, low yields contributed to the stress of drought and heat in flowering (21 per cent of low yields) and the recording of water during harvesting (29% of low yields) [47]. The type of low temperatures that often result in yield reduction is not the type often recorded in Ghana. Shallow temperatures experienced during the growing seasons in Ghana, hence its positive association with all the parameters except cocoa.

Regression estimates showed in Figure 7, extreme weather could explain almost 35.2% of the variance seen in cereal production ($R^2 = 0.352$), 45.3% of the variance seen in cocoa production ($R^2 = 0.453$), 32.6% of the variance seen in livestock production ($R^2 = 0.326$), 32.4% of the variance seen in crop production ($R^2 = 0.324$), and 32.9% of the variance seen in food production index ($R^2 = 0.328$). The whole model demonstrated an acceptable fit to the data for (APC = 0.341, $p < 0.001$), (ARS = 0.393, $p < 0.001$) and AVIF = 1.033

4.6.4. Paths Equations

As seen below, Equations (11)–(15) are the path equations for prediction of the agriculture production indexes and weather extremes.

Let X_1 = Extreme Maximum Rainfall, X_2 = Extreme Maximum Temperature

X_3 = Extreme Minimum Temperature

Thus, obtained are the following regression models for indexes prediction

$$Livestock\ production\ index = -0.184\ MaxRain + 0.305\ MaxTemp + 0.457\ MinTemp \quad (11)$$

$$Crop\ production\ index = -0.189\ MaxRain + 0.206\ MaxTemp + 0.482\ MinTemp \quad (12)$$

$$Cereal\ production\ index = -0.266\ MaxRain + 0.276\ MaxTemp + 0.455\ MinTemp \quad (13)$$

$$Cocoa\ production\ index = -0.461\ MaxRain + 0.257\ MaxTemp - 0.211\ MinTemp \quad (14)$$

$$Food\ production\ index = -0.190\ MaxRain + 0.268\ MaxTemp + 0.484\ MinTemp \quad (15)$$

5. Conclusions

In this present study, we created and examined a model that could contribute to understanding the linkage, and predictability of severe weather and agriculture production in Ghana. The model and structure outlined, tested the nature of extreme maximum rainfall, extreme maximum temperature, extreme minimum temperature and the relationship that exist on agriculture production.

The annual maximum rainfall showed a decreasing trend. However, the yearly maximum temperature and minimum temperature exhibited a significant increase. As observed, there appears no significant trend heterogeneity for each month of the yearly minimum and maximum temperatures, while the annual maximum rainfall shows homogeneity for precipitation in each month. The results show that Extreme Value Theory (EVT) is a reliable tool for climate extreme scenarios construction, where maximum likelihood method supported the evaluation of distribution parameters for weather extreme. Generalised extreme value model is found to be the most suitable model with fulfilling all statistical selection criteria. The return level for the model is constructed to predict the weather extremes for a long run in future. There is generally an increase in weather extreme as it consistently increasing from time to time for the next 100 years.

Evidence from results indicated extreme maximum rainfall adversely affects cereal and cocoa production. Cereals and cocoa thrive well when the rainfall is well distributed and not concentrated in some months and leaving other months virtually without rains.

Maximum extreme temperatures contribute positively to all the indicators under consideration. Minimum extreme temperatures also except cocoa production have a positive impact on the remaining parameters. In the case of cocoa minimum extreme temperatures result in black pod diseases which causes yield reduction. The effect of the temperature and rainfall that is maximum or minimum on food production index depends on their impact on other cereals, livestock and crop productions. Where their respective measures are positive, it results in a positive outcome for food production index. To help improve the food production index of the country there is the need to consider investing in other production sectors. Based on the results the following recommendations are proposed for consideration by policymakers.

The planting time for cereals should be considered going forward, to avoid the detrimental effects of maximum extreme rainfalls. By so doing the yields of cereals will not be affected since they will avoid the period of torrential rains, which affects yields. The diversification of cereals production will help guide against the effect of maximum extreme rainfall on the cereals sector. Some cereals can withstand the impact of maximum extreme rains; diversification into those areas will help reduce the impact of maximum extreme rains if not eliminated.

Minimum extreme temperatures are reported to have detrimental effects on cereals. We recommend the developing of resistant varieties that can withstand the minimum extreme temperatures, which are negatively affecting cereals production. In the case of developing a resistant variety for cocoa, it will help deal with the situation. Since cocoa it is a perennial crop, it will be impossible to use planting period to help deal with the effects of maximum extreme rainfalls. Developing a resistant cocoa variety that will be able to withstand both extreme conditions will be a key in mitigating extreme effects on cocoa yields.

Other research focuses on the more complex problem of catastrophic agricultural risk. To some degree, the catastrophic agricultural risk is the result of extreme weather events. However, the catastrophic agricultural risk is not the same as extreme weather risk. Factors such as environment, agricultural investment, and farmer management should be of interest. To this extent, the distribution of potential damages and losses after a particular type of extreme weather condition should be of interest.

Improving the resilience of Ghanaian agriculture sector is essential. To help do this, farmers and stakeholders in the food production chain should consider the options for adaptation. Adaptation is highly context-specific; this is important for crop, region and climatic zone to use specific adaptation strategies to help minimise the effect of weather extremes on agriculture. The ability of the agricultural sector to deal with climate events will assume a downward trend as the globe warms, and is likely to exceed or fall at specific temperatures and rainfalls. Therefore, farmers need to get used to measures for effective, sustainable, and resilient crop and animal production. Thereby enhancing farmers understanding of growing seasons, improved crop rotation systems, adaptive water management techniques, and higher quality weather forecasts.

For further study, researchers can make a long-term prediction for another weather parameter which indirectly affects the agriculture sector like, production industry on which human life is dependent.

Author Contributions: J.D., conceptualised and initiated the study, M.A.S. and T.B.U. conducted a literature review on the study, whiles A.-A.I.M. carried out analysis on extreme value distribution and SEM-regression analysis. All other authors coordinated and provided comments on this manuscript.

Funding: This research was funded by [National Science Foundation of China] under grants [71671080, 71471076, 71704066, 71501084 and 71701090] and The APC was funded by [Special Funds of the National Social Science Fund of China].

Acknowledgments: This work was supported in part by the National Science Foundation of China under grants 71671080, 71471076, 71704066, 71501084 and 71701090. This work was also supported by the Special Funds of the National Social Science Fund of China [18VSJ038] and by Jiangsu Provincial Natural Science Foundation of China (Grant No. BK20170542).

Conflicts of Interest: The authors declare no conflict of interest.

References

1. Da Cunha, D.A.; Coelho, A.B.; Féres, J.G. Irrigation as an adaptive strategy to climate change: An economic perspective on Brazilian agriculture. *Environ. Dev. Econ.* **2015**, *20*, 57–79. [CrossRef]
2. Kurukulasuriya, P.; Mendelsohn, R.; Hassan, R.; Benhin, J.; Deressa, T.; Diop, M.; Eid, H.M.; Fosu, K.Y.; Gbetibouo, G.; Jain, S. Will African agriculture survive climate change? *World Bank Econ. Rev.* **2006**, *20*, 367–388. [CrossRef]
3. Mendelsohn, R.; Dinar, A.; Williams, L. The distributional impact of climate change on rich and poor countries. *Environ. Dev. Econ.* **2006**, *11*, 159–178. [CrossRef]
4. Di Falco, S. Adaptation to climate change in Sub-Saharan agriculture: Assessing the evidence and rethinking the drivers. *Eur. Rev. Agric. Econ.* **2014**, *41*, 405–430. [CrossRef]
5. Ministry of Food and Agriculture. *Medium Term Agriculture Sector Investment Plan (METASIP)—2011–2015*; Ministry of Food and Agriculture: Accra, Ghana, 2010.
6. MOFA. *Agriculture in Ghana: Facts and Figures 2010*; MOFA: Accra, Ghana, 2011.
7. World Bank Group. *Economics of Adaptation to Climate Change-Ghana*; World Bank Group: Washington, DC, USA, 2010.
8. Asante, F.A.; Amuakwa-Mensah, F. Climate change and variability in Ghana: Stocktaking. *Climate* **2014**, *3*, 78–99. [CrossRef]
9. EPA (Environmental Protection Agency). *Ghana's Second National Communication to the UNFCCC*; Environmental Protection Agency and Ministry of Environment, Science and Technology: Accra, Ghana, 2011.
10. Owusu, K.; Waylen, P.; Qiu, Y. Changing rainfall inputs in the Volta basin: Implications for water sharing in Ghana. *GeoJournal* **2008**, *71*, 201–210. [CrossRef]

11. Bernstein, L.; Bosch, P.; Canziani, O.; Chen, Z.; Christ, R.; Davidson, O.; Hare, W.; Huq, S.; Karoly, D.; Kattsov, V.; et al. *Fourth Assessment Report*; Longer Report; IPCC: Geneva, Switzerland, 2007.

12. Al-Hassan, R.; Poulton, C. Agriculture and Social Protection in Ghana. 2009. Available online: http://opendocs.ids.ac.uk/opendocs/handle/123456789/2340 (accessed on 22 October 2018).

13. Senaratne, A.; Scarborough, H. Coping with climatic variability by rain-fed farmers in dry Zone, Sri Lanka: Towards understanding adaptation to climate change. In *AARES 2011: Australian Agricultural & Resource Economics Society 55th Annual Conference Handbook*; AARES: Reston, VA, USA, 2011; pp. 1–22.

14. FAO. The State of Food Insecurity in the World 2008: High Food Prices and Food Security–Threats and Opportunities 2008. Available online: http://www.fao.org/docrep/011/i0291e/i0291e00.htm (accessed on 22 October 2018).

15. Garcia-Aristizabal, A.; Bucchignani, E.; Palazzi, E.; D'Onofrio, D.; Gasparini, P.; Marzocchi, W. Analysis of non-stationary climate-related extreme events considering climate change scenarios: An application for multi-hazard assessment in the Dar es Salaam region, Tanzania. *Nat. Hazards* **2015**, *75*, 289–320. [CrossRef]

16. Jin, Z.; Zhuang, Q.; Wang, J.; Archontoulis, S.V.; Zobel, Z.; Kotamarthi, V.R. The combined and separate impacts of climate extremes on the current and future US rainfed maize and soybean production under elevated CO_2. *Glob. Chang. Biol.* **2017**, *23*, 2687–2704. [CrossRef] [PubMed]

17. Berezuk, A.G.; da Silva, C.A.; Lamoso, L.P.; Schneider, H. Climate and Production: The Case of the Administrative Region of Grande Dourados, Mato Grosso do Sul, Brazil. *Climate* **2017**, *5*, 49. [CrossRef]

18. Adger, W.N.; Arnell, N.W.; Tompkins, E.L. Successful adaptation to climate change across scales. *Glob. Environ. Chang.* **2005**, *15*, 77–86. [CrossRef]

19. Withanachchi, S.S.; Köpke, S.; Withanachchi, C.R.; Pathiranage, R.; Ploeger, A. Water resource management in dry zonal paddy cultivation in Mahaweli River Basin, Sri Lanka: An analysis of spatial and temporal climate change impacts and traditional knowledge. *Climate* **2014**, *2*, 329–354. [CrossRef]

20. Coles, S. *An Introduction to Statistical Modeling of Extreme Values*; Springer: Berlin, Germany, 2001; ISBN 1-85233-459-2.

21. Field, C.B.; Barros, V.; Stocker, T.F.; Qin, D.; Dokken, D.J.; Ebi, K.L.; Mastrandrea, M.D.; Mach, K.J.; Plattner, G.K.; Allen, S.K. A special report of working groups I and II of the intergovernmental panel on climate change. In *Managing the Risks of Extreme Events and Disasters to Advance Climate Change Adaptation*; IPCC: Geneva, Switzerland, 2012.

22. Mannshardt, E.; Craigmile, P.F.; Tingley, M.P. Statistical modeling of extreme value behavior in North American tree-ring density series. *Clim. Chang.* **2013**, *117*, 843–858. [CrossRef]

23. Rahimpour, V.; Zeng, Y.; Mannaerts, C.M.; Su, Z. (Bob) Attributing seasonal variation of daily extreme precipitation events across The Netherlands. *Weather Clim. Extrem.* **2016**, *14*, 56–66. [CrossRef]

24. Thiombiano, A.N.; El Adlouni, S.; St-hilaire, A.; Ouarda, T.B.M.J.; El-jabi, N. Nonstationary frequency analysis of extreme daily precipitation amounts in Southeastern Canada using a peaks-over-threshold approach. *Theor. Appl. Climatol.* **2016**. [CrossRef]

25. Minkah, R. An application of extreme value theory to the management of a hydroelectric dam. *Springerplus* **2016**, *5*, 96. [CrossRef] [PubMed]

26. Ouarda, T.; El-Adlouni, S. Bayesian nonstationary frequency analysis of hydrological variables. *JAWRA J. Am.* **2011**, *47*, 496–505.

27. Seidou, O.; Ramsay, A.; Nistor, I. Climate change impacts on extreme floods II: Improving flood future peaks simulation using non-stationary frequency analysis. *Nat. Hazards* **2012**, *60*, 715–726. [CrossRef]

28. Katz, R.W. Statistics of extremes in climate change. *Clim. Chang.* **2010**, *100*, 71–76. [CrossRef]

29. Lekina, A.; Chebana, F.; Ouarda, T.B.M.J. Weighted estimate of extreme quantile: An application to the estimation of high flood return periods. *Stoch. Environ. Res. Risk Assess.* **2014**, *28*, 147–165. [CrossRef]

30. Panthou, G.; Vischel, T.; Lebel, T. Recent trends in the regime of extreme rainfall in the Central Sahel. *Int. J. Climatol.* **2014**, *34*, 3998–4006. [CrossRef]

31. Zahiri, E.P.; Bamba, I.; Famien, A.M.; Koffi, A.K.; Ochou, A.D. Mesoscale extreme rainfall events in West Africa: The cases of Niamey (Niger) and the Upper Ouémé Valley (Benin). *Weather Clim. Extrem.* **2016**, *13*, 15–34. [CrossRef]

32. Yabi, I.; Afouda, F. Extreme rainfall years in Benin (West Africa). *Quat. Int.* **2012**, *262*, 39–43. [CrossRef]

33. Ray, D.K.; Gerber, J.S.; MacDonald, G.K.; West, P.C. Climate variation explains a third of global crop yield variability. *Nat. Commun.* **2015**, *6*, 5989. [CrossRef] [PubMed]

34. Jones, P.G.; Thornton, P.K. The potential impacts of climate change on maize production in Africa and Latin America in 2055. *Glob. Environ. Chang.* **2003**, *13*, 51–59. [CrossRef]

35. Pachauri, R.K.; Allen, M.R.; Barros, V.R.; Broome, J.; Cramer, W.; Christ, R.; Church, J.A.; Clarke, L.; Dahe, Q.; Dasgupta, P. *Climate Change 2014: Synthesis Report. Contribution of Working Groups I, II and III to the Fifth Assessment Report of the Intergovernmental Panel on Climate Change*; IPCC: Geneva, Switzerland, 2014; ISBN 9291691437.

36. Warner, K.; Afifi, T. Where the rain falls: Evidence from 8 countries on how vulnerable households use migration to manage the risk of rainfall variability and food insecurity. *Clim. Dev.* **2014**, *6*, 1–17. [CrossRef]

37. Adamgbe, E.M.; Ujoh, F. Effect of variability in rainfall characteristics on maize yield in Gboko, Nigeria. *J. Environ. Prot. (Irvine, Calif.)* **2013**, *4*, 881. [CrossRef]

38. Ogunrayi, O.A.; Akinseye, F.M.; Goldberg, V.; Bernhofer, C. Descriptive analysis of rainfall and temperature trends over Akure, Nigeria. *J. Geogr. Reg. Plan.* **2016**, *9*, 195–202.

39. Nair, K.P.P. *The Agronomy and Economy of Important Tree Crops of the Developing World*; Elsevier: Amsterdam, The Netherlands, 2010; ISBN 0123846781.

40. Anim-Kwapong, G.J.; Frimpong, E.B. Vulnerability and Adaptation Assessment Under the Netherlands Climate Change Studies Assistance Programme Phase 2 (NCCSAP2). *Cocoa Res. Inst. Ghana* **2005**, *2*, 1–30.

41. Rao, G.P. *Climate Change Adaptation Strategies in Agriculture and Allied Sectors*; Scientific Publishers: Valencia, CA, USA, 2011; ISBN 9386347474.

42. Ali, F.M. Effects of rainfall on yield of cocoa in Ghana. *Exp. Agric.* **1969**, *5*, 209–213. [CrossRef]

43. Stanturf, J.A.; Warren, M.L.; Charnley, S.; Polasky, S.C.; Goodrick, S.L.; Armah, F.; Nyako, Y.A. *Ghana Climate Change Vulnerability and Adaptation Assessment*; United States Agency for International Development: Washington, DC, USA, 2011.

44. Amundson, J.L.; Mader, T.L.; Rasby, R.J.; Hu, Q.S. Environmental effects on pregnancy rate in beef cattle 1. *J. Anim. Sci.* **2006**, *84*, 3415–3420. [CrossRef] [PubMed]

45. Sprott, L.R.; Selk, P.G.E.; Adams, D.C. Factors affecting decisions on when to calve beef females. *Prof. Anim. Sci.* **2001**, *17*, 238–246. [CrossRef]

46. Hatfield, J.L.; Boote, K.J.; Kimball, B.A.; Ziska, L.H.; Izaurralde, R.C.; Ort, D.; Thomson, A.M.; Wolfe, D. Climate impacts on agriculture: Implications for crop production. *Agron. J.* **2011**, *103*, 351–370. [CrossRef]

47. Gobin, A. Weather related risks in Belgian arable agriculture. *Agric. Syst.* **2018**, *159*, 225–236. [CrossRef]

48. Piao, S.; Ciais, P.; Huang, Y.; Shen, Z.; Peng, S.; Li, J.; Zhou, L.; Liu, H.; Ma, Y.; Ding, Y. The impacts of climate change on water resources and agriculture in China. *Nature* **2010**, *467*, 43. [CrossRef] [PubMed]

49. Adger, W.N.; Huq, S.; Brown, K.; Conway, D.; Hulme, M. Adaptation to climate change in the developing world. *Prog. Dev. Stud.* **2003**, *3*, 179–195. [CrossRef]

50. Schlenker, W.; Lobell, D.B. Robust negative impacts of climate change on African agriculture. *Environ. Res. Lett.* **2010**, *5*, 14010. [CrossRef]

51. United Nations Framework Convention on Climate Change (UNFCCC). Climate Change: Impacts, Vulnerabilities and Adaptation in Developing Countries 2007. Available online: https://www.preventionweb.net/publications/view/2759 (accessed on 22 October 2018).

52. Wealth, C. *Economics for a Crowded Planet*; Allen Lane: London, UK, 2008.

53. Organization, W.H. Protecting the health of vulnerable people from the umanitarian consequences of climate change and climate related disasters. In Proceedings of the 6th session of the Ad Hoc Working Group on Long-Term Cooperative Action under the Convention (AWG-LCA 6), Bonn, Germany, 1–12 June 2009; pp. 1–12.

54. NCCPF (National Climate Change Policy Framework); Policy, C.; Goals, M.D.; African, W. Government of Ghana Ghana Goes for Green Growth Discussion Document—Summary Climate Change in Ghana. 2015. Available online: https://cdkn.org/wp-content/uploads/2011/04/NCCPF-Summary-FINAL.pdf (accessed on 22 October 2018).

55. Intergovernmental Panel on Climate Change. IPCC Climate Change 2007: The physical science basis. *Agenda* **2007**, *6*, 333.

56. Kearney, M. Hot rocks and much-too-hot rocks: Seasonal patterns of retreat-site selection by a nocturnal ectotherm. *J. Therm. Biol.* **2002**, *27*, 205–218. [CrossRef]

57. Agyemang-Bonsu, W.K.; Minia, Z.; Dontwi, J.; Dontwi, I.K.; Buabeng, S.N.; Baffoe-Bonnie, B.; Frimpong, E.B. *Ghana Climate Change Impacts, Vulnerability and Adaptation Assessments*; Environmental Protection Agency: Accra, Ghana, 2008.

58. World Bank Group. *The Costs to Developing Countries of Adapting to Climate Change—New Methods and Estimates*; World Bank Group: Washington, DC, USA, 2010; p. 20433.

59. Cameron, C. Climate Change Financing and Aid Effectiveness: Ghana Case Study. 2011. Available online: http://www.eldis.org/document/A61721 (accessed on 22 October 2018).

60. Christensen, J.H.; Kanikicharla, K.K.; Marshall, G.; Turner, J. Climate Phenomena and Their Relevance for Future Regional Climate Change. 2013. Available online: https://www.ipcc.ch/pdf/assessment-report/ar5/wg1/WG1AR5_Chapter14_FINAL.pdf (accessed on 22 October 2018).

61. Challinor, A.J.; Watson, J.; Lobell, D.B.; Howden, S.M.; Smith, D.R.; Chhetri, N. A meta-analysis of crop yield under climate change and adaptation. *Nat. Clim. Chang.* **2014**, *4*, 287. [CrossRef]

62. Codjoe, S.N.A.; Owusu, G. Climate change/variability and food systems: Evidence from the Afram Plains, Ghana. *Reg. Environ. Chang.* **2011**, *11*, 753–765. [CrossRef]

63. Ghana Statistical Service. Ghana Multiple Indicator Cluster Survey with an Enhanced Malaria Module and Biomarker. 2011; Final Report. Available online: https://www.popline.org/node/577148 (accessed on 22 October 2018).

64. Rosenzweig, C.; Iglesius, A.; Epstein, P.R.; Chivian, E. Climate Change and Extreme Weather Events—Implications for Food Production, Plant Diseases, and Pests. 2001. Available online: https://digitalcommons.unl.edu/cgi/viewcontent.cgi?article=1023&context=nasapub (accessed on 22 October 2018).

65. Huang, C.; Lin, J.-G. Modified maximum spacings method for generalized extreme value distribution and applications in real data analysis. *Metrika* **2014**, *77*, 867–894. [CrossRef]

66. Karmakar, M.; Shukla, G.K. Managing Extreme Risk in Some Major Stock Markets: An Extreme Value Approach. *Int. Rev. Econ. Financ.* **2015**, *35*, 1–25. [CrossRef]

67. Switzer, L.N.; Wang, J.; Lee, S. Extreme risk and small investor behavior in developed markets. *J. Asset Manag.* **2017**, *18*, 457–475. [CrossRef]

68. Davison, A.C.; Padoan, S.A.; Ribatet, M. Statistical Modeling of Spatial Extremes. *Stat. Sci.* **2012**, *27*, 161–186. [CrossRef]

69. Bader, B. Automated, Efficient, and Practical Extreme Value Analysis with Environmental Applications. 2016. Available online: https://arxiv.org/pdf/1611.08261.pdf (accessed on 22 October 2018).

70. Chinhamu, K.; Huang, C.K.; Huang, C.S.; Hammujuddy, J. Empirical analyses of extreme value models for the South African mining index. *S. Afr. J. Econ.* **2015**, *83*, 41–55. [CrossRef]

71. Yee, T.W. *Vector Generalized Linear and Additive Models: With an Implementation in R*; Springer: Berlin, Germany, 2015; ISBN 9781493928187.

72. Statistics, J. Distribution Fitting 2. Pearson-Fisher, Kolmogorov-Smirnov, Anderson- Darling, Wilks-Shapiro, Cramer-von-Misses and Jarque-Bera Statistics. *Bull. UASVM Hortic.* **2009**, *66*, 691–697.

73. Hair, J.F.; Sarstedt, M.; Ringle, C.M.; Mena, J.A. An assessment of the use of partial least squares structural equation modeling in marketing research. *J. Acad. Mark. Sci.* **2012**, *40*, 414–433. [CrossRef]

74. Pohlert, T. Non-parametric trend tests and change-point detection. *CC BY-ND* **2016**, *4*. Available online: http://cran.stat.upd.edu.ph/web/packages/trend/vignettes/trend.pdf (accessed on 29 October 2018).

75. Castillo, E.; Hadi, A.; Balakrishnan, N.; Sarabia, J. *Extreme Value and Related Models with Applications in Engineering and Science*; Wiley: New York, NY, USA, 2005.

76. Byrne, B.M. *Structural Equation Modeling with AMOS: Basic Concepts, Applications, and Programming*; Routledge: London, UK, 2016; ISBN 131763313X.

77. Vinzi, V.E.; Trinchera, L.; Amato, S. PLS path modeling: From foundations to recent developments and open issues for model assessment and improvement. In *Handbook of Partial Least Squares*; Springer: Berlin, Germany, 2010; pp. 47–82.

78. Hair, J.F.; Ringle, C.M.; Sarstedt, M. Partial least squares: The better approach to structural equation modeling? *Long Range Plan.* **2012**, *45*, 312–319. [CrossRef]

79. Richter, N.F.; Sinkovics, R.R.; Ringle, C.M.; Schlaegel, C. A critical look at the use of SEM in international business research. *Int. Mark. Rev.* **2016**, *33*, 376–404. [CrossRef]

80. Lowry, P.B.; Gaskin, J. Partial least squares (PLS) structural equation modeling (SEM) for building and testing behavioral causal theory: When to choose it and how to use it. *IEEE Trans. Prof. Commun.* **2014**, *57*, 123–146. [CrossRef]

81. Field, C.B.; Barros, V.; Stocker, T.F.; Qin, D.; Dokken, D.J.; Ebi, K.L.; Mastrandrea, M.D.; Mach, K.J.; Plattner, G.-K.; Allen, S.K.; et al. *Managing the Risks of Extreme Events and Disasters to Advance Climate Change Adaptation. A Special Report of Working Groups I and II of the Intergovernmental Panel on Climate Change*; Cambridge University Press: Cambridge, UK; New York, NY, USA, 2012.

82. Montagnon, C.; Leroy, T.; Eskes, A. Amélioration variétale de# Coffea canephora#. 2: Les programmes de sélection et leurs résultats. *Plant. Rech. Dév.* **1998**, *5*, 89–98.

83. Madan, M.L.; Prakash, B.S. Reproductive endocrinology and biotechnology applications among buffaloes. *Soc. Reprod. Fertil. Suppl.* **2007**, *64*, 261–281. [CrossRef] [PubMed]

84. Gilad, E.; Meidan, R.; Berman, A.; Graber, Y.; Wolfenson, D. Effect of heat stress on tonic and GnRH-induced gonadotrophin secretion in relation to concentration of oestradiol in plasma of cyclic cows. *J. Reprod. Fertil.* **1993**, *99*, 315–321. [CrossRef] [PubMed]

85. Wise, M.E.; Armstrong, D.V.; Huber, J.T.; Hunter, R.; Wiersma, F. Hormonal Alterations in the Lactating Dairy Cow in Response to Thermal Stress1. *J. Dairy Sci.* **1988**, *71*, 2480–2485. [CrossRef]

86. Dupuis, I.; Dumas, C. Influence of temperature stress on in vitro fertilization and heat shock protein synthesis in maize (Zea mays L.) reproductive tissues. *Plant Physiol.* **1990**, *94*, 665–670. [CrossRef] [PubMed]

87. Herrero, M.P.; Johnson, R.R. High Temperature Stress and Pollen Viability of Maize 1. *Crop Sci.* **1980**, *20*, 796–800. [CrossRef]

88. Schoper, J.B.; Lambert, R.J.; Vasilas, B.L.; Westgate, M.E. Plant factors controlling seed set in maize: The influence of silk, pollen, and ear-leaf water status and tassel heat treatment at pollination. *Plant Physiol.* **1987**, *83*, 121–125. [CrossRef] [PubMed]

89. Rowhani, P.; Lobell, D.B.; Linderman, M.; Ramankutty, N. Climate variability and crop production in Tanzania. *Agric. For. Meteorol.* **2011**, *151*, 449–460. [CrossRef]

90. Schlenker, W.; Roberts, M.J. Nonlinear temperature effects indicate severe damages to U.S. crop yields under climate change. *Proc. Natl. Acad. Sci. USA* **2009**, *106*, 15594–15598. [CrossRef] [PubMed]

91. Growing Cocoa—The International Cocoa Organization. Available online: https://www.icco.org/about-cocoa/growing-cocoa.html (accessed on 5 August 2018).

Review

Geographic Information and Communication Technologies for Supporting Smallholder Agriculture and Climate Resilience

Billy Tusker Haworth [1,*], Eloise Biggs [2], John Duncan [2], Nathan Wales [3], Bryan Boruff [2] and Eleanor Bruce [4]

1   Humanitarian and Conflict Response Institute, University of Manchester, Ellen Wilkinson Building, Oxford Road, Manchester M13 9PL, UK

2   UWA School of Agriculture and Environment, University of Western Australia, Perth, WA 6009, Australia; eloise.biggs@uwa.edu.au (E.B.); john.duncan@uwa.edu.au (J.D.); bryan.boruff@uwa.edu.au (B.B.)

3   School of Geography, Earth Science & Environment, University of the South Pacific, Suva, Fiji; wales_n@usp.ac.fj

4   School of Geosciences, University of Sydney, Sydney, NSW 2006, Australia; eleanor.bruce@sydney.edu.au

*   Correspondence: billy.haworth@manchester.ac.uk; Tel.: +44-(0)161-275-6114

Received: 9 November 2018; Accepted: 5 December 2018; Published: 10 December 2018

Abstract: Multiple factors constrain smallholder agriculture and farmers' adaptive capacities under changing climates, including access to information to support context appropriate farm decision-making. Current approaches to geographic information dissemination to smallholders, such as the rural extension model, are limited, yet advancements in internet and communication technologies (ICTs) could help augment these processes through the provision of agricultural geographic information (AGI) directly to farmers. We analysed recent ICT initiatives for communicating climate and agriculture-related information to smallholders for improved livelihoods and climate change adaptation. Through the critical analysis of initiatives, we identified opportunities for the success of future AGI developments. We systematically examined 27 AGI initiatives reported in academic and grey literature (e.g., organisational databases). Important factors identified for the success of initiatives include affordability, language(s), community partnerships, user collaboration, high quality and locally-relevant information through low-tech platforms, organisational trust, clear business models, and adaptability. We propose initiatives should be better-targeted to deliver AGI to regions in most need of climate adaptation assistance, including SE Asia, the Pacific, and the Caribbean. Further assessment of the most effective technological approaches is needed. Initiatives should be independently assessed for evaluation of their uptake and success, and local communities should be better-incorporated into the development of AGI initiatives.

Keywords: climate change adaptation; livelihoods; geographic information; agriculture; resilience

---

1. Introduction

The agricultural industry is supported by 500 million smallholder farms, responsible for approximately 56% of global agricultural production [1,2]. Smallholder farmers are increasingly resource-poor and confronted by challenges associated with climate change, natural disasters, resource availability and access, and food insecurity [1,3]. Global climatic changes are influencing crop growth and yield, water balances, input availability, and agricultural system management components [4], with ensuing impacts on farming practices [5–7]. Smallholders are faced with both long-term climate stressors and short-term shocks [8]. Geographic variability in climate impacts coupled with low levels of coping and adaptive capacity results in high levels of vulnerability for marginalised farmers [9–11].

Vulnerability varies geographically (often at very local levels). This arises from the complexity of smallholder livelihoods, with multiple on-farm/off-farm activities [12], variation in asset levels and market orientation [13], local (within-farm) variability in productivity [14], gendered roles and access to resources [15], and differential capacity to manage risk [16], affecting smallholder capacity to respond and adapt to climatic challenges.

Incorporating geographic components (i.e., locational properties) into information for climate adaptation is valuable for enhancing environmental decision-making in high risk sectors, such as agriculture. Rapid advancements in geographic information technologies (e.g., geographic information systems (GIS)) and the availability of geospatial data allow for sophisticated capture, analysis, storage, dissemination and access of information across space and time. Concurrently, advancements in information communication technologies (ICTs) (e.g., short message service (SMS); smartphones; Web 2.0), have further increased the usability of geographic information derived from a diversity of sources [17].

Note, while popularity in use of the term geospatial has grown (e.g., geospatial web [18]; geospatial semantics [19]), ambiguity remains over the difference between geospatial and geographic information. Geographic describes information with a reference to Earth's surface and near-surface [20], and geospatial data has been defined as location properties (any descriptive information about the location or area of, and relationships among geographic features) related to any terrestrial feature/ phenomena [21]. We adopt the term geographic information/data, despite much of the material reviewed employing the term geospatial. We consider geographic information to be any information to which location on the Earth is a relevant feature, including both explicit and implicit [22] locational data.

Geographic information used within the agriculture sector—here termed agricultural geographic information (AGI)—is increasingly available to smallholders, yet uptake is limited. Despite a range of geographic information types, such as remote sensing, household surveys, or climate/market reports, accessibility and/or availability is often not in useful/usable formats. Traditionally, information provision to smallholders in developing countries is provided via agricultural extension organisations through farmer field schools, innovation networks and farming associations [23]. However, resource constraints and the diverse needs of smallholders limit the flow of top-down information [24]. For example, resource constraints of agricultural extension staff have been identified as a challenge under climate change in the South Pacific [25] and the lack of transparency and connectivity a constraint to information delivery in India [26].

To this end, we suggest a different or complementary model to supply smallholders with information is necessary, whereby smallholders can harness AGI to make better-informed and cost-saving decisions [27]. Using ICTs to communicate with farmers directly offers a potential for AGI to enhance sustainable agriculture [28], particularly through resources provision for increasing climate resilience at multiple landscape scales [29]. For example, access to geographic information regarding which drought-resistant crops to plant, including when and how, may increase smallholders' capacities to prepare for and withstand such long term climate stresses. Or, localised and context-specific weather forecasts delivered directly to farmers' mobile phones may allow timely decisions and mitigating actions to be taken that reduce the impacts of storms on farming livelihoods. The World Bank, African Development Bank, and African union claim that the greatest opportunities for economic growth and poverty alleviation (in Africa) are provided by ICTs in the agriculture industry [30]. Yet, the evidence base for ICT and use of AGI to support adaptive capacity of smallholders is poorly documented [31]. Baumüller [32] argues that the potential use of ICTs, such as mobile services for smallholder agriculture remains largely unfulfilled. Consequently, here we review recent trends and approaches to utilising geographic information and ICTs for agriculture, and in particular, initiatives for communicating climate and other agriculture-related information to smallholder farmers for improved livelihood security, climate change adaptation and landscape resilience. Our aim is not only to contribute to rectifying the dearth of systematically documented and analysed uses of ICTs in smallholder agriculture, but also to uncover valuable lessons for the design and application of future

AGI initiatives. We achieve this through a systematic review of multi-source literature to address the following research questions:

i. What are the key challenges that AGI initiatives aim to address?
ii. What technological approaches have been adopted to provide AGI to smallholder farmers?
iii. Who are the target users of AGI initiatives and how have initiatives been adopted?
iv. What are the factors promoting or limiting the success of AGI initiatives?

We acknowledge that earlier review works exist on related topics with similar aims and methods to those we present here. The Food and Agriculture Organisation of the United Nations (FAO) [33] reviewed a decade of ICT advancements with applications to agriculture and rural development presenting important findings, such as the significant influence of elements like quality partnerships and the digital divide on project success. But this report was largely descriptive and based on a narrow selection of projects and therefore lacks the analytical depth and rigour associated with our systematic review of AGI initiatives. The World Bank [34] also produced a report on ICT in agriculture, but a similar critique to above could be applied. Baumüller [32] systematically analysed the impact of various mobile services for smallholder agriculture, offering useful lessons for future service developments and an assessment of current shortcomings, including a lack of useful empirical evidence and limitations to current methodologies for evaluating project impact. Our work differs in that it is not constrained to examining only mobile services, but includes a broader range of ICTs used in AGI initiatives, and specifically considers delivery of information of a geographic nature. Duncombe [35] also analysed mobile phone use for agriculture in developing countries, and again, our work examines a more technologically-diverse breadth of AGI initiatives. Further, our work includes the review of AGI initiatives found and described in multiple sources, as opposed to reviews based on only practice-based literature (e.g., [34]) or academic research articles (e.g., [35]).

We first provide a brief background to geographic information and farmer information needs in agriculture, followed by a detailed methodology, presentation of results and discussion in relation to the stated research questions, with particular emphasis on lessons learned from examining a broad range of AGI initiatives. We conclude by identifying critical knowledge gaps and future opportunities.

2. Geographic Information in Agriculture

AGI encompasses a wide range of information types and can be provided through a similarly wide range of technologies. This includes any agricultural information provided through ICTs that has a geographic component, such as location-specific information delivered via SMS, telephone or the Internet, as well as geographic information produced through more sophisticated technological approaches, such as GIS mapping and spatial modelling. GIS technologies provide flexible spatially-explicit tools that support decision making for environmental and natural resource management [36]. Combined with remote sensing technologies, mapping, modelling and monitoring environmental change aids climate change adaptation and mitigation initiatives across the agriculture sector [37,38]. These technologies have contributed to advances in precision agriculture and improved crop management in commercial broad acre agriculture [39–41], yet AGI utilisation by smallholders remains limited. Reflecting on successes from other sectors, geographic information has been used to respond to natural disasters and increase community resilience across a range of environments [42,43], and resilience building in the agricultural sector, particularly in smallholder communities, has similar use potential. Such an aspiration aligns well with the concept of climate smart agriculture (CSA)—to increase food and livelihood security, and farming and landscape resilience [8,44,45]—but explicitly identifies smallholders' needs for improved information access to enable better decision making for sustainable agriculture.

2.1. Information Needs of Smallholders

Smallholder farmers require diverse information to support their livelihoods, with development in the agriculture sector dependent on success in generating, sharing, and applying knowledge [1,46]. Information can be obtained from scientists, educators, advisors, policy makers, and informal networks and smallholders themselves [31]. Information needs differ between farmers based on multiple factors, including socio-economic circumstance, literacy levels, access to resources, size of landholding, and agroclimatic conditions [28]. These factors, in conjunction with a range of socio-political conditions, such as governance structures, cultural norms and gender roles, influence how different individuals obtain and seek (applicable) information (e.g., [47]).

2.1.1. Information Availability

Availability of appropriate climate change adaptation information for smallholders often varies by geography and culture. For example, public media and personal experience form dominant information sources amongst Vietnamese farmers [48]. Conversely, in India, farmers rely on external experts such as non-governmental agricultural research for advice, despite their long histories of traditional knowledge [49]. Less formal agricultural knowledge transfer takes place through face-to-face interactions and verbal communication via mobile phones in rural communities [49]. Television, radio, agriculture offices/departments, neighbours and progressive farmers provide the most useful information sources, at least in part due to exposure and availability [50]. Further, the availability of precise and timely weather-based agro-advisory messages are useful in making informed and cost-saving decisions regarding cultivation conditions [27].

2.1.2. Information Accessibility

Information is commonly delivered to farmers through agriculture extension and advisory services [23]. Primarily top-down approaches, these transfer technologies, skills and knowledge to rural farmers and families to enhance crop/livestock production systems, household food security, and livelihoods, through increasing incomes, nutrition, education, and strengthening natural resource management [3]. However, several deficiencies of extension systems restrict their effectiveness, including limited staff, rigid organisation, poor capacity, a top-down linear culture, weak links to the research sector, and limited reach to farmers [28]. In India, for example, there are many [often duplicate] extension systems, yet the majority of farmers still suffer from inadequate information access [28]. Compounding these issues, women in rural communities bear considerable proportions of farming workloads, but have limited roles in receiving information and making decisions (see [27]). Women are often poorer with less land ownership and have difficulty accessing agricultural information from sources aside from other farmers [51]. Munyna [52] argues that women being ill-informed about technologies, markets, and other agriculture information is detrimental to agricultural development.

2.1.3. Information Applicability

Scale of agricultural systems can influence who has access to [relevant] information. For example, national information produced at the government level may not be effective for improving farming practices at more localised scales. At the local scale, farmer field schools are a variation of extension services. Small groups of farmers routinely gather to observe and evaluate potential suitability of agricultural interventions for their farms [53]. This approach also builds social capital, but often exhibits fiscal limitations [54]. Researchers have argued for an increased emphasis on local rather than global initiatives in developing countries with improved relevance and applicability of information (see [55]). This includes the exchange of knowledge in appropriate formats that respect the oral traditions of many indigenous cultures [56].

3. Methodology

To identify AGI initiatives for analysis, literature was assessed from (i) peer-reviewed academic journals, and (ii) projects listed elsewhere or in grey literature, such as through government/ non-government organisation, and other key development organisations and/or private sector agency databases. Assessing academic literature involved multiple keyword searches of the Web of Science Core Collection database, which focused on the topic areas of information, climate, and agriculture practices (in that hierarchical order) (Figure 1). Articles were constrained to include only current or recent literature (published after the year 2000; the time period considered to represent the growth of relevant geographic information, the internet, and other ICTs; when mobile technology penetration rates began to expand in developing countries [32]), those published in English language, and only items with full-text versions available. We acknowledge relevant literature will also exist in other languages, such as French, Spanish, Mandarin, or Hindi, among others, and hence incapacity to analyse non-English sources is a limitation of this study [57]. Articles which met all criteria (n = 156) were read and either entered into a spreadsheet for summarisation and analysis, or discarded if deemed not relevant. Assessment of relevance was made in relation to the research questions presented in Section 1. An article may have met all search criteria by using geographic information technologies to examine some aspect of improving agricultural practices in the context of climate change, but if the article did not describe initiatives specifically for communicating such information with farmers it was deemed not applicable to our research questions and thus was excluded. This process was performed initially by one author, and afterwards verified by another. Articles were also discarded if they only provided duplication (e.g., multiple articles describing the same initiative).

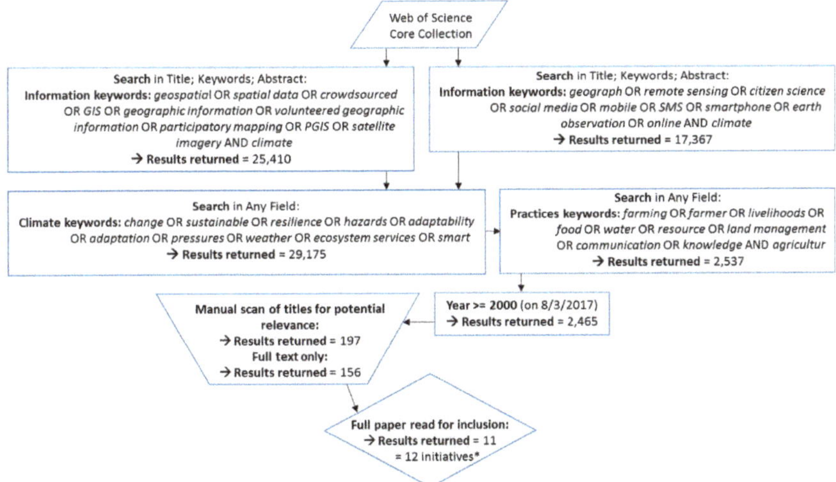

Figure 1. Flow diagram for the academic literature search resulting in 11 relevant papers (12 agricultural geographic information (AGI) initiatives) for analysis (see Table 1 for sources). * One paper described multiple initiatives.

Assessing grey literature involved identifying databases, sources, agencies, and other websites that may contain information on relevant community, agriculture and climate-related AGI initiatives. Where a database had a large number of initiatives, filtering based on keywords (in line with those presented in Figure 1) produced a subset which was manually reviewed for inclusion in a database used for summarisation and analysis. Grey literature assessment was inherently less systematic and could not be automated, and we note the limitation to findings for inclusion in this paper, as once a perceived cross-section of different types of initiatives was obtained the search was ceased. This resulted in

15 AGI initiatives identified. In total, 27 individual AGI initiatives were identified through the above scholarly and grey literature search methods (See Table 1). All initiatives were summarised and analysed in a spreadsheet according to key information relevant for answering the predefined research questions. This included descriptive information (such as initiative name, source, and year), target location and users, initiative aims and approach to achieving aims, climate-related challenges being addressed (short and long term), geographic technologies adopted, the participatory nature of each initiative, adoption and usage information, and details of if/how the initiative was evaluated and by whom.

Table 1. All AGI initiatives identified for this review, including description, target locations and source.

Initiative	Description	Targeted Country or Region	Source
Agriculture Monitoring System	Agriculture monitoring system and technologies for collecting, analysing, and disseminating information. Includes satellite remote sensing, GIS, and mobile GPS. Provides a knowledge base for government, NGOs, rural communities and other stakeholders that will aid sustainable land use and agriculture.	Afghanistan	[58]
Airtel Kilimo	Mobile phone and SMS advisory service. Dissemination of information related to crops, weather and market prices for improved farmer livelihood security.	Kenya	[59]
Avaaj Otalo	Top-down mobile phone advisory service. Delivery of weather, crop, fertiliser and other agriculture information to farmers. Addresses shortcomings of the extension system.	India	[60]
Climate Wizard Tool	Web-based system for climate change data analysis and mapping. Provides practical information for local and regional agriculture managers. Facilitates advanced statistical analyses for more technical users.	Global	[61]
CROPROTECT	Internet and smartphone application utilising GIS and Google Earth. Knowledge exchange system for farmers to acquire and share information relating to pest, weed and disease management.	United Kingdom	[62]
Digital Green	Participatory videos (local languages) used to involve local communities in sharing scientific agriculture information and local knowledge to improve livelihoods through better and more adaptive farming practices.	India, Ghana, Ethiopia	[53,63]
Farmer Decision Support System (FDSS)	Advisory information for registered farmers via SMS to assist farming decisions e.g., when and how to plant, harvest, fertilise and manage crops. 7-day weather forecasts also provided.	Philippines	[64]
Farmforce	SMS and smartphone application to link farmers with other actors in the agro-value chain to reduce transaction costs, aid compliance with food standards, and increase information exchange.	Asia, Africa, Latin America	[65]
Geospatial Information for Rice Crop Monitoring (GIRCM)	Agriculture information derived from image classification and rice crop area estimation to enhance food security. Still in proposal stage.	Afghanistan	[66]
Indian Farmers Fertiliser Cooperative (IFFCO) Kisan Agriculture App	Smartphone application to provide crop information in various formats for enhanced decision making. Aimed at farmers who are receptive to new technologies and business approaches.	India	[67]
Information Technology and Indigenous Knowledge with Intelligence (ITIKI)	Early warning system that integrates information from sensor networks and local knowledge on droughts. Communication using SMS, mobile phone calls, website posts, digital billboards and radio broadcasts to disseminate forecast information to farmers.	Kenya, Sub-Saharan Africa	[68]
iska	GPS-located weather forecasts (various time intervals) distributed via SMS to farmers to improve decision making and reduce weather-related crop losses.	West Africa	[69,70]

Table 1. *Cont.*

Initiative	Description	Targeted Country or Region	Source
Jayalaxmi Agro Tech	Crop-specific smartphone applications for access to agriculture, horticulture and animal husbandry information (English and regional languages).	India	[71]
LandCaRe DSS	Spatial simulation modelling to produce information for stakeholders and farmers involved in decision making related to land management and long-term impacts of climate change at regional and farm scales.	Germany	[72]
Mobile geospatial information for African farmers (MGIAF)	Mobile phone alerts regarding purchasing of drought-tolerant crops for farmers in remote regions. GIS maps for extension officers and community development workers for information dissemination to farmers.	Kenya	[73]
Mobile market information service (MMIS)	SMS request service for rural farmers to receive information on market information (e.g., product prices) to improve selling practices and decision making.	Papua New Guinea	[74]
Mobile soil information for African farmers (MSIAF)	Web-mapping platform for providing soil information to farmers and government workers. Accessed via the internet or mobile phone.	Kenya	[73]
(M)obile Solutions	Mobile phone voice and SMS messages (Hindi or a local language) sent to farmers. Contain information relating to weather, pests, seed varieties, climate change and climate-smart technologies. Provides recommended actions. Option for farmers to provide feedback to inform future messaging.	India	[27]
Participatory Mapping Disaster Risk Reduction Local Knowledge (PMDRRLK)	Participatory approaches and co-produced mapping to improve local resilience to climate change related hazards and increase the use of local environmental knowledge.	Switzerland	[75]
Plantwise Knowledge Bank	Online and smartphone-based knowledge bank with pest identification tools and factsheets on plant health to aid community farming.	Global	[62]
Radio Monsoon	National meteorological information and local knowledge for weather forecasts disseminated to fishermen via social media and the internet, landline and mobile phones, and loudspeakers positioned in fishing communities.	India	[76]
SmartScape	Internet and GIS tool to allow users to experiment with policy options, predict cropping system changes, and compare cropping scenarios. Produces information to be shared with stakeholders, such as policymakers, community agriculture groups, or non-government organisations.	United States of America	[77]
Sowing Application	Smartphone application and SMS used to advise registered farmers best times for sowing seeds based on soil health indicators and rainfall and weather information. Alerts issued for extreme weather conditions that may damage crops or impact farmers.	India	[78]
Tigo Kilimo	Mobile phone dissemination of information on weather, crops and markets for enhanced decision making to improve food security, livelihoods and household income for farmers.	Tanzania	[51]
Watershed Management Information System (WATMIS)	Web-based information and decision support system integrating soil, vegetation, climate and other environment information to assist agriculturalists, resource managers and the rural extension community in managing water scarcity.	India	[79]
World AgroMeteorological Information Service (WAMIS)	Web-server for disseminating agrometeorological products and information bulletins. Provides knowledge and training to large numbers of agriculture stakeholders cost effectively via the internet.	Global	[80]
Wireless Sensor Network—Decision Support System (WSN-DSS)	Wireless sensor network and web-based decision support system for irrigation scheduling. Supports farmers in restructuring agricultural land to address issues of food security and inefficient farming.	Tunisia	[81]

4. Results

4.1. AGI Initiatives

Target users of the AGI initiatives and the key challenges they seek to address are reflected in the distribution of where implementation occurred (see Table 1 for name and summary description of each initiative). Initiatives were concentrated in the global south, particularly south Asia, and east/west Africa. India and Kenya were highlighted as individual countries with the highest numbers of initiatives reviewed. Initiatives largely targeted smallholder farmers and rural communities ($n = 18$). Some AGI initiatives specifically targeted women farmers (Tigo Kilimo), farmers with low education levels (Tigo Kilimo), fishing households (Radio Monsoon), and progressive farmers more receptive to new technologies and practices (IFFCO Kisan Agriculture App). These target user groups are synonymous with those of more traditional approaches to agricultural extension and advisory services [3]. Other target users included scientists (e.g., PMDRRLK), governments (Smartscape), the agriculture extension community (WATMIS), NGOs and conservation organisations (Agriculture Monitoring System; Smartscape; LandCaRe DSS), risk management agencies (PMDRRLK), and the private sector (Agriculture Monitoring System).

Almost all initiatives adopted a top-down approach ($n = 23$), with only a few employing bottom-up practices (Digital Green, PMDRRLK and CROPROTECT). Greater emphasis was on communicating AGI to farmers, or providing a service that farmers can receive information from, rather than working with farmers to utilise AGI to support livelihoods. Of the initiatives adopting a bottom-up approach, Digital Green identified 'champions' from a local community to film and edit videos on new farming practices and topics, such as health (outputs were in local languages and topic selections were informed by scientists). Videos were then screened regularly in the community to share learnings. The localised participatory nature of Digital Green was important for people to relate to AGI information and increased adoption of sustainable livelihood practices throughout the community. IFFCO directly targeted progressive farmers, or those more likely to trial and adopt new practices based on capacity, circumstance, and interest. This assumed that farmers who receive AGI through the app, and adopt new practices, will then influence others in the community, either directly through sharing learnings or indirectly through demonstrated success.

4.1.1. Agro-Climatic Challenges Being Addressed

Many initiatives addressed climate adaptation of farmers through increasing livelihood security ($n = 19$), with some initiatives specifically aiming to increase household income or food security ($n = 15$). Several initiatives focus on addressing both long-term and short-term climate change to combat adverse impacts on livelihoods [53] and agricultural productivity [60]. In Kenya, where rainfed agriculture supports the majority of subsistence livelihoods, ITIKI sought to address the challenge of limited rainfall monitoring through the development of an integrated communication framework for indigenous knowledge and scientific drought forecast information. In Tunisia, issues of agricultural water wastage and mal-management of resources were being addressed by WSN-DSS, supporting farmers with weather information, improved irrigation scheduling and water management. In rural Africa, MSIAF aimed to mitigate the long-term stress of drought by alerting farmers to market locations to purchase drought-tolerant beans. Initiatives addressing short-term climate shocks were largely related to weather variability, including increased frequency and intensity of meteorological natural disasters (PMDRRLK; WAMIS; iska; Digital Green), extreme conditions like hailstorms and unseasonal rains (Sowing Application), and erratic weather (Radio Monsoon; (M)obile Solutions).

4.1.2. Technological Provisioning to Smallholder Farmers

Various technologies were utilised in the AGI initiatives (Figure 2). MMIS, Tigo Kilimo, Airtel Kilimo, (M)obile Solutions, and FDSS provided simple weather, crop or market information to farmers via low-tech tools, such as SMS and mobile phones, whereby farmers could either receive automatic updates (push notifications) or request information through SMS request or calling a helpline. Varying degrees of complexity were built into these basic mobile phone-based solutions. The inclusion of multiple languages and a peer-to-peer chat function were provided in the Airtel Kilimo mobile service. iska harnessed GPS technology to provide location-specific weather information via SMS. Other AGI initiatives employed internet capabilities to develop custom platforms and smartphone applications, expanding the possible information services offered in terms of both content and format, including support of images, video, animation, interactive content and maps, and hyperlinks to additional online resources. Jayalaxmi Agro Tech offered a range of crop-specific smartphone applications that aimed to enhance food and livelihood security by providing text, audio and visual content on crop information, pricing analytics, and on-demand weather to farmers in English and local languages. Similarly, IFFCO Kisan Agriculture App and Sowing Application aided farmer decision making through the provision of crop or weather information through text, voice, photo and video content. Plantwise Knowledge Bank used smartphones to augment their community-based information exchange activities by pooling information into a central resource for farmers and stakeholders to access; this is particularly useful for remote access by individuals. While GPS was explicitly stated for few AGI initiatives (WATMIS; iska; Agriculture Monitoring System), other initiatives using smart devices likely exploited this technology to provide their locational services.

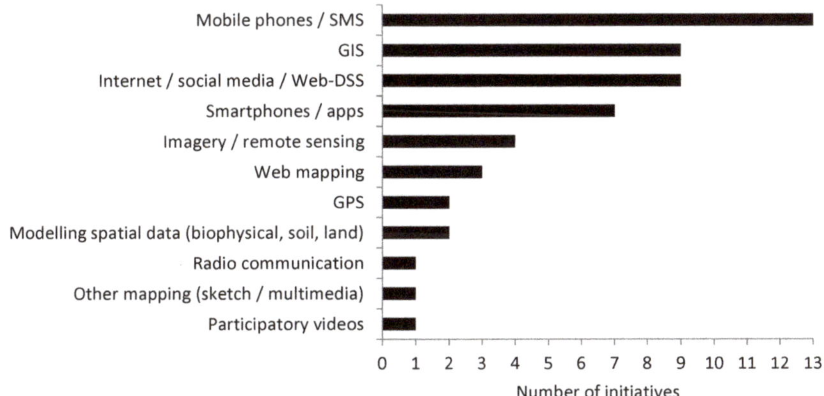

Figure 2. Technologies featured in reviewed AGI initiatives.

Some web-based platform initiatives included and disseminated more data-rich geographic information, such as fine resolution satellite imagery e.g., WATMIS and Agriculture Monitoring System. Satellite imagery and other forms of remote sensing are valuable for detailed depictions of landscape environments and remote capture of data [82]. In Afghanistan, experimentation with methods of classifying satellite imagery was undertaken to strengthen national capacity on rice crop monitoring for sustainable development and food security (GIRCM). WATMIS incorporated GIS and remote sensing data for viable and cost-effective integrated watershed and natural resource planning and management, used by agriculturalists, rural communities and extension services, and land managers. Many AGI initiatives used GIS in combination with ICTs to increase landscape resilience. For example, environmental mapping of drought extent, soils and crops were disseminated to extension workers and farmers, through mobile phones (MSIAF). Online capabilities of technologies have allowed user feedback and sharing of local knowledge for a range of applications, in particular, through social media and crowdsourcing platforms [17]. Radio Monsoon included social media through multiple AGI dissemination methods, and participatory mapping activities that harness local knowledge were used in PMDRRLK. Aside from these two initiatives, social media was absent in all other initiatives. More traditional and primitive forms of information communication, such as radio, loudspeakers and billboards in communities were utilised in some initiatives (e.g., Radio Monsoon; ITIKI).

4.1.3. Adoption of Initiatives

Our review identified limited details of AGI initiative adoption details, with a number being at the proposal, pilot or development stage ($n = 10$). For those with uptake statistics, assessments of adoption were complex. While the number of users or downloads (e.g., of a smartphone application) of an initiative seemed a standard measure of uptake, more nuanced patterns and differences between numbers of downloads, active users, repeat users, and those who implemented changes to their livelihood practices were also observed. Tigo Kilimo reported 400,000 registered users in two years. Of these, 61% were repeat users, with many trialling the service once but not returning. 30% of users reported continued use with concurrent use of new agricultural practices or growing new crops more likely. 39% were more likely to experience increased income than those not engaging with the service. An analogous service, Airtel Kilimo, reported similar adoption patterns, observing 6432 of their total 22,438 registered users (December 2014) as active, with approximately 50% of users implementing farming changes. IFFCO Kisan Agriculture App reported 170,000 users (October 2016), of which 10–20% were estimated to be active. Iska self-reported to have reached more than 80,000 farmers [70] and sent more than 8.5 million weather forecasts [71]. However, no data were provided on how farmers benefited from this service, or how weather forecasts improved livelihoods and were received/read. Digital Green claimed to have reached one million individuals across 13,592 villages through their participatory video approach, with 574,222 farmers adopting at least one of the best-practice video promotions. Yet, similar to iska, no data were available regarding individuals that have/have not implemented new practices, and why uptake has/has not occurred.

4.2. Factors Promoting or Limiting AGI Success

Given the results of the initiatives reviewed, four cross-cutting themes emerged which are important for promoting or limiting the success of AGI initiatives for climate change adaptation: Farmer capacity, delivery approach, technology used, and the organisation delivering the information (summarised in Table 2).

Table 2. Summary of factors promoting and limiting the success of AGI initiatives for addressing key agro-climatic challenges.

	Factors Promoting Success	**Factors Limiting Success**
Farmer capacity	Affordability to farmers	Participation capacity (exclusion through gender, costs, digital divide)
		Limited languages
	Available languages	Information alone often not enough for meaningful change
Approach	Partnerships with existing community groups	Methods for incorporating community knowledge into GIS
	User collaboration/sharing	Purely top-down approach—lack of interactivity
	Farmers involved in design	User registration required
Technological	High quality, locally-relevant information	Acquisition and sourcing of suitable and quality information/data
	Low tech and user friendly—ease of use	Availability and capacity of telecommunications infrastructure
	Allows participant feedback—interactivity functionality	Personal and community information security
Organisational	Organisational trust	Low user retention
	Potential for expansion—agile service	(In)ability to reach target users
	Marketing and endorsements	
	Clear business model, including funding	Funding of initiatives

4.2.1. Farmer Capacity

The most sophisticated AGI initiatives may be ineffective if target users are unable to access or utilise the information. Various socio-economic factors potentially limit accessibility for smallholder farmers, e.g., the level of disposable income required to acquire and/or access technologies like the internet, computers, smartphones, or televisions. Even relatively low-cost technologies like mobile phones may be inaccessible for many individuals, particularly in developing nations [83]. Consequently, poorer farmers are disadvantaged with increased difficulty in accessing AGI, despite often being the most in need. With reference to increasing participation in CROPROTECT, Bruce [62] described a lessening of digital divides in recent years, but poorer minorities still may lack access to ICTs. Communication technologies for enhancing knowledge access are often most beneficial for younger and more highly educated individuals [49]. Conversely, Bojovic et al. [84] demonstrated a weakening of digital divides for online participation in climate adaptation with groups that are typically excluded appearing as active participants (e.g., older or uneducated individuals). The contrasting ability of geographic information and ICTs to disproportionately benefit those who have access could be exacerbated if existing socioeconomic divisions within and across communities become greater [85].

One measure to increase farmer capacity is to incorporate local and additional languages in AGI initiatives, to ensure the usefulness of information and geographic information reach to maximise farmers benefitted. Information services provided only in English, for example, reduce the capacity of farmers who have first/only language to access the information. Producing and providing content in local languages facilitates comprehension and immediate connection with the local community (see Digital Green; [63]). However, using a local language alone reduces opportunities to expand platform use into other populations/geographical areas. Provision of information in both local/regional and national/international languages increases the probability of meeting a target user's preference [59]. Projects incorporating detailed information in multiple languages relevant to the scale of operation, including regional and local dialects (e.g., Airtel Kilimo, Jayalaxmi Agro Tech, and Digital Green) are likely to exhibit improved information dissemination and utilisation.

4.2.2. Approach

Approaches with participatory elements offer multiple potential benefits over purely top-down approaches. Where individuals can share their own information with others and/or feedback with AGI initiative developers they may feel their input is more valued and subsequently more interconnected to build community resilience [86]. Partnering with existing community groups can be a useful approach to increasing community participation. Digital Green leveraged community groups, such as women's self-help groups or farmers' groups by actively partnering with government, non-government, and private agencies with strong integration and relationships with communities, and cites these partnerships as critical to their success. Whilst having users involved in initiative development is beneficial, requiring registration for participation is seen as a limiting factor. Registering and then subscribing to content causes confusion with some users and has deterred people from using AGI services ([59]; e.g., CROPROTECT).

4.2.3. Technological

A major consideration for the successful implementation of any AGI initiative is the availability and capacity of the information and telecommunications infrastructure. This includes infrastructure for capturing and disseminating information, and for farmers to receive and use it. For example, if an initiative requires high-speed internet access to deliver high-resolution images/videos, then internet coverage is essential, as is the accessibility of affordable internet-enabled devices and data plans. Similarly, if AGI initiatives are designed to include mechanisms for user participation and feedback, then necessary ICT functionalities are required to facilitate interactivity. Many of the reviewed initiatives emphasised the importance of low-tech, user-friendly technological platforms, especially for those with low digital literacy. Additionally, the information itself is important in AGI initiatives, particularly in relation to content, quality and scale. High quality and trustworthy, locally-relevant information is most useful; sourcing and compiling such data can be technologically-challenging for the success of AGI projects [59]. Jayalaxmi Agro Tech attempted to ensure information was relevant to users by developing multiple smartphone applications specific to individual crops and livestock, whereby farmers can select an app to receive only relevant advice to their own farming practices. Attention also needs to be paid to ensuring the security and privacy of users and the data they might supply to the system, particularly in approaches that encourage public participation.

4.2.4. Organisational

Organisational factors include the organisation responsible for developing and implementing the AGI initiative and the kind of support an initiative receives. Initial funding and ongoing financial capital for maintenance, management, and information sourcing are vital for AGI initiatives. Monetary uncertainty may result in premature cessation of an initiative. Funded by a university competition prize, Radio Monsoon was received very positively by village fisherman and the local forecasters. However, the initiative ceased after two years of operation, as funding was no longer available [76]. Many of the reviewed initiatives were developed by universities and funded by external grants/agencies which resulted in uncertain or short-term initiative lifespans (<5 years) and funding unpredictability. This is problematic for climate change adaptation as climate impacts and building livelihood resilience occur over longer timeframes and multiple generations. Programs that are supported financially and in-kind by multiple sources congruently, including through local and international partnerships with the private sector, government agencies, non-government bodies, and the research sector, such as FDSS, and with a clear business model to manage these funds, appear to have greater success and longevity through decreased pressures of financial insecurity.

Reaching and maintaining users is essential for the success of any AGI initiative. Product marketing is imperative to reach users of relevance, and to raise awareness of initiative existence and accessibility. IFFCO Kisan Agriculture App utilised an existing mobile phone service with relevant potential users to target uptake. Search engine optimisation and social media sites can also provide effective and affordable marketing tools [67], but accessibility to these technologies and services is reflective of farmer socio-economic development levels. The IFFCO Kisan Agriculture App social media marketing strategy was augmented by the addition of local celebrity endorsements. GSMA [59] describe marketing and user retention challenges linked to brand identity and loyalty. Airtel Kilimo is provided to farmers through Kenyan mobile network provider Airtel, and multiple ownership, name and brand changes of Airtel have negatively impacted customer loyalty, and thus initiative uptake. Conversely, good reputation and high organisational trust can foster the success of AGI initiatives through user loyalty, sharing of positive experiences and promotion to other farmers (e.g., Tigo Kilimo).

5. Future Potential of AGI

We reflect upon the results and cross-cutting themes discussed above to recommend future avenues for ensuring successful adoption of AGI initiatives by smallholders for climate change adaptation and mitigation.

5.1. Geographical Targeting

Observational factors (Table 2) suggest that both demand- (by the need for climate adaptation solutions) and opportunity- (by the growth of populations with functional access to required ICTs) driven AGI initiatives are important. Geographical targeting of regions currently not utilising AGI initiatives could substantially benefit smallholder farmers in areas highly impacted by changing climates. Regarding regions of high climate change vulnerability and areas predicted for severe climate impacts on agriculture, various reports identify South and Southeast Asia, Africa, Caribbean nations, and small island developing states (SIDS), such as Vanuatu, Samoa and Tonga (see [4,87–89]. Nations in some of these regions already have targeted AGI initiatives (e.g., India, Afghanistan, and parts of Africa), but many other global priority areas remain untargeted. Further research is needed to expound the reasons for these geographical gaps, and for smallholders in these countries to develop appropriate AGI strategies utilising either existing or new infrastructure, technologies, or platforms that will be most effective for the populations of those regions. Vulnerable climate regions generally coincide with areas of increasing access to ICTs, with fast-growing global internet penetration rates observed in Africa, the Middle East, Latin America, and Asia (2000–2017; [90]).

5.2. Types of Information and Information Technologies

Better understanding of the types of information and technologies that are most useful is needed to target users more effectively. A detailed SWOT (strengths, weaknesses, opportunities, threats) analysis of technologies would be valuable, specifically to determine which technological approaches would most effectively deliver AGI to smallholders impacted by digital divides, for example, impoverished and uneducated farmers, women, and those in regions where access to ICT is limited. Mobile phones and SMS can be especially useful technologies for communicating AGI to smallholder farmers as necessary infrastructure is often already present, and data requirements/costs are comparatively low; in many rural regions, mobile phones are often accessible for farmers where other technologies are limited [1,59]. However, credit costs and access to electricity for charging phones can prohibit farmers' use of mobile technologies [83]. Additionally, the information disseminated via mobile phone may be limited by the text- or voice-only format. Technological, resource (cost), and skill components required to access and use AGI will present barriers for some farmers, which also impacts the inclusion of farmer feedback and local knowledge in initiatives. If technologies can be harnessed effectively, then community information sharing could promote greater peer learning and social connectedness, and contribute to increased community resilience [86].

5.3. Independent Assessment of Initiatives

Existing initiatives and future AGI projects should be independently assessed to provide robust success evaluations of their approaches. This is essential as current non-standardised, self-evaluative techniques provide no meaningful and comparable measures of AGI initiative effectiveness, and self-published usage statistics are often more aligned with marketing. The observed asymmetrical pattern of registered and active users is not unique to AGI initiatives, and transferability of assessment approaches by other online geographic information services could be investigated, e.g., OpenStreetMap has 0.5 million registered users (2011) with 38% having undertaken some mapping, and 5% classed as active contributors [91]. There is also a need to examine impacts for users with different characteristics (considering factors, such as gender, age, income, ethnicity, social status, religion and others), as usage and impacts will not be homogeneous among heterogeneous populations [32]. Furthermore, how project success is reported and marketed may have important implications for future funding and resource allocations, agriculture and climate policies, research and development directions, and the livelihoods of farmers. Thus, independent standardised approaches to evaluating AGI initiatives with an emphasis on more nuanced measures of success beyond simple user statistics are recommended. Moreover, the trust and collaboration often needed for farmers to adopt new practices and alternative ways of thinking takes time, and processes of social change can occur over generations [44], thus longitudinal assessments are also advised over raw user statistics.

5.4. Inclusivity for Multi-Level Stakeholder Communication

Ballantyne [31] argues the need for inclusive, participatory approaches to knowledge sharing, and to successfully use ICT to support farmers and rural communities, farming communities must be empowered to define their own needs. Public participation in GIS (e.g., participatory mapping by communities) to contribute their own unique spatial knowledge, often with support from government, nongovernmental, university and other organisations engaged in development and land-related planning [92], can develop community cohesion [93] and facilitate greater local engagement in land-related decision making [94]. Combining local knowledge on coping mechanisms with top-down strategies has enhanced the capacity of rural indigenous communities in SIDS to mitigate and withstand environmental pressures [95]. Additionally, enhancing smallholder social capital can provide opportunities for more effective articulation of individual and community goals/needs to policy makers, researchers and extension providers [3]. Challenges to inclusive AGI participation (e.g., education levels, household resources, local agro-ecological conditions, market access, availability of local producer organisations, and ability/willingness to collaborate and take risks) need careful consideration, particularly regarding equality for women [3]. Baumüller [32] reports for mobile services that study of behavioural factors impacting farmers' capacity and willingness to participate and/or take risks is a significant research shortfall. Technologies that are adapted to smallholders' capacity to take risks and integrated with relevant support services [28], especially to reach marginal farmers where traditional extension activities [3] or locations where reliability of traditional farming approaches [70] fall short, may prove useful in uptake of AGI to overcome cultural and socio-economic obstacles.

Underpinning each area of potential are important considerations and limitations to AGI that warrant further understanding. Adoption of AGI and any outcomes for smallholders are limited by the capacity to act on the knowledge or information gained. For example, a farmer may receive information of a locally-relevant drought-resistant crop, but may not have the financial means to acquire it. Capacity for decision making will also influence the success of AGI initiatives, and information provision alone may not result in meaningful change. Information accessibility is just one factor among many that significantly affect adaptation [96]. Improved comprehension is needed regarding how significant livelihood change occurs when farmers adopt AGI. This requires localised studies at the level of those users most affected (smallholder farmer communities). Further, as livelihood change is not a short-term process and may vary geographically, studies should be longitudinal and undertaken in a variety of climate-impacted regions. Significantly, the potential ability for AGI provision and adoption to address long-term systemic vulnerabilities requires further research attention.

6. Summary

Learning from past experiences and innovations to promote a successful climate adaptation and development research agenda for the future is crucial [97]. Under increasing livelihood pressures associated with short term, and long term, climate stressors, we advocate that smallholder farmers require diverse and locally-relevant geographic information to aid adaptation for increased food and livelihood security. As we identify, only a small percentage of targeted users of AGI initiatives we reviewed are using and acting on the information provided, which raises questions of the appropriateness of such approaches for addressing key agro-climatic challenges. Addressing these shortcomings is important for supporting smallholders to overcome global risks of extreme weather events, natural disasters, and failures of climate change mitigation and adaptation [98]. Our analysis has identified key recommendations that will serve as a valuable guide for the success of future AGI developments whereby knowledge gaps and implementation challenges should be addressed, particularly to align with the geographically varying needs of smallholder farmers (e.g., [99,100]. Use of AGI initiatives could greatly aid smallholders to move towards climate-smart agriculture [101] for sustainably increasing productivity [44], improving environmental livelihood security [102], and enhancing landscape resilience under a changing climate [103].

Author Contributions: Conceptualisation, B.T.H., J.D., E.B. (Eloise Biggs), B.B., E.B. (Eleanor Bruce) and N.W.; Methodology, B.T.H., J.D. and E.B. (Eloise Biggs); Validation, B.T.H. and E.B. (Eloise Biggs); Formal Analysis, B.T.H.; Investigation, B.T.H.; Resources, E.B. (Eloise Biggs) and B.B.; Writing—Original Draft Preparation, B.T.H.; Writing—Review and Editing, B.T.H., E.B. (Eloise Biggs), J.D., N.W., B.B. and E.B. (Eleanor Bruce); Visualisation, B.T.H. and E.B. (Eloise Biggs); Supervision, E.B. (Eloise Biggs), B.B. and E.B. (Eleanor Bruce); Project Administration, E.B. (Eloise Biggs), B.B. and E.B. (Eleanor Bruce); Funding Acquisition, E.B. (Eloise Biggs), B.B. E.B. (Eleanor Bruce) and J.D.

Funding: This research was funded under project ASEM-2016-30, an Australian Centre for International Agriculture Research (ACIAR) small research activity led by the University of Western Australia in collaboration with the University of Sydney and the University of the South Pacific.

Conflicts of Interest: The authors declare no conflict of interest.

References

1. Sylvester, G. *Success Stories on Information and Communication Technologies for Agriculture and Rural Development*, 2nd ed.; Food and Agriculture Organization of the United Nations: Rome, Italy, 2015; Available online: http://www.fao.org/3/a-i4622e.pdf (accessed on 8 May 2017).
2. Samberg, L.H.; Gerber, J.S.; Ramankutty, N.; Herrero, M.; West, P.C. Subnational distribution of average farm size and smallholder contributions to global food production. *Environ. Res. Lett.* **2016**, *11*, 124010. [CrossRef]
3. Swanson, B. *Global Review of Good Agricultural Extension and Advisory Service Practice*; Food and Agriculture Organization of the United Nations: Rome, Italy, 2008.
4. Knox, J.; Hess, T.; Daccache, A.; Wheeler, T. Climate change impacts on crop productivity in Africa and South Asia. *Environ. Res. Lett.* **2012**, *7*, 034032. [CrossRef]
5. Coumou, D.; Rahmstorf, S. A decade of weather extremes. *Nat. Clim. Chang.* **2012**, *2*, 491–496. [CrossRef]
6. Trenberth, K.E. Changes in precipitation with climate change. *Clim. Res.* **2011**, *47*, 123–138. [CrossRef]
7. Lobell, D.; Schlenker, W.; Costa-Roberts, J. Climate trends and global crop production since 1980. *Science* **2011**, *333*, 616–620. [CrossRef] [PubMed]
8. Hochman, Z.; Horan, H.; Reddy, D.R.; Sreenivas, G.; Tallapragada, C.; Adusumilli, R.; Gaydon, D.S.; Laing, A.; Kokic, P.; Singh, K.K.; et al. Smallholder farmers managing climate risk in India: 2. Is it climate-smart? *Agric. Syst.* **2017**, *151*, 61–72. [CrossRef]
9. Thornton, P.K.; Lipper, L. *How Does Climate Change Alter Agricultural Strategies to Support Food Security?* IFPRI Discussion Paper 01340; IFPRI: Washington, DC, USA, 2013.
10. Dow, K.; Berkhout, F.; Preston, B.L.; Klein, R.J.; Midgley, G.; Shaw, M.R. Limits to adaptation. *Nat. Clim. Chang.* **2013**, *3*, 305–307. [CrossRef]
11. Jayaraman, T.; Murari, K. Climate change and agriculture: Future trends and implications for India. *Rev. Agrar. Stud.* **2014**, *4*, 1–49.
12. Barrett, C.B.; Reardon, T.; Webb, P. Nonfarm income diversification and household livelihood strategies in rural Africa: Concepts, dynamics, and policy implications. *Food Policy* **2001**, *26*, 315–331. [CrossRef]
13. Jayne, T.S.; Mather, D.; Mghenyi, E. Principal challenges confronting smallholder agriculture in Sub-Saharan Africa. *World Dev.* **2010**, *38*, 1384–1398. [CrossRef]
14. Tittonell, P.; Giller, K.E. When yield gaps are poverty traps: The paradigm of ecological intensification in African smallholder agriculture. *Field Crops Res.* **2013**, *143*, 76–90. [CrossRef]
15. Fisher, M.; Carr, E.R. The influence of gendered roles and responsibilities on the adoption of technologies that mitigate drought risk: The case of drought-tolerant maize seed in eastern Uganda. *Glob. Environ. Chang.* **2015**, *35*, 82–92. [CrossRef]
16. Carter, M.R.; Little, P.D.; Mogues, T.; Negatu, W. Poverty traps and natural disasters in Ethiopia and Honduras. *World Dev.* **2007**, *35*, 835–856. [CrossRef]
17. Goodchild, M.F. Citizens as sensors: The world of volunteered geography. *GeoJournal* **2007**, *69*, 211–221. [CrossRef]
18. Elwood, S. Geographic information science: Emerging research on the societal implications of the geospatial web. *Prog. Hum. Geogr.* **2010**, *34*, 349–357. [CrossRef]
19. Kuhn, W. Geospatial Semantics: Why, of What, and How. In *Journal on Data Semantics III. Lecture Notes in Computer Science*; Spaccapietra, S., Zimányi, E., Eds.; Springer: Berlin, Germany, 2005; Volume 3534.
20. Longley, P.A.; Goodchild, M.F.; Maguire, D.J.; Rhind, D.W. *Geographic Information Systems and Science*, 2nd ed.; John Wiley and Sons Ltd.: Chichester, UK, 2006.
21. Open Geospatial Consortium (OGC). Glossary of Terms—G. 2017. Available online: http://www.opengeospatial.org/ogc/glossary/g (accessed on 5 May 2017).
22. Craglia, M.; Ostermann, F.; Spinsanti, L. Digital Earth from vision to practice: Making sense of citizen-generated content. *Int. J. Digit. Earth* **2012**, *5*, 398–416. [CrossRef]

23. Benson, A.; Jafry, T. The state of agricultural extension: An overview and new caveats for the future. *J. Agric. Educ. Ext.* **2013**, *19*, 381–393. [CrossRef]

24. Anderson, J.R.; Feder, G. Agricultural extension: Good intentions and hard realities. *World Bank Res. Obs.* **2004**, *19*, 41–60. [CrossRef]

25. Taylor, M.; McGregor, A.; Dawson, B. *Vulnerability of Pacific Island Agriculture and Forestry to Climate Change*; Pacific Community: New Caledonia, 2016; Available online: https://spccfpstore1.blob.core.windows.net/digitallibrary-docs/files/6f/6fdef19c8085874a0406d7e1f64897bd.pdf?sv=2015-12-11andsr=bandsig=qUFufXOhh7WrD9g5SQuNcg89y1d4QUHTmUiUPqK0hEo%3Dandse=2018-05-20T04%3A14%3A47Zandsp=randrscc=public%2C%20max-age%3D864000%2C%20max-stale%3D86400andrsct=application%2Fpdfandrscd=inline%3B%20filename%3D%22Vulnerability_Pacific_agriculture_climate_change.pdf%22 (accessed on 12 December 2017).

26. Duncan, J.M.; Tompkins, E.L.; Dash, J.; Tripathy, B. Resilience to hazards: Rice farmers in the Mahanadi Delta, India. *Ecol. Soc.* **2017**, *22*, 3. [CrossRef]

27. Mittal, S. Role of Mobile Phone-enabled Climate Information Services in Gender-inclusive Agriculture. *Gend. Technol. Dev.* **2016**, *20*, 200–217. [CrossRef]

28. Glendenning, C.J.; Babu, S.; Asenso-Okyere, K. Review of Agricultural Extension in India. IFPRI Discussion Paper 01048. International Food Policy Research Institute, 2010. Available online: http://cdm15738.contentdm.oclc.org/utils/getfile/collection/p15738coll2/id/7280/filename/7281.pdf (accessed on 16 June 2017).

29. Balaghi, R.; Badjeck, M.C.; Bakari, D.; De Pauw, E.; De Wit, A.; Defourny, P.; Donato, S.; Gommes, R.; Jlibene, M.; Ravelo, A.C.; et al. Managing Climatic Risks for Enhanced Food Security: Key Information Capabilities. *Procedia Environ. Sci.* **2010**, *1*, 313–323. [CrossRef]

30. Yonazi, E.; Kelly, T.; Halewood, N.; Blackman, C. *eTransform Africa 2012: The Transformational Use of Information and Communication Technologies in Africa*; World Bank, African Development Bank and African Union: Washington, DC, USA; Tunis, Tunisia, 2012.

31. Ballantyne, P. Accessing, sharing and communicating agricultural information for development: Emerging trends and issues. *Inf. Dev.* **2009**, *25*, 260–271. [CrossRef]

32. Baumüller, H. The little we know: An exploratory literature review on the utility of mobile phone-enabled services for smallholder farmers. *J. Int. Dev.* **2018**, *30*, 134–154. [CrossRef]

33. Food and Agriculture Organization of the United Nations (FAO). *e-Agriculture 10 Year Review Report*; Food and Agriculture Organization of the United Nations: Rome, Italy, 2015.

34. World Bank. *ICT in Agriculture: Connecting Smallholders to Knowledge, Networks, and Institutions*; World Bank: Washington, DC, USA, 2011.

35. Duncombe, R. Mobile phones for agricultural and rural development: A literature review and suggestions for future research. *Eur. J. Dev. Res.* **2016**, *28*, 213–235. [CrossRef]

36. Wang, J.; Chen, J.; Ju, W.; Li, M. IA-SDSS: A GIS-based land use decision support system with consideration of carbon sequestration. *Environ. Model. Softw.* **2010**, *25*, 539–553. [CrossRef]

37. Adenle, A.A.; Azadi, H.; Arbiol, J. Global assessment of technological innovation for climate change adaptation and mitigation in developing world. *J. Environ. Manag.* **2015**, *161*, 261–275. [CrossRef]

38. Kroschel, J.; Sporleder, M.; Tonnang, H.E.Z.; Juarez, H.; Carhuapoma, P.; Gonzales, J.C.; Simon, R. Predicting climate-change-caused changes in global temperature on potato tuber moth Phthorimaea operculella (Zeller) distribution and abundance using phenology modeling and GIS mapping. *Agric. For. Meteorol.* **2013**, *170*, 228–241. [CrossRef]

39. Aubert, B.A.; Schroeder, A.; Grimaudo, J. IT as enabler of sustainable farming: An empirical analysis of farmers' adoption decision of precision agriculture technology. *Decis. Support Syst.* **2012**, *54*, 510–520. [CrossRef]

40. Bongiovanni, R.; Lowenberg-DeBoer, J. Precision agriculture and sustainability. *Precis. Agric.* **2004**, *5*, 359–387. [CrossRef]

41. Zhang, N.; Runquist, E.; Schrock, M.; Havlin, J.; Kluitenburg, G.; Redulla, C. Making GIS a versatile analytical tool for research in precision farming. *Comput. Electron. Agric.* **1999**, *22*, 221–223. [CrossRef]

42. Meier, P. Crisis mapping in action: How open source software and global volunteer networks are changing the world, one map at a time. *J. Map Geogr. Libr.* **2012**, *8*, 89–100. [CrossRef]

43. Haworth, B.T.; Bruce, E.; Whittaker, J.; Read, R. The good, the bad, and the uncertain: Contributions of volunteered geographic information to community disaster resilience. *Front. Earth Sci.* **2018**, *6*, 183. [CrossRef]

44. FAO. *Climate-Smart Agriculture Sourcebook*; FAO: Rome, Italy, 2013; Available online: http://www.fao.org/docrep/018/i3325e/i3325e00.htm (accessed on 4 December 2018).

45. Thornton, P.K.; Aggarwal, P.; Parsons, D. Editorial: Prioritising climate-smart agricultural interventions at different scales. *Agric. Syst.* **2017**, *151*, 149–152. [CrossRef]

46. World Bank. *Enhancing Agricultural Innovation: How to Go beyond the Strengthening of Research Systems*; World Bank: Washington, DC, USA, 2007; Available online: http://siteresources.worldbank.org/INTARD/Resources/EnhancingAgInnovationebook.pdf (accessed on 9 May 2017).

47. Obayelu, A.; Ogunlade, I. Analysis of the uses of information communication technology (ICT) for gender empowerment and sustainable poverty alleviation in Nigeria. *Int. J. Educ. Dev. Using ICT* **2006**, *2*, 45–69.

48. Dang, H.L.; Li, E.; Bruwer, J.; Nuberg, I. Farmers' perceptions of climate variability and barriers to adaptation: Lessons learned from an exploratory study in Vietnam. *Mitig. Adapt. Strateg. Glob. Chang.* **2014**, *19*, 531–548. [CrossRef]

49. Hudson, S.; Krogman, N.; Beckie, M. Social practices of knowledge mobilization for sustainable food production: Nutrition gardening and fish farming in the Kolli hills of India. *Food Secur.* **2016**, *8*, 523–533. [CrossRef]

50. Kumar, K.R.; Nain, M.S.; Singh, R.; Bana, R.S. Analysis of farmers' communication network and factors of knowledge regarding agro meteorological parameters. *Indian J. Agric. Sci.* **2015**, *85*, 1592–1596.

51. Palmer, T.; Pshenichnaya, N. *Tigo Kilimo Impact Evaluation*; GSMA: London, UK, 2015; Available online: http://www.gsma.com/mobilefordevelopment/programme/magri/assessing-the-impact-of-tigo-kilimo (accessed on 15 August 2017).

52. Munyna, H. Application of ICTs in Africa's agricultural sector: A gender perspective. In *Gender and the information revolution in Africa*; Rathgeber, E.M., Adera, E.O., Eds.; International Development Research Centre: Ottawa, ON, Canada, 2000; pp. 85–124.

53. Gandhi, R.; Veeraraghavan, R.; Toyama, K.; Ramprasad, V. Digital Green: Participatory video for agricultural extension. In Proceedings of the 2007 International Conference on Information and Communication Technologies and Development, Bangalore, India, 15–16 December 2007.

54. Rola, A.C.; Quizon, J.B.; Jamias, S.B. Do Farmer Field School Graduates Retain and Share What They Learn? An Investigation in Iloilo, Philippines. *J. Int. Agric. Ext. Educ.* **2002**, *5*, 65–75. [CrossRef]

55. Walsham, G.; Sahay, S. Research on information systems in developing countries: Current landscape and future prospects. *Inf. Technol. Dev.* **2006**, *12*, 7–24. [CrossRef]

56. Metcalfe, M.; Joham, C. The "ear" and "eye" digital divide. In *Organizational Information Systems in the Context of Globalization*; Korpela, M., Montealegre, R., Poulymenakou, A., Eds.; Springer: Boston, MA, USA, 2003; pp. 419–434.

57. Livoreil, B.; Glanville, J.; Haddaway, N.R.; Bayliss, H.; Bethel, A.; de la Chapelle, F.F.; Robalino, S.; Savilaakso, S.; Zhou, W.; Petrokofsky, G.; et al. Systematic searching for environmental evidence using multiple tools and sources. *Environ. Evid.* **2017**, *6*, 23. [CrossRef]

58. FAO. Enhancing Agriculture Monitoring System Based on Geospatial Technology in Afghanistan. 2016. Available online: http://www.fao.org/3/a-i5569e.pdf (accessed on 15 August 2017).

59. Groupe Spéciale Mobile Association (GSMA). *Case Study: Airtel Kilimo, Kenya*; GSMA: London, UK, 2015; Available online: http://www.gsma.com/mobilefordevelopment/wp-content/uploads/2015/02/GSMA_Case_Airtel_FinalProof02.pdf (accessed on 15 August 2017).

60. Cole, S.; Fernando, A.N. Mobile'izing Agricultural Advice: Technology Adoption, Diffusion and Sustainability. Harvard Business School Working Paper. Cambridge, UK, April 2016. Available online: https://papers.ssrn.com/sol3/papers.cfm?abstract_id=2179008anddownload=yes (accessed on 10 October 2017).

61. Girvetz, E.H.; Zganjar, C.; Raber, G.T.; Maurer, E.P.; Kareiva, P.; Lawler, J.L. Applied Climate-Change Analysis: The Climate Wizard Tool. *PLoS ONE* **2009**, *4*, e8320. [CrossRef] [PubMed]

62. Bruce, T.J.A. The CROPROTECT project and wider opportunities to improve farm productivity through web- based knowledge exchange. *Food Energy Secur.* **2016**, *5*, 89–96. [CrossRef] [PubMed]

63. Gandhi, R. Case Study 1: Digital Green: Leveraging social networks for agricultural extension. In *Success Stories on Information and Communication Technologies for Agriculture and Rural Development*, 2nd ed.; Sylvester, G., Ed.; Food and Agriculture Organization of the United Nations: Rome, Italy, 2015; Available online: http://www.fao.org/3/a-i4622e.pdf (accessed on 8 May 2017).

64. Trogo, R.; Ebardaloza, J.B.; Sabido, D.J.; Bagtasa, G.; Tongson, E.; Balderama, O. SMS-based Smarter Agriculture Decision Support System for Yellow Corn Farmers in Isabela. In Proceedings of the IEEE Canada International Humanitarian Technology Conference (Ihtc2015), Ottawa, ON, Canada, 31 May–4 June 2015.

65. GSMA. *Farmforce*; GSMA: London, UK, 2013; Available online: http://www.gsma.com/mobilefordevelopment/programme/magri/farmforce (accessed on 16 October 2017).

66. FAO. Monitoring of Rice Crop Using Satellite Remote Sensing and GIS Technologies in Northern and Eastern Afghanistan. 2016. Available online: http://www.fao.org/3/a-i6146e.pdf (accessed on 22 August 2017).

67. Darabian, N. Case Study IFFCO Kisan Agriculture App: Evolution to Data Driven Services in Agriculture. GSMA, London. 2016. Available online: http://www.gsma.com/mobilefordevelopment/programme/magri/iffco-kisan-agricultural-app-evolution-to-data-driven-services-in-agriculture (accessed on 15 August 2017).

68. Masinde, M.; Bagula, A.; Muthama, N. Implementation Roadmap for Downscaling Drought Forecasts in Mbeere Using Itiki. In Proceedings of the 2013 Itu Kaleidoscope Academic Conference: Building Sustainable Communities (K-2013), Kyoto, Japan, 22–24 April 2013; pp. 63–70.

69. Ignitia. Ignitia, Tropical Weather Forecasting. 2017. Available online: www.ignitia.se (accessed on 16 June 2017).

70. United Nations Development Program (UNDP). Using SMS Texts to Provide Weather Forecasts for Small Farmers in West Africa. December 2015. Available online: http://www.undp.org/content/undp/en/home/presscenter/pressreleases/2015/12/22/using-sms-texts-to-provide-weather-forecasts-for-small-farmers-in-west-africa.html (accessed on 16 June 2017).

71. Jayalaxmi Agro Tech. Welcome to Jayalaxmi Agro Tech. 2014. Available online: http://www.jayalaxmiagrotech.com/ (accessed on 21 August 2017).

72. Wenkel, K.O.; Berg, M.; Mirschel, W.; Wieland, R.; Nendel, C.; Kostner, B. LandCaRe DSS—An interactive decision support system for climate change impact assessment and the analysis of potential agricultural land use adaptation strategies. *J. Environ. Manag.* **2013**, *127*, S168–S183. [CrossRef] [PubMed]

73. Ogodo, O. African Farmers Get Geospatial Info on Their Phones. SciDevNet, 24 April 2009. Available online: http://www.scidev.net/global/farming/news/african-farmers-get-geospatial-info-on-their-phone.html (accessed on 19 August 2017).

74. Laraki, J. Case Study 6: Mobile market information service: A pilot project of ICT use for smallholder farmers in Papua New Guinea. In *Success Stories on Information and Communication Technologies for Agriculture and Rural Development*, 2nd ed.; Sylvester, G., Ed.; Food and Agriculture Organization of the United Nations: Rome, Italy, 2015; Available online: http://www.fao.org/3/a-i4622e.pdf (accessed on 8 May 2017).

75. Reichel, C.; Frömming, U.U. Participatory mapping of local disaster risk reduction knowledge: An example from Switzerland. *Int. J. Disaster Risk Sci.* **2014**, *5*, 41–54. [CrossRef]

76. Slawson, N. Radio Monsoon attempts to ensure safety reigns among fisherman in south India. The Guardian, 24 April 2017. Available online: https://www.theguardian.com/global-development/2017/apr/24/radio-monsoon-safety-fishermen-south-india-kerala (accessed on 15 August 2017).

77. Tayyebi, A.; Meehan, T.D.; Dischler, J.; Radloff, G.; Ferris, M.; Gratton, C. SmartScape™: A web-based decision support system for assessing the tradeoffs among multiple ecosystem services under crop-change scenarios. *Comput. Electron. Agric.* **2016**, *121*, 108–121. [CrossRef]

78. Agarwal, V. New App Promises to Tell Indian Farmers When to Sow Crops. Available online: https://blogs.wsj.com/indiarealtime/2016/06/17/new-app-promises-to-tell-indian-farmers-when-to-sow-crops/ (accessed on 17 June 2016).

79. Aher, P.D.; Adinarayana, J.; Gorantiwar, S.D.; Sawant, S.A. Information System for Integrated Watershed Management Using Remote Sensing and GIS. *Remote Sens. Appl. Environ. Res.* **2014**, 17–34. [CrossRef]

80. Aggarwal, P.K.; Baethegan, W.E.; Cooper, P.; Gommes, R.; Lee, B.; Meinkef, H.; Rathoreg, L.S.; Sivakumarh, M.V.K. Managing Climatic Risks to Combat Land Degradation and Enhance Food security: Key Information Needs. *Procedia Environ. Sci.* **2010**, *1*, 305–312. [CrossRef]

81. Fourati, M.A.; Chebbi, W.; Kamoun, A. Development of a Web-based weather station for irrigation scheduling. In Proceedings of the 2014 Third IEEE Colloquium in Information Science and Technology, Tetouan, Morocco, 20–22 October 2014; pp. 37–41. [CrossRef]

82. Lillesand, T.; Kiefer, R.W.; Chipman, J. *Remote Sensing and Image Interpretation*; John Wiley and Sons: New York, NY, USA, 2014.

83. Duncan, J.M.; Haworth, B.; Biggs, E.; Boruff, B.; Wales, N.; Bruce, E. Managing multifunctional landscapes: Local insights from a Pacific Island country context. *J. Environ. Manag.* in review.

84. Bojovic, D.; Bonzanigo, L.; Giupponi, C.; Maziotis, A. Online participation in climate change adaptation: A case study of agricultural adaptation measures in Northern Italy. *J. Environ. Manag.* **2015**, *157*, 8–19. [CrossRef]

85. Haworth, B.T. Implications of Volunteered Geographic Information for Disaster Management and GIScience: A More Complex World of Volunteered Geography. *Ann. Am. Assoc. Geogr.* **2018**, *108*, 226–240. [CrossRef]

86. Haworth, B.; Whittaker, J.; Bruce, E. Assessing the application and value of participatory mapping for community bushfire preparation. *Appl. Geogr.* **2016**, *76*, 115–127. [CrossRef]

87. Wheeler, D. *Quantifying Vulnerability to Climate Change: Implications for Adaptation Assistance*; Center for Global Development Working Paper No. 240; Center for Global Development: Washington, DC, USA, 2011.

88. Kreft, S.; Eckstein, D.; Dorsch, L.; Fischer, L. Global Climate Risk Index 2016: Who Suffers Most from Extreme Weather Events? Weather-Related Loss Events in 2014 and 1995 to 2014. Germanwatch Nord-Süd Initiative eV. 2015. Available online: https://germanwatch.org/fr/download/13503.pdf (accessed on 15 August 2017).

89. Notre Dame Global Adaptation Initiative (NDGAI). Notre Dame Global Adaptation Index. 2017. Available online: http://index.gain.org (accessed on 14 August 2017).

90. Internet Usage Statistics. Internet World Stats. N.p., 9 June 2017. Available online: http://www.internetworldstats.com/stats.htm (accessed on 12 June 2017).

91. Neis, P.; Zipf, A. Analyzing the contributor activity of a volunteered geographic information project—The case of OpenStreetMap. *Int. J. Geo-Inf.* **2012**, *1*, 146–165. [CrossRef]

92. IFAD. *Good Practices in Participatory Mapping*; Prepared by Corbett, J.M.; The International Fund for Agricultural Development: Rome, Italy, 2009.

93. Corbett, J.M.; Keller, C.P. An analytical framework to examine empowerment associated with participatory geographic information systems (PGIS). *Cartogr. Int. J. Geogr. Inf. Geovis.* **2005**, *40*, 91–102. [CrossRef]

94. Corbett, J. "I Don't Come from Anywhere": Exploring the Role of the Geoweb and Volunteered Geographic Information in Rediscovering a Sense of Place in a Dispersed Aboriginal Community. In *Crowdsourcing Geographic Knowledge: Volunteered Geographic Information (VGI) in Theory and Practice*; Sui, D.Z., Elwood, S., Goodchild, M.F., Eds.; Springer: Berlin, Germany, 2013; pp. 223–241.

95. Mercer, J.; Dominey-Howes, D.; Kelman, I.; Lloyd, K. The potential for combining indigenous and western knowledge in reducing vulnerability to environmental hazards in small island developing states. *Environ. Hazards* **2007**, *7*, 245–256. [CrossRef]

96. Menike, L.; Arachchi, K.K. Adaptation to climate change by smallholder farmers in rural communities: Evidence from Sri Lanka. *Procedia Food Sci.* **2016**, *6*, 282–292. [CrossRef]

97. Jones, L.; Harvey, B.; Cochrane, L.; Cantin, B.; Conway, D.; Cornforth, R.J.; De Souza, K.; Kirbyshire, A. Designing the next generation of climate adaptation research or development. *Reg. Environ. Chang.* **2018**, *18*, 297–304. [CrossRef]

98. World Economic Forum (WEF). *The Global Risks Report 2017*, 12th ed.; World Economic Forum: Geneva, Switzerland, 2017; Available online: http://reports.weforum.org/global-risks-2017/ (accessed on 14 August 2017).

99. Harvey, C.A.; Saborio-Rodriguez, M.; Martinez-Rodriguez, M.R.; Viguera, B.; Chain-Guadarrama, A.; Vignola, R.; Alpizar, F. Climate change impacts and adaptation among smallholder farmers in Central America. *Agric. Food Secur.* **2018**, *7*, 57. [CrossRef]

100. Abdul-Razak, M.; Kruse, S. The adaptive capacity of smallholder farmers to climate change in Northern Region of Ghana. *Clim. Risk Manag.* **2017**, *17*, 104–122. [CrossRef]

101. Martinez-Baron, D.; Orjuela, G.; Renzoni, G.; Loboguerrero Rodriguez, A.M.; Prager, S.D. Small-scale farmers in a 1.5 °C future: The importance of local social dynamics as an enabling factor for implementation and scaling of climate-smart agriculture. *Curr. Opin. Environ. Sustain.* **2018**, *31*, 112–119. [CrossRef]

102. Biggs, E.M.; Bruce, E.; Boruff, B.; Duncan, J.M.A.; Horsley, J.; Pauli, N.; McNeil, K.; Neef, A.; Van Ogtrop, F.; Curnow, J.; et al. Sustainable development and the water-energy-food nexus: A perspective on livelihoods. *Environ. Sci. Policy* **2015**, *54*, 389–397. [CrossRef]
103. Kremen, C.; Merenlender, A.M. Landscapes that work for biodiversity and people. *Science* **2018**, *362*, eaau6020. [CrossRef] [PubMed]

Article

Opportunities for Green Energy through Emerging Crops: Biogas Valorization of *Cannabis sativa* L. Residues

Carla Asquer [1], Emanuela Melis [1,*], Efisio Antonio Scano [1] and Gianluca Carboni [2]

1   Biofuels and Biomass Laboratory, Renewable Energies Facility, Sardegna Ricerche – VI strada ovest Z.I. Macchiareddu, 09010 Uta, Italy; asquer@sardegnaricerche.it (C.A.); efisioas@tin.it (E.A.S.)
2   Agris Sardegna, Viale Trieste 111, 09123 Cagliari, Italy; gcarboni@agrisricerca.it
*   Correspondence: emanuela.melis@sardegnaricerche.it or emymelis@hotmail.com; Tel.: +39-070-9243-2708

Received: 31 October 2019; Accepted: 11 December 2019; Published: 13 December 2019

Abstract: The present work shows the experimental evidence carried out on a pilot scale and demonstrating the potential of *Cannabis sativa* L. by-products for biogas production through anaerobic digestion. While the current state-of-the-art tests on anaerobic digestion feasibility are carried out at the laboratory scale, the here described tests were carried out at a pilot-to-large scale. An experimental campaign was carried out on hemp straw residues to assess the effective performance of this feedstock in biogas production by reproducing the real operating conditions of an industrial plant. An organic loading rate was applied according to two different amounts of hemp straw residues (3% wt/wt and 5% wt/wt). Also, specific bioenhancers were used to maximize biogas production. When an enzymatic treatment was not applied, a higher amount of hemp straw residues determined an increase of the median values of the gas production rate of biogas of 92.1%. This reached 116.6% when bioenhancers were applied. The increase of the specific gas production of biogas due to an increment of the organic loading rate (5% wt/wt) was +77.9% without enzymatic treatment and it was +129.8% when enzymes were used. The best management of the biodigester was found in the combination of higher values of hemp straw residues coupled with the enzymatic treatment, reaching 0.248 $\text{Nm}^3 \cdot \text{kg}_{\text{volatile solids}}^{-1}$ of specific biogas production. Comparisons were made between the biogas performance obtained within the present study and those found in the literature review coming from studies on a laboratory scale, as well as those related to the most common energy crops. The hemp straw performance was similar to those provided by previous studies on a laboratory scale. Values reported in the literature for other lignocellulosic crops are close to those of this work. Based on the findings, biogas production can be improved by using bioenhancers. Results suggest an integration of industrial hemp straw residues as complementary biomass for cleaner production and to contribute to the fight against climate change.

Keywords: renewable energy technologies; sustainability; clean energy; bioenergy; biogas; industrial hemp; anaerobic digestion

---

1. Introduction

Industrial hemp (*Cannabis sativa* L.) is a valuable crop, and all parts of the plant can be used in many ways. Recent surveys carried out in the past few years (e.g., [1,2]) suggested that industrial hemp is a niche crop of increasing interest for its properties and versatility. New uses and innovative products appear on the market (more than 25,000 products have been discerned [3]), thus *Cannabis sativa* L. is becoming a very attractive crop on a global scale.

In Europe, hemp cultivation is mainly a multi-purpose crop. The market interest for hemp seeds and the need for attaining maximum economic viability of the related supply chains are stimulating a progressive shift of interest from traditional stem fiber use (textile, pulp or paper) towards multi-purpose cultivations. Indeed, in recent years, an increasing interest for new products obtainable as food or feed from seeds and for phyto-based cosmetics from inflorescence is emerging [4].

Hemp is a crop with fast growth, high biomass production at low inputs (fertilisers/pesticides), good CO_2 capture per hectare (about 2.5 t/ha), and soil protection due to the length of its roots, suitable for many industrial processes [5,6]. Appropriate soils for hemp are deep, show pH between 6.0 and 7.5, and have a good availability of nutrients and water holding capacity [7]. Moreover, hemp requires proper preparation of the seedbed, especially on clay soils, for a homogenous emergence due to its particular sensitivity to waterlogging. Sandy soils are less suitable for this cultivation, because of its poor water holding capacity determining greater water requirements [8]. It depends on climatic conditions. Indeed, in the South Mediterranean environment, higher irrigation volumes are required, with respect to the North Mediterranean one [9,10], but hemp water requirements are lower [11] compared to other specialized and common crops, such as maize, which are also cultivated for biogas production in Europe.

Industrial hemp cultivation is growing over time. A great increase was recorded from 2013 to 2017 in Europe [2], because of the introduction of policies and local incentives to the hemp industry [12].

As a result of the Italian Regulation [13], industrial hemp cultivation and processing assumed an increased national relevance. The regulation supports (also by including economic incentives) and promotes the development of integrated supply chains valuing research findings and pursuing local integration, as well as effective environmental and economic sustainability.

During the first decades of the 20th century, Italy was one of the most important producers on a global scale. In 1940, cultivated areas exceeded 100,000 ha, corresponding to more than 80,000 tons of hemp fibers [14].

The extension of the cultivated areas in Italy from 1961 to 2017 are reported in Figure 1 (sources of data: [15–17]).

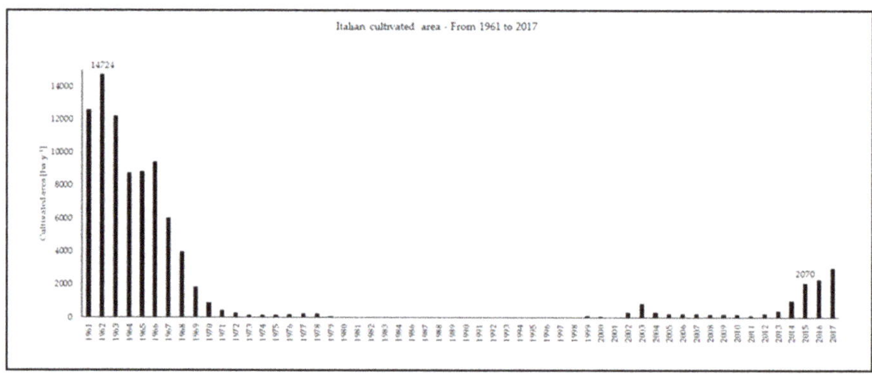

Figure 1. Hemp harvested areas (hectares by year) in Italy, from 1961 to 2017 (source of data: [15–17]).

Cultivated areas were significant during the 1960s and 1970s, and cultivations stopped in the 1980s and 1990s, mainly due to strict policies and regulations against the use of narcotic and psychotropic drugs. From 1999 to 2018, a new interest in hemp cultivation was developed, supported by national and European funding (the Italian Ministry of Agricultural Food and Forestry Policies financed a project to promote industrial hemp supply chains; during the same period, the European Union funded 3-year projects to reintroduce this feedstock for multiple purposes and to differentiate crops).

Due to the global resurgence of hemp cultivation needed to meet the requirements of the hemp sectors widespread today (building construction, food/animal feed, pharmaceutical, paper, textile, etc.), the recovery of hemp fiber and hurd residues should be addressed.

In this regard, some research and development projects funded by the European Union, such as MultiHemp [18] and GRACE [19], were already developed to demonstrate the sustainability of hemp-derived products according to the biorefinery concept.

However, another possible recycling path, aimed at increasing the economic and environmental benefits in a circular economy perspective, is the conversion of the agro-industry by-products into energy carriers.

Despite its thousands of uses, hemp by-products' potential as an energy feedstock is yet to be examined in depth. To date, few works have identified industrial hemp as an energy crop (for instance, a potential energy crop to produce bioenergy in [20]; ethanol production in [21,22]; methyl ester production in [23]; pyrolysis feedstock in [24]; biomass for thermochemical processes in [5]; combustion in [25,26]; co-firing in coal and peat power stations in [27]; and gasification or co-firing in [28]).

A few scientific works related to industrial hemp as a potential energy crop for biogas production can be found in the literature, going from 1990 to 2014. Rehman et al. [29] give an overall perspective of using hemp as a bioenergy crop in Pakistan, including biogas production.

Kreuger et al. [30], Heiermann et al. [31], Adamovics et al. [32], Mallik et al. [33] and Kaiser [34] provided data from anaerobic digestion trials carried out on a laboratory scale (a few liters-capacity reactors) (Table 1). As a pre-treatment, hemp was ground to a few mm or powder size. Since the grinding size influences the digestion kinetics [35] and, more generally, particle size affects the hydrolytic phase of the biodegradation of lignocellulosic feedstock [36], this factor should be considered when analyzing and comparing studies based on a laboratory scale. However, fine grinding is not reasonably achievable and economically affordable in industrial applications processing huge amounts of biomass.

Table 1. Main experimental conditions and information related to the scientific works found in the literature on anaerobic digestion of hemp, based on a laboratory scale.

Experiment	Hemp Cultivar	Country	Thermal Conditions	HRT	Specific Biogas Yield	Specific Methane Yield	Methane Content
[30]	Futura75	Sweden	50 °C	30 days	-	234 ± 35 m$^3 \cdot$t^{-1} VS (mean $\pm$ std.dev. [1])	-
[31]	Fedora19	Germany	35 °C	35 days	$453 \div 567$ L$_N \cdot$kg^{-1} VS	$259 \div 301$ L$_N \cdot$kg^{-1} VS [2]	$53 \div 57$ (%vol)
[32]	Futura75 (among other cultivars)	Latvia	38 ± 1 °C	53 days	$0.357 \div 0.370$ L$\cdot$g^{-1} VS (coarse particles); $0.470 \div 0.530$ L$\cdot$g^{-1} VS (fine particles)	$0.172 \div 0.185$ L$\cdot$g^{-1} VS (coarse particles); $0.240 \div 0.270$ L$\cdot$g^{-1} VS (fine particles)	-

[1] std.dev. means "standard deviation"; [2] N means "normal", VS means "volatile solids".

For the present work, the study of Mallik et al. [33] was excluded from the comparison reported in Table 1, because of a lack of information about biogas production/methane production performance. In [34], experiments conducted in a batch digester were presented, where industrial hemp was co-digested with other vegetable wastes and poultry litter. These three types of biomass, fed to 10-L reactors, had the same size (2 cm), which did not allow the authors to consider the relationships between chemical composition and size, and how they influenced anaerobic digestion. It was difficult to assign biogas/methane production performances to each biomass making up the admixtures (with attention to hemp) to make comparisons with the results obtained in the present study.

The present work goes beyond the past research approaches to anaerobic digestion of hemp. It focuses on assessing actual opportunities of a large-scale use of industrial hemp straw residues. Indeed, in the south of Italy, hemp is primarily cultivated for seed production while hemp straw residues are ordinarily left in the field, due to their scarce economic value as well as to the limited industrial interest and knowledge about this by-product. This study aimed to provide a more comprehensive knowledge of this residual lignocellulosic biomass.

The screening carried out on hemp straw residues for biogas generation completes the major gaps identified in the related state-of-the-art, by offering in-depth knowledge of the effective performance of *Cannabis sativa* L. residues in biogas production.

This work considers an alternative use of hemp straw residues with respect to the already developed market sectors (see, for instance, [29]) and, consequently, it suggests new market opportunities for hemp-derived products. The outcomes show the effective potential of developing a new supply chain, based on an emerging lignocellulosic crop for biofuel production. These aspects will be economically relevant both for farmers and contractors in biogas/biomethane sectors.

As pointed out by [37], lignocellulosic crops are not a common source of biomass for biogas production. The authors emphasize that the most significant constraints to hemicellulose/cellulose digestion are related to the lignin content, crystallinity of cellulose, and particle size. These limits may be reduced through optimization of the methodologies and technologies supporting biogas and subsequent biomethane production, for more sustainable use of crop residues for energy purposes. Among other techniques, the use of specific enzyme systems should be considered to reduce the lignocellulose's recalcitrance to anaerobic digestion. An in-depth presentation of the chemical and biological mechanisms between recalcitrant biomass and enzymes was provided by [38].

As reported by [39], commercial bioenhancers are not thoroughly characterized, but the positive results provided by the recent literature (+30% increase circa, as reported by [40]) related to biogas production from biomass with a complex lignocellulosic structure stimulate further applications and studies.

This work includes treatment with a commercial preparation of bioenhancers developed to improve biogas production through anaerobic digestion of cellulose and hemicellulose in lignocellulosic crops, like *Cannabis sativa* L. residues, to contribute to an advance in the field.

Additionally, by assessing the current state of industrial hemp usage and deployment, it emerged that a synergistic approach along the entire supply chain should be adopted, by integrating high-value components of hemp and other parts of the plant into a well-designed biorefinery, in order to support the local economy in a more sustainable way.

2. Materials and Methods

In 2017, an experimental hemp crop on the *Cannabis sativa* L. cultivar "Futura75" was carried out at San Giovanni Suergiu (pilot site located in the south-western side of the Sardinian Island, Italy). This trial is part of the CANOPAES project (the Italian acronym for "CANapa: OPportunità Ambientali ed Economiche in Sardegna", focused on the environmental and economic opportunities of hemp in the Sardinian Island).

Futura75 was chosen for its diffusion in Europe and its ability to produce both seeds and biomass.

According to the available long-term data, the San Giovanni Suergiu's climate is typically the Mediterranean. During the crop cycle, the thermopluviometric trend was characterized by maximum temperatures above the average, and rainfall was equal to about one-third of the seasonal average. Hemp was sown at a density of 120 plants·m^{-2} and at a depth of 0.02 m.

In addition, 60 kg·ha^{-1} as urea were top-dressed at about one month after emergence. Irrigation was performed by sprinklers with 75% ETm (maximum evapotranspiration) restitution and no weed control was required. Hempseeds were harvested by an ordinary combine. After that, the by-product straw naturally dried on the field (moisture <15% on a wet basis). Then, straw was raked and baled for transportation to the pilot plant. The green biomass yield was about 20 t·ha^{-1} while naturally dried straw was about 3.7 t·ha^{-1}.

Based on the assumption that different uses of hemp straw can coexist, though the specific features of the local market drive the types of use (as stated by [6] and [7], the dual-purpose oil-fiber of *Cannabis sativa* L. is dominant in the European territory), this work assumed a hypothetical scenario made of a dual-purpose supply chain: Hempseeds were harvested by a combine, to be used for oil extraction, while residues were processed for energy carrier generation (specifically, biogas).

Then, single-step digestion was performed in the pilot plant described below. The duration of the experiment was of 423 days (from March 2018 to June 2019).

2.1. Feedstock Characterization and Pre-Treatment, Admixture Preparation, and Pilot Plant

Since the chemical composition and physical characteristics (e.g., moisture content M) were used to define the admixtures proportion, to manage the process stability and to optimize anaerobic digestion, proximate analysis and ultimate analysis of hemp straw residues were performed.

Samples were prepared by drying hemp straw at 105 ± 2 °C in a thermostatic oven (Memmert GmbH, Schwabach, Germany), and by shredding and mixing the material through a cutter.

The proximate analysis was conducted using a thermogravimetric analyzer (TGA701, LECO Corporation, St. Joseph, MI, USA) following [41], to determine the moisture content (M), volatile solids on a dry basis (VS$_{d.b.}$), ash content, and fixed carbon (FC) (reported as percentage by mass [%wt]).

Total carbon, hydrogen, total nitrogen, and sulphur were determined by conducting the ultimate analysis through a CHNS analyzer (Truspec, LECO Corporation, St. Joseph, MI, USA) in accordance with [42].

Fiber composition (ADF: Acid detergent fiber, NDF: Neutral detergent fiber, ADL: Acid detergent lignin) of the lignocellulosic feedstock was used to determine the daily intake of enzymes (see Section 2.2). Values were obtained by using a fiber analyzer ANKOM 2000 (ANKOM, Macedon NY, USA), following the Van Sœst methodology [43–46]. Concerning the hemicellulose and cellulose contents, those values were estimated by subtracting ADF from NDF and ADL from ADF [47,48].

Chemical and physical characteristics (with their standard deviations) of hemp residues are listed in Tables 2–4.

Table 2. Proximate analysis of hemp residues. The cultivar "Futura 75".

Proximate Analysis	
[%wt]	
M^1$_{d.b.}$	7.71 ± 0.01
VS^2$_{d.b.}$	81.37 ± 0.08
Ash$_{d.b.}$	2.50 ± 0.25
FC^3$_{d.b.}$	16.13 ± 0.35

[1] M means "moisture"; [2] VS means "Volatile Solids"; [3] FC means "Fixed Carbon".

Table 3. Ultimate analysis of hemp residues. The cultivar "Futura 75".

Ultimate Analysis	
[%wt]	
Carbon$_{d.b.}$	47.41 ± 0.04
Hydrogen$_{d.b.}$	6.52 ± 0.10
Nitrogen$_{d.b.}$	1.64 ± 0.02
Sulphur$_{d.b.}$	0.18 ± 0.00

Table 4. Fiber composition (mean values, [% dry matter]) of hemp residues. The cultivar "Futura 75".

Chopped Hemp, Reproductive Stage		
ADL [1] [%wt]	NDF [2] [%wt]	ADF [3] [%wt]
7.87	59.16	44.40

[1] ADL means "Acid detergent lignin"; [2] NDF means "neutral detergent fiber; [3] ADF means "acid detergent fiber"

The feedstock characterization did not include parameters, such as starch and sugar contents, because of the composition of hemp straw residues mainly characterized by the lignocellulosic structure.

The anaerobic digester used in the present work was a tubular, horizontal reactor of 1.13 m^3 total volume. It is 2.25 m long and its external diameter is 779 mm. It was radially mixed using a mechanical stirrer. The reactor was fed via a pneumatic pump, conveying the substrate previously introduced into a 250-kg-capacity feeding hopper. The reactor was tested by filling it with 960 L of digestate (corresponding to about 85% of the total volume).

The digestate produced during the process was discharged into a 200-kg-capacity tank, by using a pneumatic pump. The reactor was heated by three electrical resistances located in its center, loading, and discharging sides.

Sampling operations for the reactor sludge were performed using two valves located in the loading and discharging sides of the reactor.

Operations and parameters settings were managed and controlled by a programmable logic controller (PLC).

The feedstock pre-treatment consisted in mechanical milling, by shredding hemp straw residues using a 20-L-capacity cutter (dry cut). Then, coarse particles (maximum size: 1 cm) were mixed with the recirculated digestate in a 40-L-capacity cutter. When necessary, different amounts of water were added.

The operative settings were changed during the experimental period to investigate different process conditions (see Section 2.2).

2.2. Feeding Phases

The reactor was filled with digestate coming from an anaerobic digestion industrial plant treating corn silage and triticale. The digestate was used as received from the industrial plant. Then, a start-up phase was performed. During this phase, the temperature was increased by 2 °C daily until a constant value of 39 °C (mesophilic conditions) was reached.

Subsequently, the daily feeding rate of admixtures (Q, $[m^3_{substrate} \cdot d^{-1}]$) was increased during the first phase to reach an adaptation of the bacterial consortium to the specific substrate. After the start-up phase, the hemp to digestate proportion (hereinafter: percentage of new hemp straw in the admixture C, [% wt/wt]), Q, and digestate recirculation ratio (R, adimensional) were changed during the experimental period.

C and R values were chosen keeping in account the fluid-dynamics behavior of the admixtures. An increase of the hemp straw amount, indeed, could make the pumping of the admixtures itself difficult.

Two reference values of C (C1: 3% wt/wt; C2: 5% wt/wt) were set to evaluate the process. These conditions were tested by considering the presence/absence of specific bioenhancers. Treatments were randomly applied during the entire experiment.

Concerning the ranges of the organic loading rate (OLR, $[kg_{VS} \cdot (m^3_{reactor} \cdot d)^{-1}]$), hydraulic retention time (HRT, [d], dependent on Q), R, and C, different regimes were defined.

The abovementioned variables were analyzed along with the specific gas production (SGP, $[m^3_{biogas\ or\ methane} \cdot kg_{VS}^{-1}]$) and the gas production rate (GPR, $[m^3_{biogas\ or\ methane} \cdot (m^3_{reactor} \cdot d)^{-1}]$).

A commercial enzymatic preparation (Micropan Biogas ®from Eurovix, IT) was applied to reduce the current supply of hemp straw residues and to maximize biogas production at the same time. It is made of microbial enzymes containing cellulase, lipase, xylanase, active principles of *Fucus Laminariae*, algae *Lithothamnium calcareum*, natural nutrients/grow factors, selected yeast, mineral biological catalysts rich in oligo elements, and selected microorganisms (facultative anaerobic bacteria, such as: *Bacillus subtilis*, *Bacillus macerans*; strictly anaerobic bacteria genus *Methanobacterium*). The specific gravity was 0.8 t/m^3.

The daily intake was introduced into the reactor by dissolving the powder into water (1:4 wt/v).

The dosage was divided into two parts, depending on the fiber composition of lignocellulosic feedstock (Table 4), C, and Q. The daily intake was defined by multiplying the percentage of cellulose and hemicellulose of hemp residues (see Table 4) with the hemp mass in Q and a coefficient of 0.05, as suggested by the producers. Thus, 20 g per day were obtained.

The first enzymatic inoculum of the reactor sludge was calculated by dividing the working volume of the reactor by Q, and by multiplying this value with the daily intake (500 g of bioenhancers were introduced into the reactor).

2.3. Management of the Reactor and Process Stability

Both the process stability and reactor features were controlled and managed by using a set of parameters.

Management of the reactor: As regards the reactor, the following were considered:

- TS (total solids), $VS_{d.b.}$, determined via proximate analysis on a weekly/sub-weekly basis for the new admixture introduced into the feeding hopper, the material in the hopper/in the digestate tank, and the sludge inside the reactor;
- HRT [d], calculated as:

$$HRT = \frac{V}{Q},\qquad(1)$$

where V is the total digester volume [$m^3_{reactor}$] and Q is the daily feeding rate [$m^3_{substrate} \cdot d^{-1}$];
- OLR ([$kg_{VS} \cdot (m^3_{reactor} \cdot d)^{-1}$]), calculated as:

$$OLR = \frac{V \cdot S}{Q},\qquad(2)$$

where S is the VS concentration on a wet basis in the feeding admixtures [$\%wt_{w.b.}$].

Management of process stability: The process stability was monitored by considering the below listed parameters:

- pH and FOS/TAC ratio (volatile fatty acids content/buffer capacity) of the reactor sludge (daily measures by means of, respectively, a multi-parametric analyzer Orion Versa Star (ThermoScientific Inc., Waltham, MA, USA) and an automatic titrator T70 (Mettler Toledo International Inc., Columbus, OH, USA));
- Biogas production [$m^3_{biogas} \cdot d^{-1}$] (daily values provided by a biogas flow meter);
- Biogas composition daily (CH_4, CO_2, O_2 [%wt]; NH_3, and H_2S [ppm]), determined using a portable gas analyzer GA2000 (Geotechnical Instruments UK Ltd., Coventry, UK). Biogas composition is strictly related to SGP and GPR;
- Temperature of the reactor sludge, measured through three temperature probes located in the center, the loading, and discharging sides of the reactor, and monitored through the PLC.

Performance parameters: The two main performance parameters considered in the anaerobic digestion trials carried out on hemp straw residues are:

- SGP [$Nm^3_{biogas\ or\ methane} \cdot kg_{VS}^{-1}$], calculated as:

$$SGP = \frac{G}{Q \cdot S},\qquad(3)$$

where Q and S were already described, G is the daily production of biogas/methane [$m^3_{biogas\ or\ methane} \cdot d^{-1}$];
- GPR ([$Nm^3_{biogas\ or\ methane} \cdot (m^3_{reactor} \cdot d)^{-1}$]), calculated as the daily production of biogas/methane per m^3 of sludge accumulated in the reactor.

The abovementioned parameters are related to:

- C (percentage of new hemp in the admixture, [% wt/wt]), calculated as:

$$C = \frac{Mass_{hemp}}{Mass_{hemp} + Mass_{digestate} + Mass_{water}} \cdot 100,\qquad(4)$$

where: $Mass_{hemp}$, $Mass_{digestate}$, and $Mass_{water}$ are the mass of the hemp, digestate, and water composing the admixtures;

- R (digestate recirculation ratio, adimentional), following Equation (5):

$$R = \frac{\Sigma_i \, Mass_{digestate}}{\Sigma_i \, Mass_{hemp}}, \tag{5}$$

where $\Sigma_i Mass_{hemp}$ and $\Sigma_i Mass_{digestate}$ [g·10^3] are the cumulative amounts of new hemp or digestate composing hemp-digestate admixtures in a specific time window.

2.4. Statistical Analyses

Statistical analyses were performed on biogas composition, SGP, and GPR using Statgraphic Centurion XVI [49]. The mean, standard deviation, maximum and minimum values, and median were calculated for each feeding phase composing the experimental period. Also, the skewness and kurtosis were calculated to determine which kind of statistical analysis should be performed.

Depending on this first data analysis and with a specific focus on the skewness and kurtosis, any evenness emerged among variances. Consequently, non-parametric tests were applied.

Two non-parametric tests were considered: Mann–Whitney [50,51] and Kruskal–Wallis [52]. The former was used to determine if two ordinal and independent random samples (feeding phases) were part of the same population. The latter was useful to compare median values of the different groups to identify whether they belonged to a population characterized by the same median.

Statistical analyses were developed by considering $p < 0.05$. Graphical representations were provided by using box-and-whisker diagrams.

3. Results and Discussion

3.1. Feeding Phases

Based on the outcomes of the experiment, it was divided into phases. The main characteristics of the feeding phases, except for the start-up phase, are reported in Table 5. Indeed, the start-up phase was characterized by high instability of the main process parameters as well as by a rapid variation of admixture feeding rates over time. Thus, the remaining feeding phases were labeled from 1 to 7.

Table 5. Main parameters (with their standard deviations) of anaerobic digestion trials on hemp residues.

Phase [-]	Description [-]	Duration [d]	OLR [1] [kg$_{VS}$·m^{-3}·d^{-1}]	HRT [2] [d]	C [3] [% wt/wt]	R [4] [-]
1	No enzymatic treatment	45 (day 98–day 143)	2.8 ± 0.6	29 ± 2	2.3 ± 0.0	16.3 ± 1.1
2	No enzymatic treatment	27 (day 144–day 171)	1.3 ± 0.2	34 ± 4	3.0 ± 0.9	18.9 ± 0.4
3	No enzymatic treatment	55 (day 172–day 227)	2.9 ± 0.8	34 ± 7	2.9 ± 1.8	20.7 ± 0.8
4	Enzymatic treatment	34 (day 228–day 262)	3.8 ± 0.8	31 ± 7	2.5 ± 1.7	22.0 ± 0.3
5	Enzymatic treatment	35 (day 263–day 298)	3.2 ± 0.8	30 ± 6	5.1 ± 0.2	22.0 ± 0.4
6	No enzymatic treatment	36 (day 299–day 335)	3.1 ± 0.9	33 ± 9	5.2 ± 1.0	20.9 ± 0.3
7	Enzymatic treatment	87 (day 336–day 423)	3.1 ± 1.0	29 ± 3	4.4 ± 2.0	20.3 ± 0.2

[1] OLR means "organic loading rate"; [2] HRT means "hydraulic retention time"; [3] C means "percentage of new hemp in the admixture"; [4] R means "digestate recirculation ratio".

Feeding phases were classified according to enzymatic treatment, OLR, HRT, C, and R (Table 5).

The two C reference values applied in this study are close to the organic loadings used in the industrial plants of anaerobic digestion.

3.2. Management of the Reactor and Process Stability

Management of the reactor -TS and $VS_{d.b.}$ trends of the material in the feeding hopper, in the reactor sludge, and the digestate tank are reported in Figures 2 and 3.

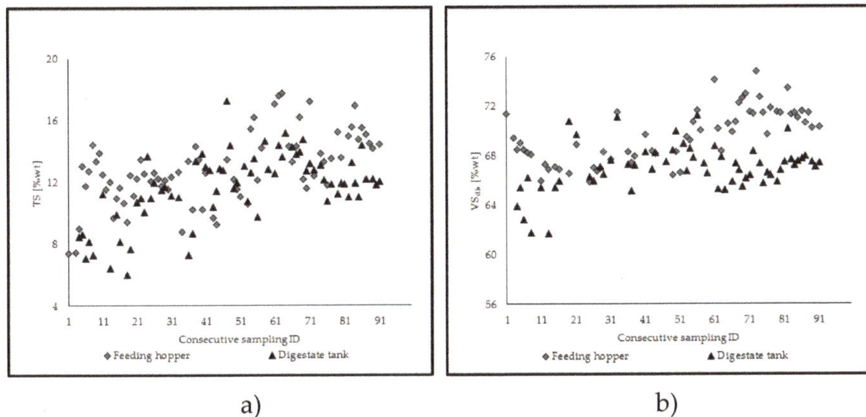

a)   b)

Figure 2. TS (total solids) (**a**) and $VS_{d.b.}$ (volatile solids on a dry basis); (**b**) trends of the material in the feeding hopper and the digestate tank, concerning consecutive sampling.

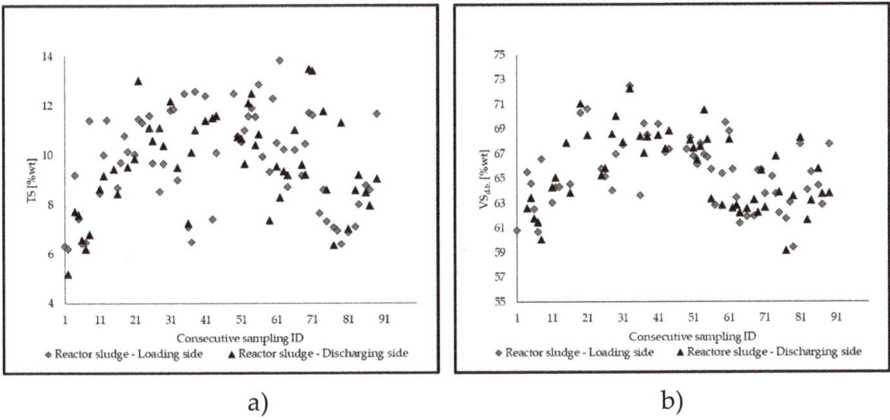

a)   b)

Figure 3. TS (total solids) (**a**) and $VS_{d.b.}$ (volatile solids on a dry basis); (**b**) trends of the reactor sludge, concerning consecutive sampling.

By comparing the TS and $VS_{d.b.}$ trends in the feeding hopper and digestate, it can be seen that from the 70th sampling, these parameters were notably lower in the discharged sludge than in the corresponding feeding mixtures. The 70th sampling corresponds to the beginning of the 5th feeding phase and this occurrence emerged from the observations related to the 6th and 7th phases as well. In the previous feeding phases, a distinction between the TS and $VS_{d.b.}$ evolution in the fed slurry and the corresponding digestate cannot be seen. This result is related to the higher reference value of C (C2), which was better than the other one (C1).

With regard to Figure 3, the reactor sludge did not show relevant differences in terms of TS and $VS_{d.b.}$ by comparing the loading side to the discharging side, mainly due to a certain mixing of the sludge along the longitudinal section of the reactor. TS and $VS_{d.b.}$ seemed to increase from phase "1" to phase "4" and to decrease in the subsequent phases (characterized by the reference value, C2).

Management of process stability: Concerning the main design and operation process parameters of the reactor, the trends of HRT and OLR are shown in Figures 4 and 5.

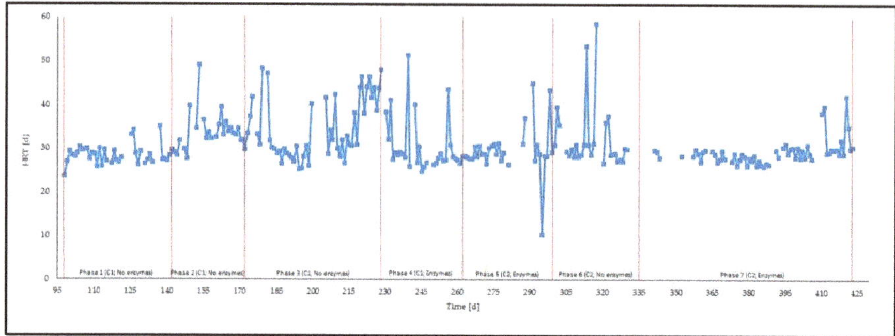

Figure 4. Hydraulic retention time (HRT) trend.

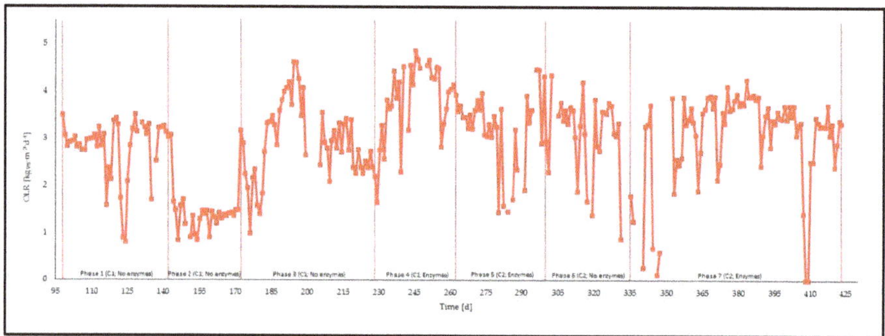

Figure 5. Organic loading rate (OLR) trend.

HRT did not show significant variations during the entire experiment (Figure 4) and it was excluded from the statistical analysis.

OLR (Figure 5) was monitored through the daily determination of TS and $VS_{d.b.}$ in the feeding admixture and adjusted by setting the hemp share in the feeding admixture and the loading flow rate (Figure 2).

Regarding the process stability parameters, the trends of FOS/TAC and pH, biogas production, and biogas composition (CH_4, CO_2, NH_3, H_2S) are reported in Figures 6–9.

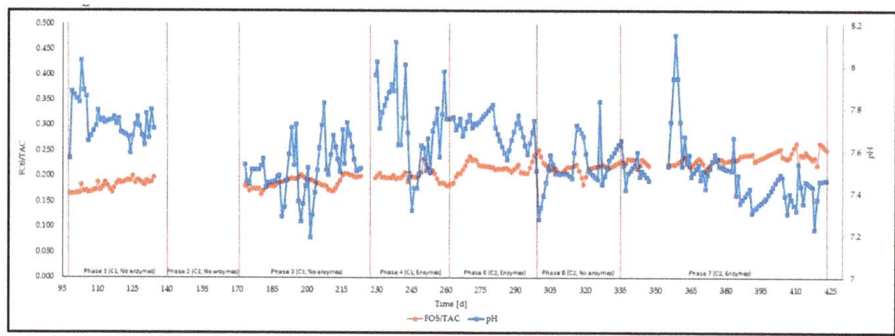

Figure 6. FOS/TAC (volatile fatty acids/buffer capacity) and pH trends.

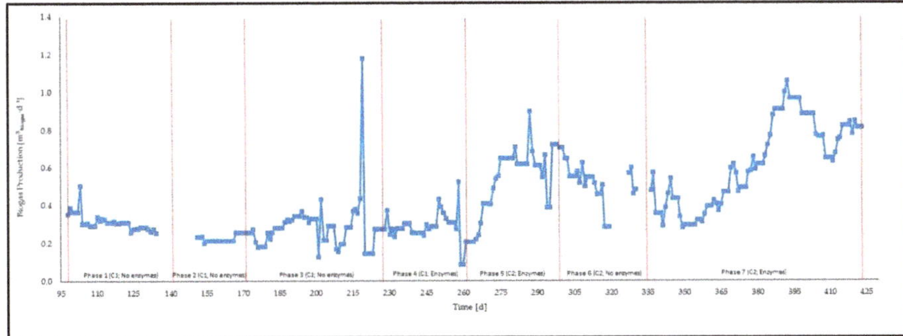

Figure 7. Biogas production trends (GPR) (via biogas metering).

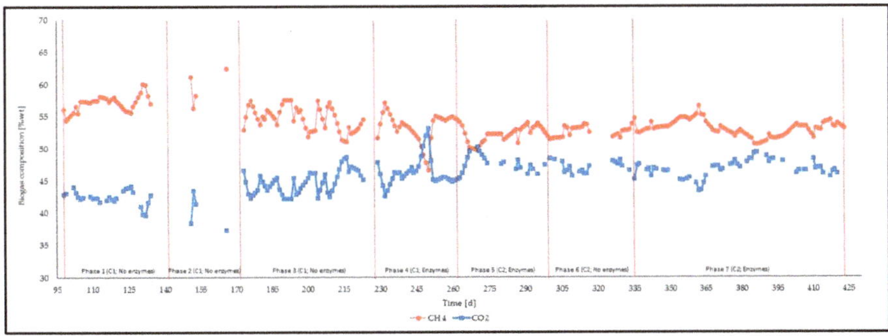

Figure 8. Biogas composition (CH_4 and CO_2, [%wt]) detected by the portable gas analyzer during the experimental period.

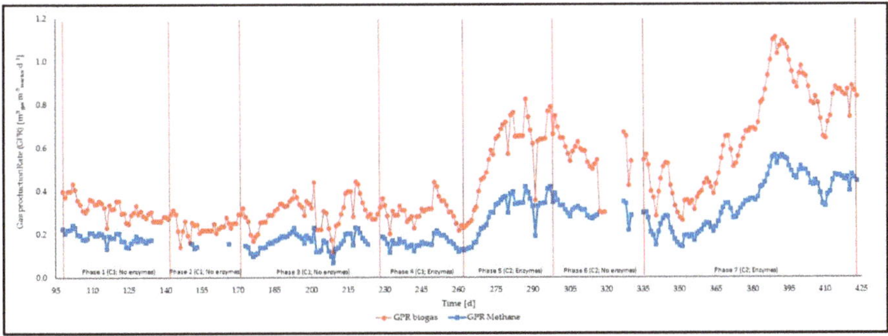

Figure 9. Gas production rate (GPR) trends of biogas and methane.

By considering the entire experimental campaign, pH varied between 7.2 and 8.0, accompanied by higher values during the first phase.

The FOS/TAC ratio increased over time, from about 0.170 to values close to 0.270. These values are very similar to the typical ones of the industrial plants treating lignocellulosic biomass, such as corn silage.

Overall, the FOS/TAC trend is consistent with the increase in the percentage of new hemp in the feeding admixtures (C2 treatment) (see Table 5) from the fifth feeding phase to the seventh one. It can be supposed that the introduction of higher amounts of hemp provoked a shift of the anaerobic digestion microbial dynamics towards the predominance of acidogenic reactions.

Daily biogas production (Figure 7) was characterized by high variability during the experimental campaign. It showed a significant increase of the biogas produced during the experimental phases from "5" to "7" (C2), accompanied by a rapid increase during the fifth phase and a decrease over the sixth one (without enzymes). The last feeding phase showed rising values for most of the days (to reach about 1.1 m^3 biogas per day), corresponding to the coupling of higher C and enzymatic treatment. It can be considered as the most suitable among all the treatments applied. On average, the CH$_4$ content [%wt] in the biogas produced during the entire experimental campaign was 53.8 ± 2.2. CO$_2$ content [%wt] was 45.6 ± 2.4. CH$_4$ concentration showed a slow reduction from the phase "1" to the phase "4". From the phase "5", a gradual increase of its concentration was detected. CH$_4$ and CO$_2$ trends (Figure 8) did not show any peak attributed to organic overload.

Performance parameters: The performance parameters, SGP and GPR, of the anaerobic digestion trials are shown in Figures 9 and 10.

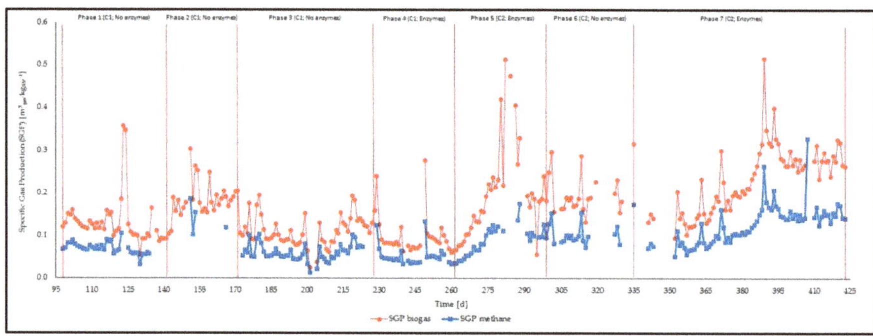

Figure 10. Specific gas production (SGP) trends of biogas and methane.

SGP of biogas/methane is the energy yield of an anaerobic digestion system regardless of OLR, thus it plays an important role when assessing the energy yield of the considered process.

Both trends showed an increase starting in the fifth feeding phase. It was more evident during the seventh period.

By comparing these outcomes with those obtained on a laboratory scale [30,32], it can be pointed out that the present work provided higher SGP values both for biogas and methane than the previous literature on anaerobic digestion trials of hemp straw. Overall, the management of the reactor and the process stability was enhanced with respect to biogas production, especially in the last part of the experiment, characterized by the combination of higher values of C and the application of enzymes.

SGPs of the present work are lower compared to those obtained in the same pilot plant, using vegetable feedstock characterized by high degradability (lower lignin contents and higher starch content). Due to the lignocellulosic nature of this crop, the specific biogas/methane production of hemp straw was lower than those related to raw potatoes (0.68 Nm^3 of biogas ·kgVS^{-1} and 0.37 Nm^3 of methane ·kgVS^{-1}) [53], potato chips (0.81 Nm^3 of biogas ·kgVS^{-1} and 0.47 Nm^3 of methane ·kgVS^{-1}) [54], and fruit and vegetable wastes (0.78 Nm^3 of biogas ·kgVS^{-1} and 0.43 Nm^3 of methane ·kgVS^{-1}) [55].

Experiments carried out on the same pilot digester using admixtures made of different kinds of feedstock (admixture composition: 30%wt of shredded corn, the remaining part consisting in whey, vegetable water, pomace pitted, and manure to maintain the OLR between 2.5 and 3.5 kgVS·m^3$_{reactor}$·d^{-1}) reported SGP$_{biogas}$ from 0.623 ± 0.212 Nm^3·kgVS^{-1} to 0.768 ± 0.227 Nm^3·kgVS^{-1} and SGP$_{methane}$ from 0.281 ± 0.160 Nm^3·kgVS^{-1} to 0.438 ± 0.096 Nm^3·kgVS^{-1} [56]. Also, hemp straw residues' performance in anaerobic digestion was found to be similar to other lignocellulosic crops [57], the specific methane production of which is between 0.17 and 0.39 Nm^3·kgVS^{-1}. Results are comparable to those reported by [58] for other agricultural crops, such as oats, flax, and sorghum, but lower than the hemp energy yields considered there.

Thus, hemp straw residues used in the same pilot plant showed lower SGP of biogas and methane probably because of the lignocellulosic composition. The results obtained by the residues considered in this work could be affected by the harvesting time, which is the reproductive stage when the lignin content is higher than in the vegetative stage. By considering that the reproductive stage is the core of the supply chain scenario assumed in this study, related to the extraction of oil from seeds, the results already discussed about the potential of biogas production from straw residues suggest the recovery of this low-value by-product to energy conversion. By considering the ultimate analysis, the carbon:nitrogen ratio is useful to define the biomass suitability for biochemical (ratio <30) or thermochemical processes (ratio >30). The ratio of hemp straw residues is 28.9 (Table 3), thus it can be considered for both, but with slightly higher suitability for biochemical conversion.

3.3. Statistical Analysis

Results of the data analysis (mean, standard deviation, maximum and minimum, skewness, kurtosis, and median values) carried out on the most significant process parameters and outcomes are reported in Tables 6 and 7.

Table 6. Main statistical variables of the biogas composition and biogas and methane gas production rate (GPR) and specific gas production (SGP).

Process Parameter	Variable	Unit	Feeding Phase	No. of Values	Mean ± Std.dev.	Minimum	Maximum
Biogas composition	CH_4	[%wt]	1	37	57.1 ± 1.2	54.4	60.0
			2	4	59.5 ± 2.8	56.3	62.4
			3	51	54.7 ± 1.9	51.0	57.6
			4	34	53.3 ± 2.3	46.6	57.1
			5	36	52.1 ± 1.2	49.7	54.1
			6	28	52.6 ± 0.9	51.4	54.7
			7	88	53.0 ± 1.2	50.5	56.6
	CO_2	[%wt]	1	27	42.4 ± 1.1	39.6	44.2
			2	3	41.1 ± 2.6	38.4	43.5
			3	51	44.7 ± 1.8	42.2	48.7
			4	34	46.2 ± 2.2	42.6	53.1
			5	20	47.7 ± 1.3	45.4	50.3
			6	19	47.0 ± 1.0	45.2	48.5
			7	52	46.8 ± 1.3	43.4	49.4
GPR	GPR Biogas	$[Nm^3{\cdot}d^{-1}]$	1	46	0.317 ± 0.047	0.228	0.433
			2	28	0.232 ± 0.035	0.140	0.291
			3	56	0.299 ± 0.070	0.119	0.441
			4	35	0.301 ± 0.052	0.199	0.438
			5	36	0.581 ± 0.162	0.238	0.238
			6	26	0.552 ± 0.115	0.297	0.297
			7	88	0.668 ± 0.036	0.262	1.109
	GPR CH_4	$[Nm^3{\cdot}d^{-1}]$	1	37	0.186 ± 0.025	0.132	0.241
			2	4	0.147 ± 0.010	0.135	0.155
			3	51	0.164 ± 0.038	0.068	0.231
			4	34	0.160 ± 0.025	0.114	0.215
			5	36	0.303 ± 0.086	0.129	0.418
			6	23	0.307 ± 0.036	0.218	0.388
			7	88	0.353 ± 0.123	0.141	0.424
SGP	SGP Biogas	$[Nm^3{\cdot}kgVS^{-1}]$	1	45	0.129 ± 0.054	0.054	0.358
			2	27	0.191 ± 0.037	0.148	0.304
			3	54	0.110 ± 0.035	0.025	0.194
			4	33	0.097 ± 0.047	0.059	0.277
			5	32	0.207 ± 0.113	0.055	0.514
			6	23	0.198 ± 0.048	0.132	0.316
			7	75	0.250 ± 0.119	0.095	0.825
	SGP CH_4	$[Nm^3{\cdot}kgVS^{-1}]$	1	35	0.069 ± 0.013	0.032	0.105
			2	4	0.140 ± 0.037	0.103	0.186
			3	49	0.060 ± 0.019	0.013	0.104
			4	31	0.050 ± 0.023	0.033	0.133
			5	27	0.092 ± 0.033	0.041	0.174
			6	20	0.104 ± 0.027	0.071	0.173
			7	75	0.132 ± 0.062	0.052	0.439

Table 7. Skewness, kurtosis, and median values of the biogas composition, and biogas and methane gas production rate (GPR) and specific gas production (SGP).

Process Parameter	Variable	Unit	Feeding Phase	No. of Values	Skewness	Kurtosis	Median
Biogas composition	CH_4	[%wt]	1	37	0.041	0.571	57.3
			2	4	−0.167	−1.191	–
			3	51	−0.297	−1.401	44.7
			4	34	−3.061	2.19	53.8
			5	36	−1.043	−0.722	52.2
			6	28	0.773	−0.717	52.5
			7	88	0.543	0.326	53.1
	CO_2	[%wt]	1	27	−1.652	1.25	42.3
			2	3	−0.383	–	–
			3	51	0.934	−1.069	43.6
			4	34	3.808	3.712	45.7
			5	20	0.478	−0.446	48.2
			6	19	−0.257	−0.962	46.6
			7	52	−1.431	0.653	46.8
GPR	GPR Biogas	[Nm^3·d^{-1}]	1	46	1.231	−0.350	0.311
			2	28	−1.027	1.151	0.232
			3	56	−0.473	−0.079	0.295
			4	35	1.028	0.835	0.303
			5	36	−2.012	−0.395	0.639
			6	26	−2.098	0.995	0.582
			7	88	0.161	−2.276	0.674
	GPR CH_4	[Nm^3·d^{-1}]	1	37	−0.197	0.073	0.188
			2	4	−0.444	−1.202	–
			3	51	−0.761	−0.559	0.168
			4	34	0.588	−0.33	0.157
			5	36	−1.944	−0.643	0.339
			6	23	−0.172	1.138	0.310
			7	88	0.000	−2.399	0.351
SGP	SGP Biogas	[Nm^3·kgVS^{-1}]	1	45	8.717	16.069	0.118
			2	27	3.400	2.619	0.182
			3	54	1.225	0.748	0.104
			4	33	6.655	9.930	0.084
			5	32	2.940	1.566	0.193
			6	23	2.462	0.869	0.185
			7	75	8.592	16.019	0.248
	SGP CH_4	[Nm^3·kgVS^{-1}]	1	35	0.240	1.828	0.069
			2	4	0.375	−0.860	–
			3	49	1.195	0.743	0.054
			4	31	6.431	9.402	0.046
			5	27	0.570	0.032	0.080
			6	20	2.450	0.980	0.095
			7	75	9.078	17.583	0.130

As already mentioned in Section 2.4, skewness and kurtosis indices show that, except for SGP of biogas, the considered parameters can be attributed to a normal distribution, but variances of the two groups differ. Thus, comparisons were made on medians instead of mean values.

The outcomes of the Kruskal–Wallis test performed on the feeding phases "3", "4", "5", and "6", which are the most representative phases with respect to the treatments applied in this work (C1 and C2; presence/absence of enzymes), are shown in Figure 11.

The main outcomes of the two comparisons, "3" and "6", and "4" and "5", are described below.

Methane content [%wt] showed a tendency to lower median values when C is higher (C2) and higher values for the lower C regime (C1). This is confirmed by the Kruskal–Wallis test. Conversely, the behavior of the median value of the CO_2 content in biogas is opposite to the methane content, and it was higher for the higher C regime (C2). This is due to the addition of new raw material that modifies the reactions towards acidogenic conditions and promotes an increase of the CO_2 content in biogas with respect to the effect of the digestate recirculated, containing an amount of undigested substrate that is less reactive.

An increase of C led to an increase of GPR (median values: Phase "3": 0.295 Nm^3·d^{-1}; phase "4": 0.303 Nm^3·d^{-1}; phase "5": 0.639 Nm^3·d^{-1}; phase "6": 0.582 Nm^3·d^{-1}). When the enzymatic treatment was not applied (phase "3" and phase "6"), the higher C regime (C2) corresponded to an increase of the GPR_{biogas} median values of 92.1%. This was about 116.6% when bioenhancers were applied.

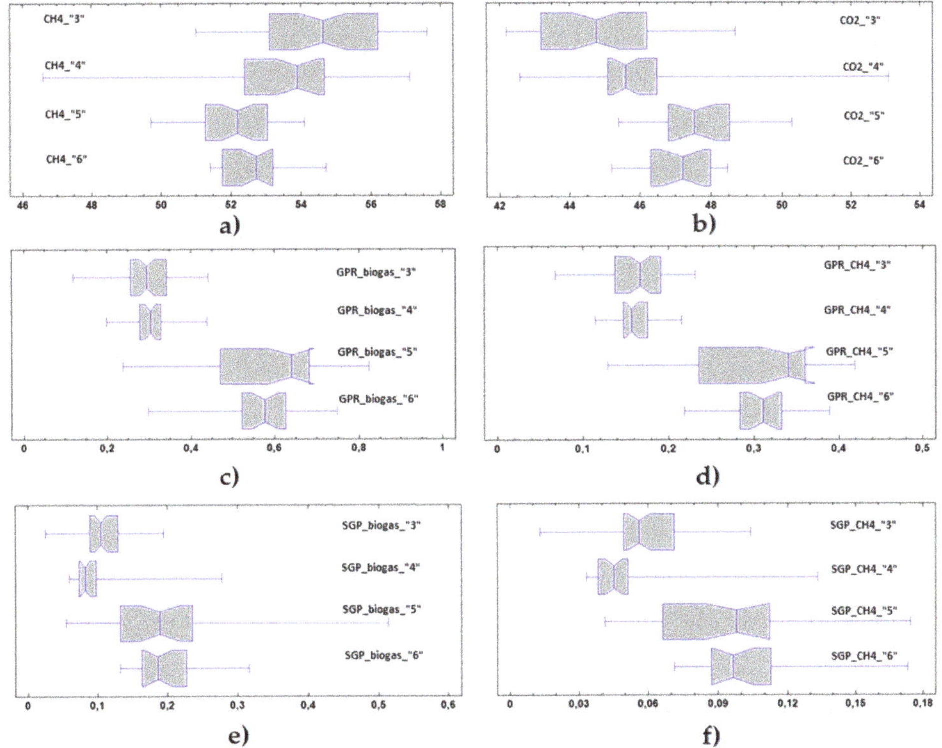

Figure 11. Box-and-whisker plots of the feeding-phases "3"–"6" and "4"–"5". Kruskal–Wallis test on the statistic variables: CH_4 content in biogas (**a**), CO_2 content in biogas (**b**), gas production rate (GPR) biogas (**c**), GPR methane (**d**), specific gas production (SGP) biogas (**e**) and SGP methane (**f**).

$GPR_{methane}$ showed similar trends observed for GPR_{biogas}: Higher values in the phases characterized by the C2 regime (median values: Phase "5": 0.339 $Nm^3 \cdot d^{-1}$; phase "6": 0.310 $Nm^3 \cdot d^{-1}$) than in those related to the C1 treatment (median values: Phase "3": 0.168 $Nm^3 \cdot d^{-1}$; phase "4": 0.157 $Nm^3 \cdot d^{-1}$). Hence, the increase due to the higher C regime was, respectively, +115.9% and +84.5% with and without enzymes.

With regard to the SGP of biogas/methane, an increase of C values led to higher energy yields.

The increase of SGP_{biogas} due to the increment of C was +77.9% without enzymatic treatment and +129.8% with enzymatic treatment. Thus, the coupling of a higher C regime with the addition of enzymes allowed the best management of the pilot plant to be obtained.

Essentially, similar behavior was observed for $SGP_{methane}$: The increasing of C promoted $SGP_{methane}$ (+165.9%) when enzymes were not applied. The increase associated with the enzymatic treatment was +73.9.

Statistically significant differences were found between phases "5" and "6" and phases "3" and "4", for all the parameters considered in the statistical analysis. Thus, it is reasonable to assert that variations of C influence all the parameters contributing to energy yields (SGP, GPR). The enzymatic treatment, instead, showed statistically significant differences in SGP in phases characterized by lower C values ("3" and "4").

The results of the Mann–Whitney test on the feeding phases "6" and "7", performed in order to assess the effect of the enzymatic treatment when the biodigester is managed by applying higher values of C, are reported in the box-and-whisker plot of Figure 12.

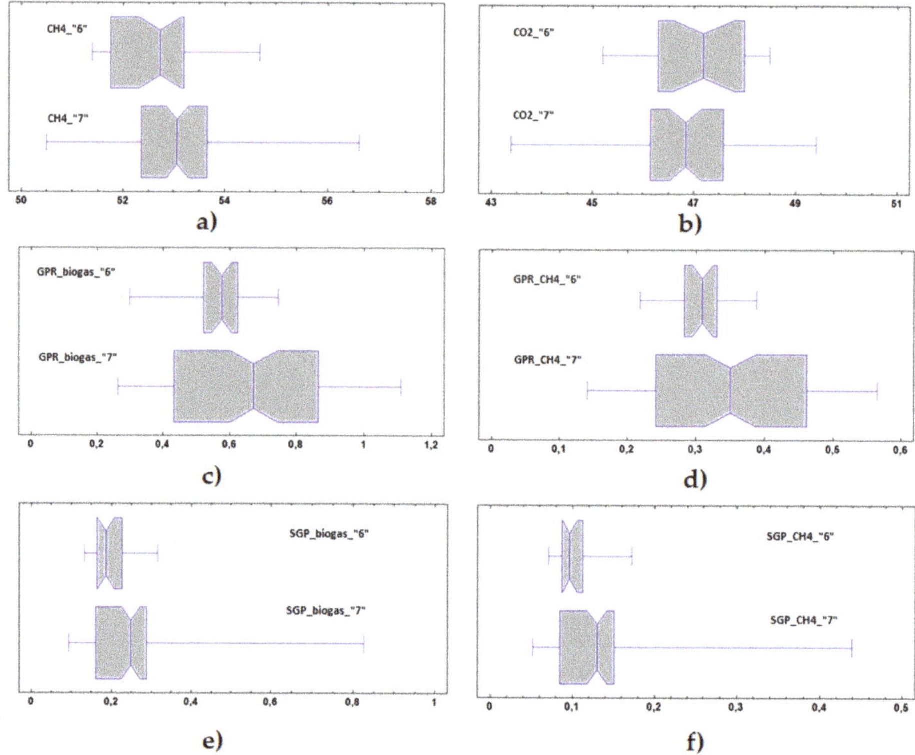

Figure 12. Box-and-whisker plots of the feeding-phases "6" and "7". Mann–Whitney test on the statistic variables: CH_4 content in biogas (**a**), CO_2 content in biogas (**b**), Gas Production Rate (GPR) biogas (**c**), GPR methane (**d**), Specific Gas Production (SGP) biogas (**e**) and SGP methane (**f**).

Skewness and kurtosis (Table 7) showed normal distributions of CH_4 and CO_2 contents in biogas, but SGP and GPR deviated from normality.

The Mann–Whitney test did not show any statistically significant difference in CH_4 content. Similar results were obtained for the CO_2 content in biogas (median of phase "6": 47.2% wt, median of phase "7": 46.8% wt).

SGP and GPR of biogas and methane were higher in the last phase of the experiment (with enzymatic treatment) than the second-last phase (without enzymes). More specifically, GPR and SGP of biogas and methane in the seventh phase reached the maximum values of the entire experimental campaign.

Concerning $GPR_{methane}$, any statistically significant difference was found between the two feeding phases (median value of phase "6": 0.310 $Nm^3 \cdot d^{-1}$, median value of phase "7": 0.351 $Nm^3 \cdot d^{-1}$).

More generally, in the last feeding phase, characterized by the enzymes, SGP and GPR of biogas and methane were higher than in the previous periods. Thus, the best management of the biodigester was characterized by this combination: A higher percentage of new hemp straw in the admixtures (C2) coupled with enzymatic addition.

4. Conclusions

The experimental campaign carried out on the cultivar "Futura 75" grown on the pilot site "San Giovanni Suergiu" (Sardinia, Italy) allowed an assessment of the effectiveness of using *Cannabis sativa* L. straw residues as a substrate for anaerobic digestion at an industrial scale and to enhance the management of the biodigester fed with hemp straw residues.

In this work, the feasibility of using this substrate in anaerobic digestion (which is currently often underutilized) was evaluated.

Results in terms of GPR and SGP of biogas/methane are promising, especially if compared to other vegetable feedstocks commonly used in anaerobic digestion and by considering that industrial hemp is characterized by higher values of lignin, which leads to high recalcitrance [59,60]. However, the SGP of biogas/methane is lower compared to corn silage, commonly used in industrial plants of anaerobic digestion (common values of about 0.7 to 0.85 $Nm^3 \cdot kg_{VS}^{-1}$), but the low cost of hemp straw residues and their behavior in the anaerobic digestion contribute to the definition of this by-product as a good process moderator when using other types of biomass leading easily to process instability.

The comparison between the findings of this work and the literature related to previous experiments carried out on a laboratory scale led to the assertion that biogas and methane yields provided by the trials carried out on hemp straw residues in the Sardinian pilot digester are similar or higher than those provided by the previous studies based on few liters-capacity reactors. It should be considered that differences in the energy yields reported may depend on environmental conditions (climate, soil type, crop management, etc.), the different cultivars used, hemp stage (vegetative versus reproductive stage), and hence, chemical and physical characteristics (such as TS, $VS_{d.b.}$, carbon:nitrogen ratio, etc.).

Results of SGP are close to those of other lignocellulosic crops but lower than those produced by highly degradable vegetable feedstocks studied through the same pilot plant. It suggests conducting additional experimental studies on hemp straws residues as a co-substrate in anaerobic digestion involving one or more easily digestible types of biomass.

Energy yields of anaerobic digestion carried out on hemp straw residues are influenced by different operating conditions: Increased feeding admixture composition (depending on C and R) produced a statistically significant increase in terms of methane content in biogas and of the parameters influencing GPR and SGP. Enzymatic treatments tended to enhance the SGP of biogas/methane.

The fluid dynamics of hemp-digestate admixtures play an important role in digestion kinetics, affected by solid–liquid separation and solid particles' tendency to sedimentation. Thus, further research should pay attention to this topic, to define relationships between the reactor and specific characteristics of admixtures.

These prodromal studies based on pilot-scale experiments on *Cannabis sativa* L. residues should be continued by analyzing more extensive conditions for factors inhibiting anaerobic digestion of hemp (e.g., heavy metals absorbed by roots, straw, leaves, and seeds during plant growth), and to different daily intakes of enzymatic preparations. Further investigations should pay attention to the enzymatic or other chemical additives' effects on energy yields (GPR, SGP, etc.).

The sustainability of hemp straw residues' biogas conversion should be evaluated as well, to define achievable costs and economic benefits. This hypothetical chain based on this emerging crop must be compared to the most commonly used energy feedstock. The main advantage in the energy conversion of hemp straws residues is to use a by-product of a cultivation carried out to obtain seeds as the main product: The consumption of water and fertilizers, however limited, is necessary to obtain seeds and no other input is spent, except for harvest and transport operations from the field to the plant.

In addition, with respect to the hemp-related supply chain, it has to be considered that relevant constraints to industrial hemp market development are fewer innovations in harvesting technologies and processes or processing facilities, as well as transportation/distribution issues (mainly due to the high low bulk density of this type of biomass) [61]. New research should overcome these current limits of industrial hemp exploitation and valorization to ensure more effective development of sustainable supply chains.

The results of this study produce a baseline to stimulate new perspectives of using hemp straw residues in the biogas sector and to inspire its consideration in the biorefinery thinking.

Author Contributions: Conceptualization, C.A., E.M. and E.A.S.; Data curation, C.A., E.M. and E.A.S.; Formal analysis: C.A.; Methodology, C.A., E.M. and E.A.S.; Project administration, G.C.; Supervision, E.A.S.; Validation, C.A., E.M. and E.A.S.; Writing—original draft preparation, E.M.; Writing—review and editing, C.A., E.M., G.C., E.A.S.

Funding: This research was funded by Agris Sardegna under a three-year agreement with Sardegna Ricerche.

Acknowledgments: Gratefully acknowledges Agris Sardegna (Regional Agency for Research in Agriculture) of the Sardinia Autonomous Region for the providing of hemp fibre composition data (Table 4), as part of the project agreement "CANOPAES", under the Regional Law No. 15/2015 of the Sardinia Autonomous Region.

Conflicts of Interest: The authors declare no conflict of interest.

References

1. Carus, M.; Sarmento, L. *The European Hemp Industry: Cultivation, Processing and Applications for Fibres, Shivs, Seeds and Flowers; Report 2016–05*; European Industrial Hemp Association: Brussels, Belgium, 2016; p. 9. Available online: eiha.org/media/2016/05/16-05-17-European-Hemp-Industry-2013.pdf (accessed on 1 April 2019).

2. Carus, M. *The European Hemp Industry: Cultivation, Processing and Applications for Fibres, Shivs, Seeds and Flowers*; Report 2017–03-26; European Industrial Hemp Association: Brussels, Belgium, 2017; p. 9. Available online: eiha.org/media/2017/12/17-03_European_Hemp_Industry.pdf (accessed on 1 April 2019).

3. Johnson, R. *Hemp as an Agricultural Commodity*; CRS Report; Congressional Research Service: Washington, DC, USA, 2014; p. 34.

4. Carus, M.; Karst, S.; Kauffmann, A.; Hobson, J.; Bertucelli, S. *The European Hemp Industry: Cultivation, Processing and Applications for Fibres, Shivs and Seeds*; European Hemp Industry Association: Brussels, Belgium, 2013; Available online: www.votehemp.com/wp-content/uploads/2018/09/13-03_European_Hemp_Industry.pdf (accessed on 1 April 2019).

5. Żuk-Gołaszewska, K.; Gołaszewski, J. *Cannabis sativa* L.—Cultivation and quality of raw material. *J. Elem.* **2018**, *23*, 971–984. [CrossRef]

6. Tang, K.; Struik, P.C.; Yin, X.; Thouminot, C.; Bjelková, M.; Stramkale, V.; Amaducci, S. Comparing hemp (*Cannabis sativa* L.) cultivars for dual-purpose production under contrasting environments. *Ind. Crops Prod.* **2016**, *87*, 33–44. [CrossRef]

7. Amaducci, S.; Scordia, D.; Liuc, F.H.; Zhang, Q.; Guo, H.; Testa, G.; Cosentino, S.L. Key cultivation techniques for hemp in Europe and China. *Ind. Crops Prod.* **2015**, *68*, 2–16. [CrossRef]

8. Struik, P.C.; Amaducci, S.; Bullard, M.J.; Stutterheim, N.C.; Venturi, G.; Cromack, H. Agronomy of fibre hemp (*Cannabis sativa* L.) in Europe. *Ind. Crops Prod.* **2000**, *11*, 107–118. [CrossRef]

9. Amaducci, S.; Amaducci, M.T.; Benati, R.; Venturi, G. Crop yield and quality parameters of four annual fibre crops (hemp, kenaf, maize and sorghum) in the North of Italy. *Ind. Crops Prod.* **2000**, *11*, 179–186. [CrossRef]

10. Cosentino, S.L.; Riggi, E.; Testa, G.; Scordia, D.; Copani, V. Evaluation of European developed fibre hemp genotypes (*Cannabis sativa* L.) in semi-arid Mediterranean environment. *Ind. Crops Prod.* **2013**, *50*, 312–324. [CrossRef]

11. Di Bari, V.; Campi, P.; Colucci, R.; Mastrorilli, M. Potential productivity of fibre hemp in southern Europe. *Euphytica* **2004**, *140*, 25–32. [CrossRef]

12. Vantreese, V.L. Hemp Support. *J. Ind. Hemp* **2002**, *7*, 17–31. [CrossRef]

13. Italian Republic. *Law n. 242, 2 December 2016. Disposizioni Per la promozione della Coltivazione e della Filiera Agroindustriale della Canapa*; General Series n. 304; Gazzetta Ufficiale della Repubblica Italiana: Rome, Italy, 30 December 2016.

14. Di Candilo, M.; Ranalli, P.; Liberalato, D. Gli interventi necessari per la reintroduzione della canapa in Italia. *Agroindustria* **2003**, *2*, 27–36.

15. FAOSTAT. Available online: www.fao.org/faostat (accessed on 23 April 2019).

16. ISTAT. Available online: www.agri.istat.it (accessed on 23 April 2019).

17. European Industrial Hemp Association (EIHA). Available online: www.eiha.org (accessed on 29 November 2019).

18. MultiHemp Project. Available online: www.multihemp.eu (accessed on 29 November 2019).

19. GRACE Project. Available online: www.grace-bbi.eu (accessed on 29 November 2019).

20. Tedeschi, A.; Tedeschi, P. The potential of hemp to produce bioenergy. In Proceedings of the 2nd World Conference on Biomass for Energy, Industry and Climate Protection, Rome, Italy, 10–14 May 2004; pp. 148–152.

21. González-García, S.; Luo, L.; Moreira, M.T.; Feijoo, G.; Huppes, G. Life cycle assessment of hemp hurds use in second generation ethanol production. *Biomass Bioenergy* **2012**, *36*, 268–279. [CrossRef]

22. Kuglarz, M.; Gunnarsson, I.B.; Svensson, S.-E.; Prade, T.; Johansson, E.; Angelidaki, I. Ethanol production from industrial hemp: Effect of combined dilute acid/steam pretreatment and economic aspects. *Bioresour. Technol.* **2014**, *163*, 236–243. [CrossRef]

23. Ragit, S.S.; Mohapatra, S.K.; Gill, P.; Kundu, K. Brown hemp methyl ester: Transesterification process and evaluation of fuel properties. *Biomass Bioenergy* **2012**, *41*, 14–20. [CrossRef]

24. Branca, C.; Di Blasi, C.; Galgano, A. Experimental analysis about the exploitation of industrial hemp (*Cannabis sativa*) in pyrolysis. *Fuel Process. Technol.* **2017**, *162*, 20–29. [CrossRef]

25. Rice, B. Hemp as a feedstock for biomass-to-energy conversion. *J. Ind. Hemp* **2008**, *13*, 145–156. [CrossRef]

26. Burczyk, H.; Grabowska, L.; Kołodziej, J.; Strybe, M. Industrial Hemp as a Raw Material for Energy Production. *J. Ind. Hemp* **2008**, *13*, 37–48. [CrossRef]

27. Finnan, J.; Styles, D. Hemp: A more sustainable annual energy crop for climate and energy policy. *Energy Policy* **2013**, *58*, 152–162. [CrossRef]

28. Hanegraaf, M.C.; Biewinga, E.E.; van der bijl, G. Assessing the ecological and economic sustainability of energy crops. *Biomass Bioenergy* **1998**, *15*, 345–355. [CrossRef]

29. Rehman, M.S.U.; Rashid, N.; Saif, A.; Mahmood, T.; Han, J.-I. Potential of bioenergy production from industrial hemp (*Cannabis sativa*): Pakistan perspective. *Renew. Sustain. Energy Rev.* **2013**, *18*, 154–164. [CrossRef]

30. Kreuger, E.; Prade, T.; Escobar, F.; Svensson, S.-E.; Englund, J.-E.; Björnsson, L. Anaerobic digestion of industrial hemp–Effect of harvest time on methane energy yield per hectare. *Biomass Bioenergy* **2011**, *35*, 893–900. [CrossRef]

31. Heiermann, M.; Ploechl, M.; Linke, B.; Schelle, H.; Herrmann, C. Biogas Crops-Part I: Specifications and Suitability of Field Crops for Anaerobic Digestion. *Agric. Eng. Int. CIGR J.* **2009**, *11*, 1–17.

32. Adamovics, A.; Dubrovskis, V.; Platace, R. Productivity of industrial hemp and its utilisation for anaerobic digestion. In Energy Production and Management in the 21st Century, Vol. 2. *WIT Trans. Ecol. Environ.* **2014**, *190*, 1045–1055.

33. Mallik, M.K.; Singh, U.K.; Ahmad, N. Batch digester studies on biogas production from Cannabis sativa, water hyacinth and crop wastes mixed with dung and poultry litter. *Biol. Wastes* **1990**, *31*, 315–319. [CrossRef]

34. Kaiser, F.; Diepolder, M.; Eder, J.; Hartmann, S.; Prestele, H.; Gerlach, R.; Ziehfreund, G.; Gronauer, A. Biogas yields from various renewable raw materials. In Proceedings of the 7th FAO/SREEN Workshop, Uppsala, Sweden, 30 November–2 December 2005.

35. Dumas, C.; Silva Ghizzi Damasceno, G.; Barakat, A.; Carrere, H.; Steyer, J.-P.; Rouau, X. Effects of grinding processes on anaerobic digestion of wheat straw. *Ind. Crops Prod.* **2015**, *74*, 450–456. [CrossRef]

36. Lynd, L.R.; Weimer, P.J.; Van Zyl, W.H.; Pretorius, I.S. Microbial cellulose utilization: Fundamentals and biotechnology. *Microbiol. Mol. Biol. Rev.* **2002**, *66*, 506–577. [CrossRef] [PubMed]

37. Merlin Christy, P.; Gopinath, L.R.; Divya, D. A review on anaerobic decomposition and enhancement of biogas production through enzymes and microorganisms. *Renew. Sustain. Energy Rev.* **2014**, *34*, 167–173. [CrossRef]

38. Xu, N.; Liu, S.; Xin, F.; Zhou, J.; Jia, H.; Xu, J.; Jiang, M.; Dong, W. Biomethane production from lignocellulose: Biomass recalcitrance and its impacts on anaerobic digestion. *Front. Bioeng. Biotechnol.* **2019**, 1–12. [CrossRef]

39. Čater, M.; Zorec, M.; Marinšek Logar, R. Methods for improving anaerobic lignocellulosic substrates degradation for enhanced biogasp. *Springer Sci. Rev.* **2014**, *2*, 51–61. [CrossRef]

40. Herrero Garcia, N.; Benedetti, M.; Bolzonella, D. Effects of enzymes addition on biogas production from anaerobic digestion of agricultural biomasses. *Waste Biomass Valor.* **2019**, *10*, 3711–3722. [CrossRef]

41. ASTM D7582-15. Standard Test Methods for Proximate Analysis of Coal and Coke by Macro Thermogravimetric Analysis. Available online: https://www.astm.org/Standards/D7582.htm (accessed on 30 June 2017).

42. ASTM D5373-16. Standard Test Methods for Determination of Carbon, Hydrogen and Nitrogen in Analysis Samples of Coal and Carbon in Analysis Samples of Coal and Coke. Available online: https://www.astm.org/Standards/D5373.htm (accessed on 2 January 2018).

43. Van Sœst, P.J.; Robertson, J.B.; Lewis, B.A. methods for dietary fiber, Neutral Detergent Fiber, and nonstarch polysaccharides in relation to animal nutrition. *J. Dairy Sci.* **1991**, *74*, 3583–3597. [CrossRef]

44. ANKOM Technologies. *Acid Detergent Fiber in Feeds—Filter Bag Technique (for A200 and A200I)*; ANKOM Technologies: Macedon, NY, USA, 2011.

45. ANKOM Technologies. *Method 8—Determining Acid Detergent Lignin in Beakers*; ANKOM Technologies: Macedon, NY, USA, 2005.

46. ANKOM Technologies. *Neutral Detergent Fiber in Feeds—Filter Bag Technique (for A200 and A200I)*; ANKOM Technologies: Macedon, NY, USA; Available online: www.ankom.com/sites/default/files/document-files/Method_6_NDF_A200.pdf (accessed on 29 November 2019).

47. Jung, H.-J.G. Analysis of forage fiber and cell walls in ruminant nutrition. *J. Nutr.* **1997**, *127*, 810S–813S. [CrossRef]

48. Theander, O.; Aman, P.; Westerlund, E.; Andersson, R.; Pettersson, D. Total dietary fiber determined as neutral sugar residues, uronic acid residues, and Klason lignin (the Uppsala method). *J. Assoc. Anal. Chem. Int.* **1995**, *78*, 1030–1044.

49. Statgraphics. Available online: www.statgraphics.com (accessed on 15 September 2019).

50. Kruskal, W.H. Historical Notes on the Wilcoxon Unpaired Two-Sample Test. *J. Am. Stat. Assoc.* **1957**, *52*, 356–360. [CrossRef]

51. Neuhäuser, M. Wilcoxon–Mann–Whitney Test. In *International Encyclopedia of Statistical Science*; Springer: Berlin/Heidelberg, Germany, 2011.

52. Kruskal, W.H.; Wallis, W.A. use of ranks in one-criterion variance analysis. *J. Am. Stat. Assoc.* **1952**, *47*, 583–621. [CrossRef]

53. Pistis, A.; Asquer, C.; Scano, E.A. Anaerobic digestion of potato industry by-products on a pilot-scale plant under thermophilic conditions. *Environ. Eng. Manag. J.* **2013**, *12*, 93–96.

54. Asquer, C.; Pistis, A.; Scano, E.A.; Cocco, D. Energy-oriented optimization of an anaerobic digestion plant for the combined treatment of solid and liquid wastes in a potato chips industrial plant. In Proceedings of the 22nd EUBCE, Hamburg, Germany, 23–26 June 2014.

55. Scano, E.A.; Asquer, C.; Pistis, A.; Ortu, L.; Demontis, V.; Cocco, D. Biogas from anaerobic digestion of fruit and vegetable wastes: Experimental results on pilot-scale and design of a full-scale power plant. *Energy Convers. Manag.* **2014**, *77*, 22–30. [CrossRef]

56. Scano, E.A. Trattamento di Biomasse Vegetali e Algali Finalizzato All'Ottenimento di Energia. Potenziali Sviluppi in Sardegna. Ph.D. Thesis, University of Cagliari, Cagliari, Italy, 2016. Available online: http://hdl.handle.net/11584/266883 (accessed on 29 November 2019).

57. Frigon, J.-C.; Guiot, S. Biomethane production from starch and lignocellulosic crops: A comparative review. *Biofuels Bioprod. Bioref.* **2010**, *4*, 447–458. [CrossRef]

58. International Energy Agency. Biogas from Crop Digestion. Bioenergy Task 32. 2011. Available online: http://www.ieabioenergy.com/publications/biogas-from-energy-crop-digestion/ (accessed on 27 September 2019).

59. Ghosh, S.; Henry, M.P.; Christopher, R.W. Hemicellulose conversion by anaerobic digestion. *Biomass* **1985**, *6*, 257–269. [CrossRef]

60. Brodeur, G.; Yau, E.; Badal, K.; Collier, J.; Ramachandran, K.B.; Ramakrishnan, S. Chemical and physicochemical pretreatment of lignocellulosic biomass: A review. *Enzyme Res.* **2011**, *2011*, 17. [CrossRef]

61. Fortenbery, T.R.; Bennett, M. Opportunities for Commercial Hemp Production. *Rev. Agric. Econ.* **2004**, *26*, 97–117. [CrossRef]

Article

Impact of Climate Change on Twenty-First Century Crop Yields in the U.S.

Lillian Kay Petersen

Los Alamos High School, Los Alamos, NM 87544, USA; lilliankay.petersen@gmail.com; Tel.: +1-505-709-0687

Received: 3 February 2019; Accepted: 8 March 2019; Published: 14 March 2019

Abstract: Crop yields are strongly dependent on the average climate, extreme temperatures, and carbon dioxide concentrations, all of which are projected to increase in the coming century. In this study, a statistical model was created to predict US yields to 2100 for three crops using low and high-emissions future scenarios (RCP 4.5 and 8.5). The model is based on linear regressions between historical crop yields and daily weather observations since 1970 for every county in the US. Yields were found to be most strongly dependent on heat waves, summer average temperatures, and killing degree days; these relationships were hence used to predict future yields. The model shows that warming temperatures will significantly decrease corn and soybean yields, but will not have as strong of an influence on rice. Before accounting for CO_2 fertilization, crops in the high-emissions scenario are predicted to produce 77%, 85%, and 96% of their expected yield without climate change for corn, soybeans, and rice, respectively. When a simple CO_2 fertilization factor is included, corn, a C4 plant, increases slightly, while the yields of the C3 plants (soybeans and rice) are actually predicted to increase compared to today's yields. This study exhibits the wide range of possible impacts of climate change on crop yields in the coming century, and emphasizes the need for field research on the combined effects of CO_2 fertilization and heat extremes.

Keywords: future crop yields; climate change impacts; CO_2 fertilization; corn; rice; soybeans

1. Introduction

1.1. Impacts of Climate Change and the Social Cost of Carbon

On 8 October 2018, the Intergovernmental Panel on Climate Change (IPCC) released a new report titled *Global Warming of 1.5 °C*, which concludes that drastic action must be taken to limit global temperature rise and avoid serious negative impacts. It finds that natural, managed and human systems have a high risk of permanent damage as the climate warms. Extreme weather events will occur more often, including droughts, floods, coastal storms, and heat waves, increasing mortality and property damage. As warmer temperatures combine with more extreme weather events, cereal yields will decrease [1,2]. If we fail to limit fossil fuel combustion, all of the effects will increase [3].

Also on 8 October, the Nobel Prize in Economic Science was awarded to Paul Romer and William Nordhaus for their research on using economics as a driving factor to reduce greenhouse gas emissions [4]. Nordhaus, recognized as the founder of climate change economics, developed economic models to weigh the cost of reducing carbon footprints today against future costs of current emissions [5]. Romer focused on how market factors influence technological growth. Both advocate carbon taxes to employ market forces to reduce emissions and spur innovation in energy efficiency.

The correct amount to tax carbon may be found through the social cost of carbon, or the external cost of carbon emissions. The social cost of carbon is calculated by integrating all future economic losses due to climate change discussed in the IPCC. In total, warming temperatures cause a loss of annual national average gross domestic product (GDP) of 1.0% to 3.0% at the end of the century [6]. Agriculture is a substantial portion of the economy, and crop yields are highly dependent on temperature. Other impacts are more difficult to put a price on, such as biodiversity and ecosystem loss. The currently accepted cost of carbon dioxide when considering these externalities varies between $37 and $220 per ton emitted [7].

1.2. Warming Temperatures Impact Agriculture

The United States produces 41% of the world's corn and 38% of the world's soybeans, two of the four largest crop sources of caloric energy [8]. These crops are thus crucial to food security, and understanding how their yields will change in the next century could help drive more informed policy decisions.

The growing world population requires a larger food supply. Historical improvements in crop yields from agricultural technology (e.g., pesticides, fertilizers, farm machinery, gene modification, and shifting of production to large corporations [9]) have kept up with increasing demand for several decades, but it is doubtful that yields will continue to grow at the same rate as they have since 1970. Population is unlikely to stop growing this century, and by 2100 there will be between 9.6 and 12.3 billion people on earth [10]. Research has shown that yields are projected to drop in coming decades due to warming temperatures and the potential emergence of virulent crop diseases [11,12].

Crop yields are strongly dependent on the weather and may be predicted from observed weather events during the growing season [13–16]. Over 60% of yield variability in global breadbaskets can be explained by climate variation [17], particularly temperature extremes during crucial phases of the growing season, such as the grain fill stages [18]. Some research suggests that yields decrease exponentially as temperatures warm [8]. Therefore, a warming climate could harm crop yields and global food security. In fact, corn and wheat yields have already decreased by 1–2% per decade since 1980 relative to the expected harvest without warming [19].

There are many different approaches to identifying the the impacts of climate change on crop yields. Statistical models, including this study, use historical correlations from observations to develop empirical relationships between yields and weather. These relationships are then applied to climate model output to predict future crop yields. Process models are based on the mechanisms of an individual plant's physiology and then are scaled up to large domains. Each type of model has its own advantages and disadvantages. Statistical models are accurate for the specific locations and conditions of their training data sets, and are a direct way to model yields within those constraints [20,21]. Process models offer a deeper understanding of the cause and effect of the environmental impacts on yields, and they can potentially model future yields outside of historical observations. Process models have become more sophisticated in recent years [22], but still have difficulty reproducing historical yields in certain circumstances [23].

Many previous studies have analyzed the relationships between crop yields and emperature [13,24,25], precipitation [14,26,27], or radiation measurements [28,29], and have predicted future yields based on these relationships [13,26,30]. These models may be statistical [17,31], process-based [30,32], or both [33], and have focused on US [14,24,33], China [23,25,29,31,32], Europe [28,30], or global bread baskets [15,17,27]. All of these studies conclude that climate change will have a negative impact on future crop yields, and bread basket failures could pose a threat to food security.

The purpose of this study is to evaluate the future economic losses or gains of three crops through 2100 for different climate scenarios. It examines the historical relationship between crop yields and extreme weather to better understand which factors affect yields, and then projects crop yields into the future for every county in the United States. This study also incorporates CO_2 fertilization to show the range of possible future impacts from these processes. Finally, a monetary value of changing crop yields is calculated, an integral part to the social cost of carbon.

2. Methods

A statistical model was created to evaluate historical weather and crop data, compute correlations and linear regressions for each county, and project crop yields to 2100 based on two future climate model scenarios. Annual crop yield data of corn, soybeans, and rice was obtained for every US county for 1970 through 2015 from the United States Department of Agriculture (USDA, [34]). In addition, 1970 was chosen as a start date because yields were more variable and the farming practices were not as standardized before then (e.g., irrigation, pesticides, fertilizers). Daily weather station observations, provided by the Daily Global Historical Climatology Network [35], were downloaded for all weather stations in the US with data since 1970. Daily minimum and maximum air temperature and precipitation were computed for each county from the average of the two weather stations closest to the center of that county. This provided redundancy—if one station was missing data, the other station's data was used.

Next, various means and extremes were computed for each county and year. Most of these are standard measures reported by the Intergovernmental Panel on Climate Change (IPCC, [36], Box 2.4 p. 221). Table 1 lists all extremes computed from the daily temperature. Values of the 10th and 90th percentile for each variable and county were computed from the daily data from the years 1970 to 1990. The extreme measures were computed over the growing season, which varies for each crop and state, and were obtained from the USDA [37].

Table 1. Temperature measures computed to find correlations to crop yields. Summer average temperature, heat waves, and killing degree days had the highest correlation, and thus were used as predictors of crop yields. All statistics after the first four are summed over the growing season for each crop and location. Here, highs and lows refer to the recorded daily high and low temperature at each site.

Measurement	Definition	Units
Average Yearly high	Average of all daily highs in a year	°C
Average Yearly low	Average of all daily lows in a year	°C
Summer Average	Average of all daily max temps over months June, July, and August	°C
Warmest Day	The warmest high in the growing season	°C
Coldest Day	The coldest high in the growing season	°C
Warmest Night	The warmest low in the growing season	°C
Coldest Night	The coldest low in the growing season	°C
Heat Waves of highs	Frequency of 3 daily highs in a row >90th percentile	#/year
Heat Waves of lows	Frequency of 3 daily lows in a row >90th percentile	#/year
Cold Spells of lows	Frequency of 3 daily lows in a row <10th percentile	#/year
Cold Spells of highs	Frequency of 3 daily highs in a row <10th percentile	#/year
Warm Days	Days when daily high >90th percentile	days/year
Cold Days	Days when daily high <10th percentile	days/year
Warm Nights	Days when daily low >90th percentile	days/year
Cold Nights	Days when daily low <10th percentile	days/year
Tropical Nights	Frequency of daily lows >20 °C (68 °F)	days/year
Frost Nights	Frequency of daily lows <0 °C (32 °F)	days/year
Growing Degree Days	Summation of daily highs above 10 °C (50 °F)	°C*days
Killing Degree Days	Summation of daily highs above 29 °C (68 °F) [13]	°C
Precipitation	Total precipitation	mm/year

Correlations between the detrended crop yield and each of the weather statistics were then computed for each county and crop. In order to account for increasing crop yields due to improvements in agricultural technology, the yields were first detrended for each county. Figure 1a shows the increase in corn yields since 1970 in an example county of Champaign, Illinois, and Figure 1b shows the correlation between the detrended corn yield and summer average temperature for that same county. Correlations with a *p*-value less than 0.05 are considered significant and ones under 0.01 are highly significant [38]. The three with the highest correlations were summer average temperature, heat waves, and killing degree days. Thus, these three statistics were used to predict future yields. Although regressions were also computed with precipitation, it was found to have little to no correlations with yields.

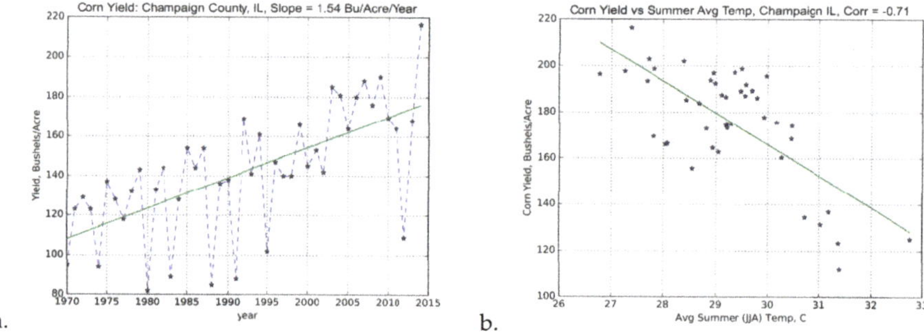

Figure 1. The corn yield over time for an example county (**a**) and detrended corn yield plotted against summer average temperature (**b**). The correlation of -0.71 is highly significant. Data is from the USDA [34].

Future climate model data was obtained from the Coupled Model Intercomparison Project Version 5 (CMIP5), using the Community Climate System Model (CCSM4) [39], and was obtained from the MACA data portal. I downloaded datasets for two IPCC scenarios: a high emission future with a Representative Concentration Pathway (RCP) that induces an extra 8.5 W/m^2 of radiative forcing (RCP 8.5) and a low emissions scenario with an RCP of 4.5 W/m^2 (RCP 4.5). Climate forcings in the MACAv2-METDATA were drawn from a statistical downscaling of global climate model data from CMIP5 [40] utilizing a modification of the Multivariate Adaptive Constructed Analogs (MACA) [41] method with the METDATA [42] observational dataset as training data. The climate data had high resolution in space (one-tenth of a degree) and time (daily). Summer average temperature, heat waves, and killing degree days were computed for each county for every year until 2100, using data from the closest model grid-cell to the center of each county. The histograms displaying these future heat measurements (Figure 2) were computed using a latitude/longitude rectangle around the dominant corn-growing region, with corners at (40N, 100W), (44N, 85W).

Crop yields were then predicted by applying the historical linear regressions to the future projections of summer average temperature, heat waves, and killing degree days. For each county, the crop predictions from the three statistics were averaged, as each measure predicted the yields slightly differently. National averages of crop yields were computed by averaging all counties that either consistently grew their crop over the past 10 years or grew at least 10% as much as the highest-producing county for that crop.

After future crop yields were predicted from temperature projections, the yield was multiplied by the expected yield factor from CO$_2$ fertilization. Future carbon dioxide concentrations to 2100 for RCP8.5 and RCP4.5 scenarios were obtained from [43]. The yield factor is the change in yield due to carbon dioxide fertilization, where 0.8 is a 20% reduction and 1.2 is a 20% increase in yield. The yield factor for C3 and C4 crops under different CO$_2$ concentrations (Figure 3) was acquired from [44], in which results from the DSSAT4 models [45] were interpolated based on CO$_2$ enrichment experiments, and were then normalized at 2015 CO$_2$ concentrations. Figure 4 displays the yield factor due to carbon dioxide fertilization to year 2100 for C3 and C4 crops under low and high-emissions scenarios.

Crop yields in the US have improved in recent decades due to better technologies. In fact, corn yields have doubled since 1970. It is not known whether these trends will continue to hold in the future or if biological constraints will impose maximum achievable crop yields. In this study, best and worst case scenarios were computed as a proxy for all possible futures. For the best case scenario of continuous technology improvement, the historical trend was added to the future yield predictions. For the worst case scenario of no future technology improvement, no trend was added to the yield predictions.

Ultimately, future yield forecasts were computed based on three sets of indicators: (1) only temperature changes, (2) only CO_2 fertilization, and (3) both combined. Each of these predictions were then split into two technology scenarios: (1) no technology improvement and (2) continuous technology improvement, resulting in six forecasts for each crop.

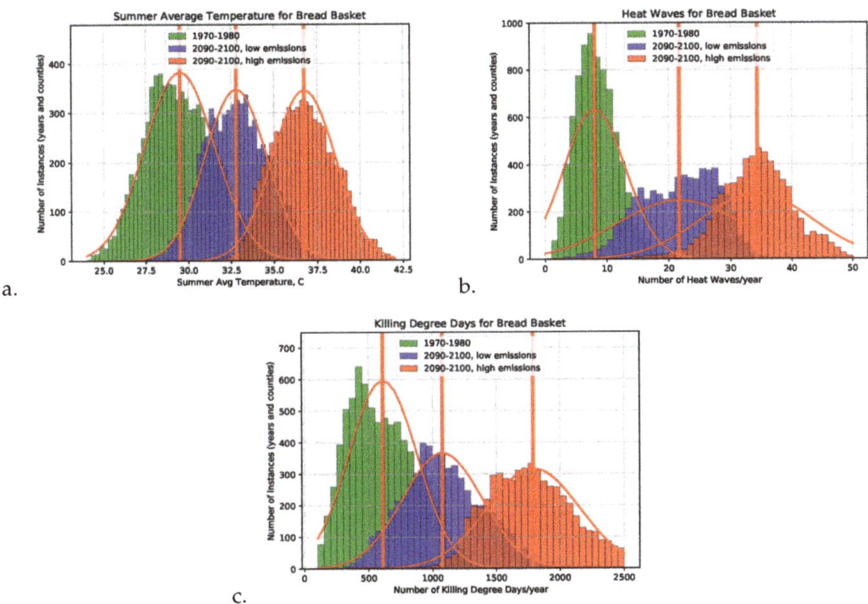

Figure 2. Distributions of summer average temperature (**a**), heat waves (**b**), and killing degree days (**c**) for historical (green), future low-emissions RCP 4.5 scenario (blue) and future high-emissions RCP 8.5 scenario (red). Results are for the US corn growing region. Historical data from [35] and future projections from [41].

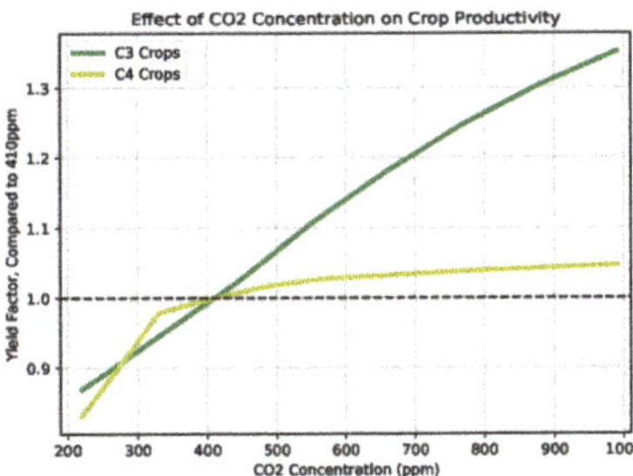

Figure 3. Yield factor for C3 (green) and C4 (yellow) crops versus CO_2 concentration. Crop productivity was acquired from [44], in which results from the DSSAT4 models [45] were interpolated based on CO_2 enrichment experiments.

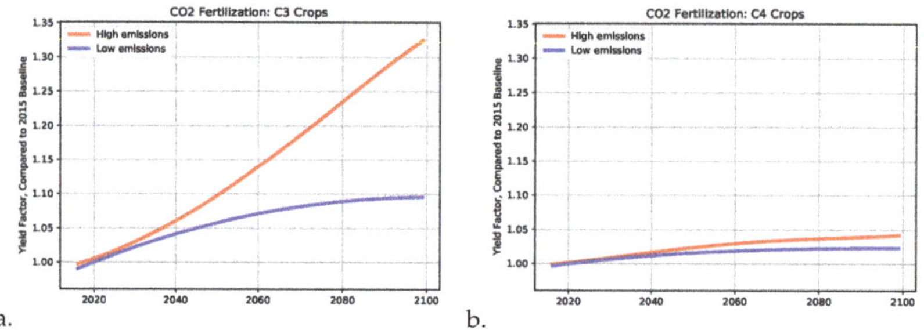

Figure 4. Yield factor for C3 (**a**) and C4 (**b**) crops through 2100 for a low (blue) and high (red) emissions scenario. Crop productivity under CO_2 concentrations was obtained from [44], and CO_2 concentrations for RCP4.5 and RCP8.5 are from [43].

3. Results

3.1. Historical Correlations and Regressions

The slopes of the linear regressions computed between crop yields and weather statistics for every county in the US can be seen in Figure 5. Almost all of the slopes are negative, meaning that higher temperatures result in lower crop yields. Corn has average correlations of -0.44, -0.46, and -0.41 to summer average temperature, heat waves, and killing degree days, respectively (Figure 6), and three in every five counties have a significant correlation. All temperature indices have similar correlations, indicating that all three have similar predictive power. Corn and soybeans observe large geographical distributions in the eastern and central US, while rice is mostly grown along the Mississippi River. Spatial variations of slopes and correlations can be examined for corn and soybeans. In southern growing regions (Missouri, southern Illinois, and Indiana), slopes are very negative with highly significant correlations (Figures 5 and 6). Crops here are thus extremely sensitive to heat extremes. In contrast, slopes and correlations farther north (Minnesota and South Dakota) are about zero, indicating that the yields are not affected by temperatures classified as extreme events in these states. The correlations with heat waves, summer average temperature, and killing degree days all follow similar geographic distributions. These results indicate that the places where crops are grown will likely shift north over time, where average temperatures are cooler.

Heat waves have the highest impact on all three crops. Corn has the strongest correlation to heat measurements, soybeans have a slighly weaker correlation, and rice has little to no correlation to temperatures. When averaged across crop-growing counties, soybeans have a correlation of -0.37 with heat waves, and about half of the counties have significant correlations. Rice has an average correlation -0.22 with heat waves and no counties have significant correlations (Figure 6).

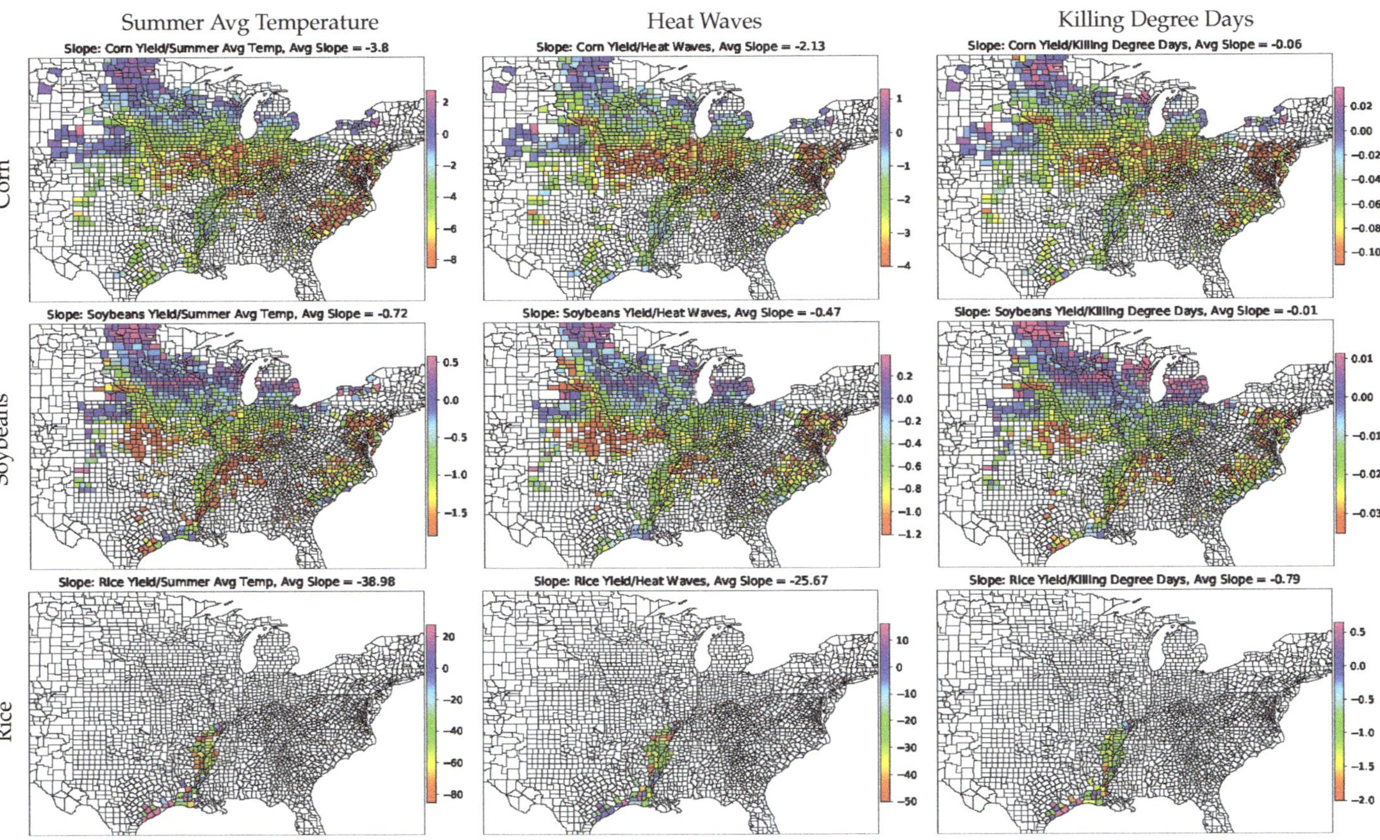

Figure 5. The slopes of the linear regressions between corn (top), soybean (middle), and rice (bottom) yields and summer average temperature (left), heat waves (center), and killing degree days (right), in each US county with reported crop yields.

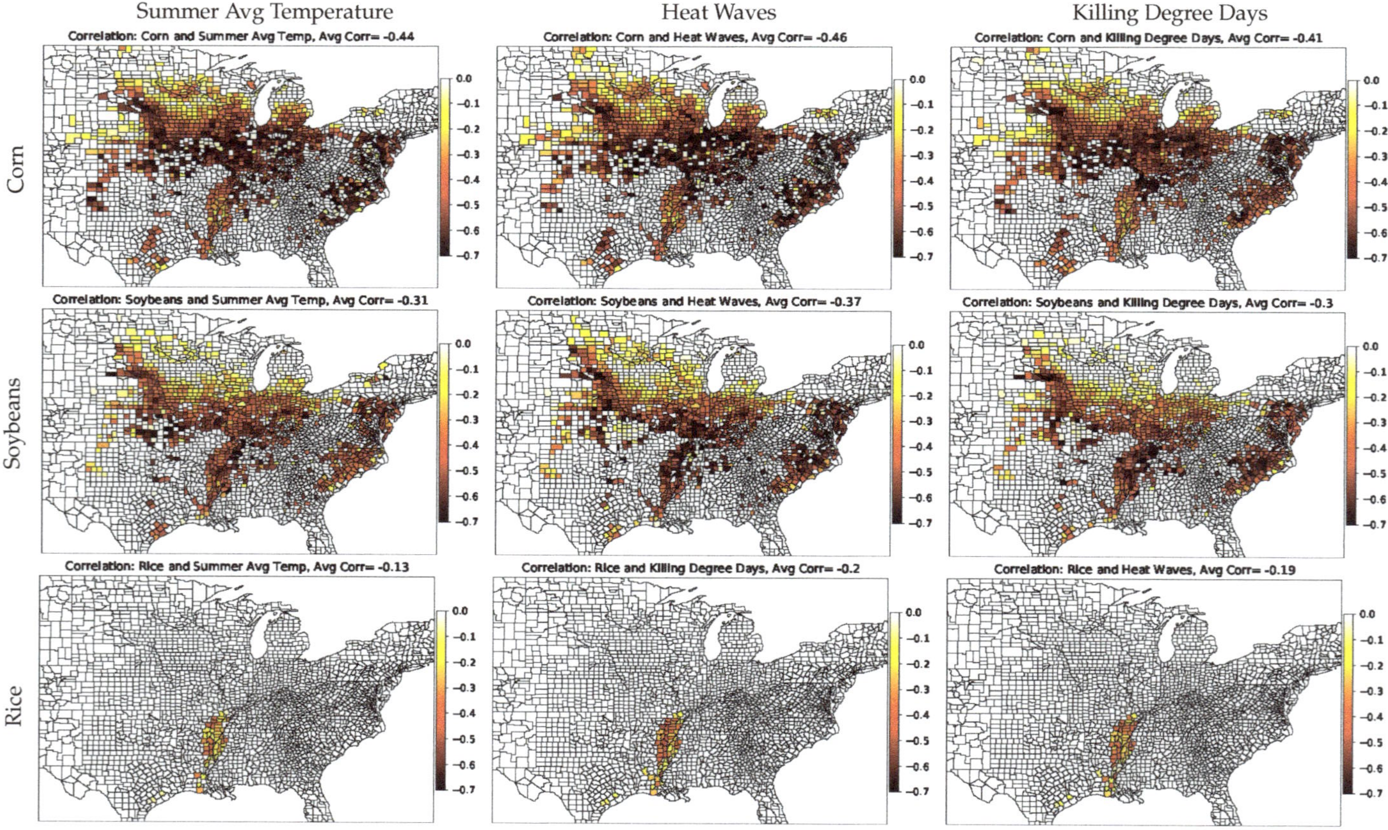

Figure 6. The same as Figure 5, but for correlations between the crop yield and the temperature metric. Correlations below −0.49 are significant (*p* < 0.05) and below −0.59 are highly significant (*p* < 0.01).

3.2. Prediction of Future Crop Yield

Heat extremes are expected to increase in the coming century. Histograms of the three heat measurements are shown for three different times and scenarios: 1970–1980 observed, 2090–2100 low emissions, and 2090–2100 high emissions, over the corn growing counties (Figure 2). Both mean and extreme temperatures dramatically increase in the future, with high emission scenarios increasing more than low emission. For example, the average summer daily high temperature was 29 °C (85 °F) in 1970 to 1980. In 2090 to 2100, the summer average temperature is projected to be 33 °C (91 °F) for RCP 4.5 and 36 °C (97 °F) for RCP 8.5. Heat waves and killing degree days are also expected to increase dramatically (Figure 2b,c). Interestingly, the tails of histograms in the future are much wider, indicating higher probabilities of extreme weather events.

Historical regressions and future climate extremes were used to predict future yields to 2100 for each county, year, and crop. First, results without CO_2 fertilization will be presented, followed by those including CO_2 fertilization.

Before accounting for CO_2 fertilization, crop yields are projected to dramatically drop in the coming century. Forecasted crop yields, with and without future technology improvements, can be seen in Figure 7. In addition to having the highest correlations, corn is also affected the most by the warming climate. Average US corn yields doubled from 80 bushels/acre in 1970 to 170 bushels/acre in 2015. Predicted yields in 2096 through 2100 drop to 76% (86%) of expected yields without warming for a high (low) emissions scenario. This translates to a 3.8% decrease in corn yields per decade for a high emissions scenario and a 1.8% decrease per decade for a low emissions scenario, compared to a historical 24% increase in corn yields per decade due to agricultural technology improvements. For more details, refer to Table 2.

Table 2. Statistics on future yield predictions, for forecasts without and with CO_2 fertilization. Future yield change per decade are in comparison to expected yields without climate change. Monetary losses are calculated from the acres harvested and the crop prices in 2016 [46,47].

	Corn		Soybeans		Rice	
Historical yield change/decade (%)	23.7		17.7		17.4	
Future: low emissions	no CO_2	with CO_2	no CO_2	with CO_2	no CO_2	with CO_2
Yield change/decade (%)	−1.77	−1.50	−1.20	+0.002	−0.367	0.860
Projected yield diff to steady climate (%)	87.9	90.0	92.3	101	98.6	108
Monetary loss or gain (billion 2019 US$)	−12.7	−10.5	−5.00	+0.717	−0.343	+1.99
Future: high emissions	no CO_2	with CO_2	no CO_2	with CO_2	no CO_2	with CO_2
Yield change/decade (%)	−3.78	−3.41	−2.40	0.96	−0.830	3.09
Projected yield diff to steady climate (%)	76.8	80.0	84.6	111	95.7	126
Monetary loss or gain (billion 2019 US$)	−24.3	−21.0	−10.0	+7.29	−1.06	+6.42

Even with the optimistic conditions of continuous technology improvement, there is a huge loss in yields below expected yields with a steady climate. In 2100, there is a loss of $24 billion per year for high emissions and $13 billion per year for a low emissions scenario. This estimate assumes the acres harvested and the cost of corn in 2016, and does not account for inflation [46,47]. Soybeans are affected by temperature extremes slightly less than corn, with losses of $5 and $10 billion per year in 2100 for high and low emission scenarios. Rice, being least sensitive to climate change, only has losses of $0.34 and $1.06 billion per year (Table 2). Rice is the least affected by heat, likely because it is grown in flooded conditions. These computations assume the current market prices of crops and do not include CO_2 fertilization.

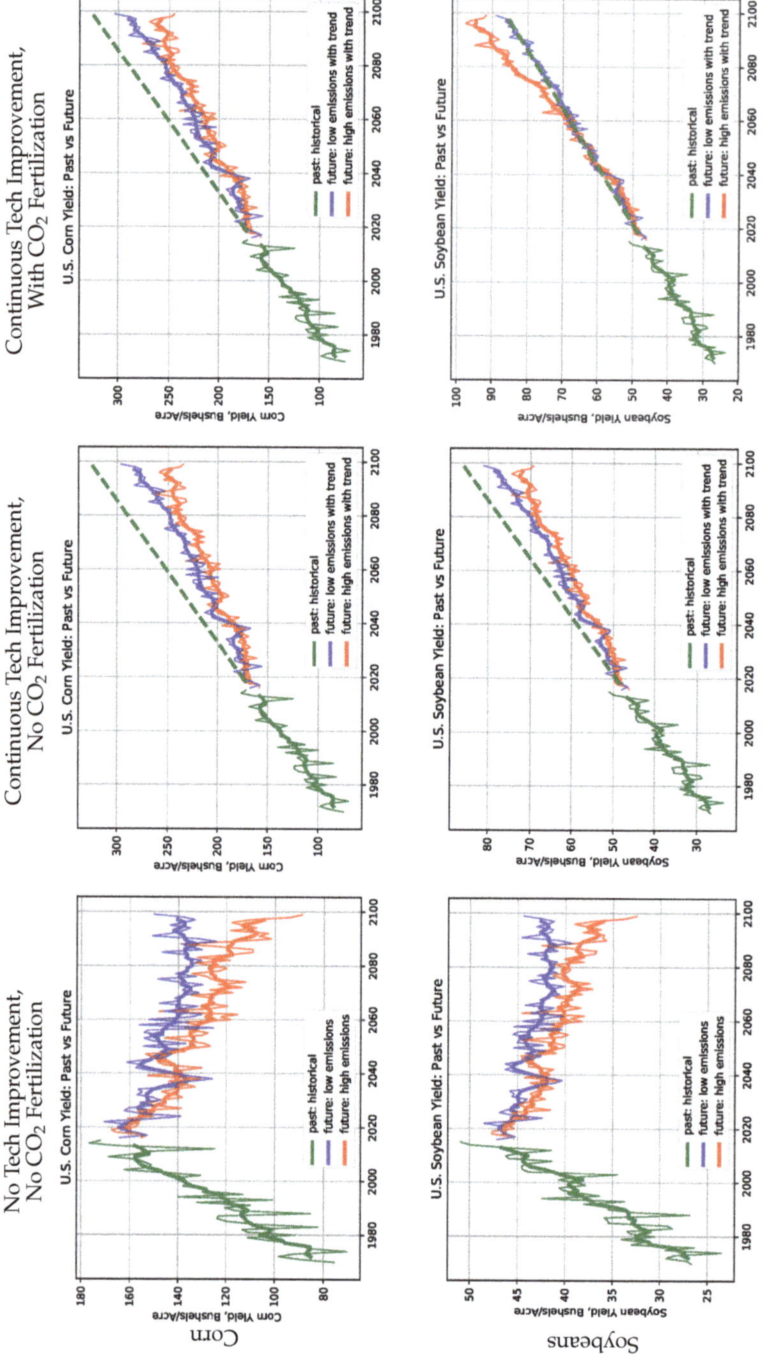

Figure 7. *Cont.*

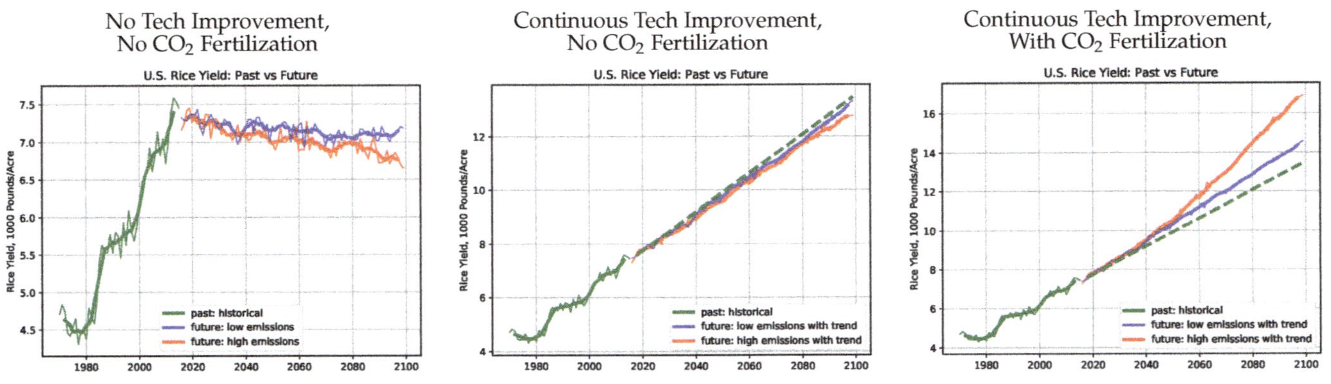

Figure 7. Projected US corn (top), soybean (middle), and rice (bottom) yields to 2100. The prediction scenarios include: (left) no technology improvement and no CO_2 fertilization; (center) continuous technology improvement and no CO_2 fertilization; and (right) continuous technology improvement with CO_2 fertilization. The green dashed line is a linear extension of the 1970–2015 trend. Thin lines are yearly data, solid is the five-year running average.

Although these results display a dismal future, the story changes when we account for CO_2 fertilization. In RCP 8.5, carbon dioxide concentrations break 900 ppm by 2100. That translates to almost 135% productivity for C3 crops (here: soybeans and rice) compared to productivity in 2015. Even for a low emissions scenario, C3 crops gain 10% productivity by the end of the century. CO_2 concentrations have a much smaller effect on C4 crops (here: corn), which reach 104% and 102% productivity in 2100 for a high and low-emissions scenario.

The projections of corn, a C4 crop, are very similar before and after accounting for CO_2 fertilization (Figure 7). Projections of soybeans and rice with CO_2 fertilization, however, are considerably higher. In fact, rice reaches 126% expected productivity by 2100 with a high emissions scenario. A summary of future crop yield estimates is shown in Figure 8.

The spatial distributions of projected crop yields may be examined (Figure 9). In 2005 through 2015, corn and soybean yields are relatively uniform across the US. In contrast, their yields are spatially disparate in 2100, with very low yields in the south and much higher yields in the northern Midwest. This distribution likely is a product of cooler climates farther north, and holds true with and without CO_2 fertilization. Few spatial differences are predicted in rice yields, as rice is grown in a relatively small geographic range and has weaker correlations to temperature.

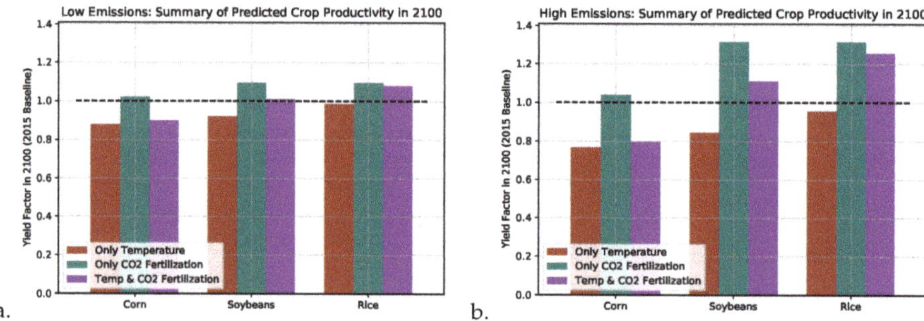

Figure 8. Crop productivity averaged over 2096–2100, compared to a 2011 through 2015 baseline for a low (**a**) and high (**b**) emissions scenario. Estimates are based on only temperatures (red), only CO_2 fertilization (teal), and both assuming compounding effects (purple).

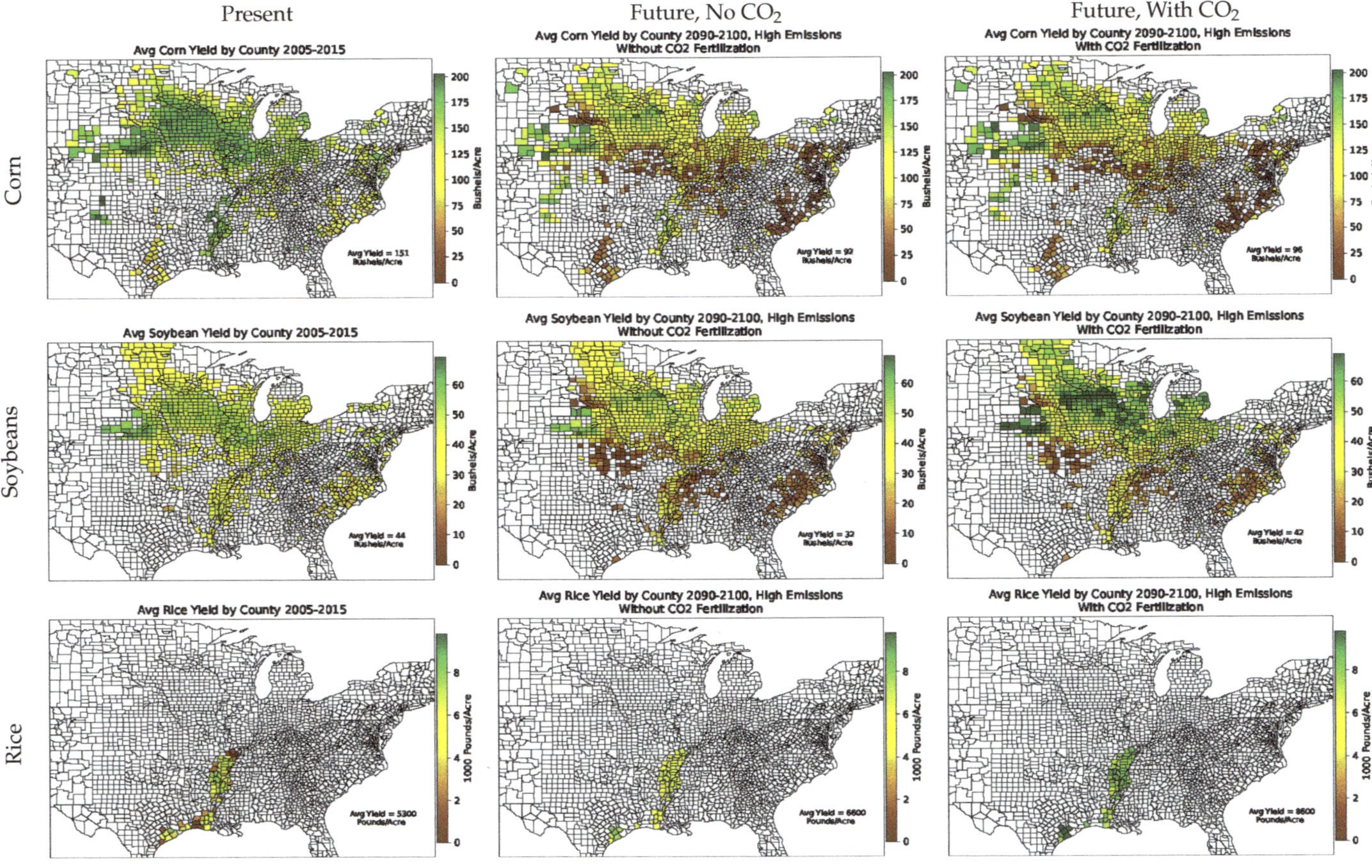

Figure 9. Crop yield of corn (top), soybeans (middle), and rice (bottom) for 2005–2015 (left), 2090–2100 without CO$_2$ fertilization (center), and 2090–2100 with CO$_2$ fertilization (right). Crop predictions are for a high-emissions scenario without future technology improvements.

4. Discussion

The purpose of this study was to analyze the potential economic losses or gains of US crop yields due to climate change by applying historical relationships between yields and heat extremes to multiple future climate scenarios, while accounting for CO_2 fertilization. Historical records give insights into factors that affect yields. The most dominant influence on crop yields since 1970 is the secular trend due to improving farming practices and technologies, where yields nearly double over that period. On top of this trend, there is year-to-year variability that can be explained by local weather. Corn, soybean, and rice yields were correlated to several measures of mean and extreme weather, and all were most strongly dependent on heat waves, summer average temperature, and killing degree days. This indicates that hot temperatures have the strongest effect on crop growth, while moderate or cold temperatures have little effect. Interestingly, precipitation had insignificant correlations with crop yields, possibly because of the prevalence of drought-resistant breeds in the US or the confounding influence of irrigation.

This study evaluates future crop yields with and without CO_2 fertilization. Without CO_2 fertilization, increasing temperatures significantly decrease crop production, with yields reaching as low as 77% of their expected productivity without climate change. In all cases, corn is most sensitive to heat, soybeans slightly less so, and rice the least, likely because rice is grown in flooded conditions. When carbon dioxide fertilization is added into the future projections, crop yields increase dramatically. Carbon dioxide concentrations alone are expected to increase C3 crop yields by 35% by the end of the century for a high emissions scenario. When the effects of CO_2 and temperatures are combined in a simple way, rice and soybean yields are actually shown to increase over the next century.

In this study, the effects of temperature extremes on crop yields were measured during historical CO_2 concentrations, while the effects of increased CO_2 concentrations were measured under laboratory conditions, independently of heat extremes. To project the effects of both together, the two factors were multiplied, but this may in fact not be how plants respond. It is unknown whether CO_2 fertilization and temperatures will have compounding physiological effects: the positive influence of CO_2 fertilization may be severely curbed at sufficient heat. Thus, more data on the effects of these combined influences is required for more accurate yield forecasts.

Projected yield losses due to climate change may be compared to past studies, shown in Table 7-2 of the IPCC Working Group II [36]. Numbers in this study compare well with the past corn and rice studies (10 and 13 studies, respectively). The soybean losses here are much greater than the 10 past publications when only accounting for temperatures, but are comparative when including CO_2 fertilization. These comparisons are complicated by the mixture of scenarios and model types in the IPCC summary, and most previous studies do not include CO_2 fertilization.

This study is very similar to [44], which predicts global crop yields for different low-emission climate scenarios, with and without CO_2 fertilization. They find that accounting for CO_2 fertilization mitigates the effect of warming temperatures. The results of this study are comparable to those of [44], since both studies find that C3 crops benefit strongly from higher carbon dioxide concentrations. The results of this study are also similar to those of [15], who use temperature, precipitation, and CO_2 fertilization to predict global corn and wheat yields. They also find clear benefits of reducing greenhouse gas emissions for corn yields, but fewer for wheat yields, as wheat is a C3 crop and benefits from CO_2 fertilization [48].

The statistical model developed here includes several assumptions. The three seasonal climate statistics only involve temperature, but crop yields may also be correlated with other conditions, such as precipitation, soil moisture, and radiation. Some other statistical models include these, but this study found much higher correlations with temperature than precipitation. Soil moisture and radiation were not available from weather station data. In order to project into the future, the model assumes that temperature continues to influence crop yields as they have historically, despite potential changes in other factors such as precipitation, soil conditions, or more advanced technologies. Another assumption is that linear regressions may be used as a predictive model for temperatures much hotter than those recorded historically, even though previous papers have found crop growth to respond nonlinearly to climate [8,49]. Despite these shortcomings, the high correlations in Figure 6 show that the climate statistics used for prediction are reasonable predictors of crop yields.

This study indicates that land in the northern United States may be more suitable for crop production as temperatures rise. Crops in the northern counties are less susceptible to events classified as extreme temperatures in those counties (Figure 6), and are forecasted to have much higher yields in 2100 (Figure 9). A limitation of a statistical model is that it can only predict crop yields in areas with sufficient crop data. Thus, this model cannot judge whether crops will have high yields in counties further north where crops have not been historically grown. A process model would be required to predict crop performance in these counties. A large unknown in this study is whether agricultural technology will continue to improve or whether crop yields will hit a fundamental biological limit. Given the difficulties of predicting future technologies, this study has instead projected a best case and a worst case scenario. Most likely, technology will have decreasing effects on crop yields due to biological constraints on production, and crop yields will increase at slower rates than have been in the last 50 years. In order to prepare for climate change, we should develop farming practices and crop breeds that are resistant to stronger and more frequent heat extremes.

This study highlights the trade-offs of climate change, where CO_2 fertilization is a potential benefit even while average temperatures and heat waves are increasing. Research to date has shown that the positive influences of CO_2 fertilization will increase yields for much of the 21st century, but will be countered by increasingly hot and dry conditions [50–52]. The results in this study exhibit the wide range of possible future impacts of climate change in the next century, and emphasize the need for continued research on the compounding effects of carbon dioxide fertilization and heat extremes.

Funding: This research received no external funding

Acknowledgments: L.K.P. is grateful to Mark Petersen and Phillip Wolfram for discussions on statistical methods and scientific presentation.

Conflicts of Interest: The author declares no conflict of interest.

References

1. Shrestha, S.; Deb, P.; Bui, T.T.T. Adaptation strategies for rice cultivation under climate change in Central Vietnam. *Mitig. Adapt. Strateg. Glob. Chang.* **2016**, *21*, 15–37. [CrossRef]
2. Deb, P.; Shrestha, S.; Babel, M.S. Forecasting climate change impacts and evaluation of adaptation options for maize cropping in the hilly terrain of Himalayas: Sikkim, India. *Theor. Appl. Climatol.* **2015**, *121*, 649–667. [CrossRef]
3. IPCC. Summary for Policymakers. In *Global Warming of 1.5 °C*; An IPCC Special Report on the Impacts of Global Warming of 1.5 °C above Pre-Industrial Levels and Related Global Greenhouse Gas Emission Pathways, In the Context of Strengthening the Global Response to the Threat of Climate Change, Sustainable Development, and Efforts to Eradicate Poverty; Intergovernmental Panel on Climate Change (IPCC): Geneva, Switzerland, 2018.
4. Schiermeier, Q. Economists Who Changed Thinking on Climate Change Win Nobel Prize. *Nature*, 8 October 2018. [CrossRef]
5. Nordhaus, W.D. Economic aspects of global warming in a post-Copenhagen environment. *Proc. Natl. Acad. Sci. USA* **2010**, *107*, 11721–11726. [CrossRef] [PubMed]
6. Hsiang, S.; Kopp, R.; Jina, A.; Rising, J.; Delgado, M.; Mohan, S.; Rasmussen, D.J.; Muir-Wood, R.; Wilson, P.; Oppenheimer, M.; et al. Estimating economic damage from climate change in the United States. *Science* **2017**, *356*, 1362–1369. [CrossRef] [PubMed]
7. Moore, F.C.; Diaz, D.B. Temperature impacts on economic growth warrant stringent mitigation policy. *Nat. Clim. Chang.* **2015**, *5*, 127–131. [CrossRef]
8. Schlenker, W.; Roberts, M.J. Nonlinear temperature effects indicate severe damages to U.S. crop yields under climate change. *Proc. Natl. Acad. Sci. USA* **2009**, *106*, 15594–15598. [CrossRef] [PubMed]

9. Edgerton, M.D. Increasing Crop Productivity to Meet Global Needs for Feed, Food, and Fuel. *Plant Physiol.* **2009**, *149*, 7–13. [CrossRef] [PubMed]

10. Gerland, P.; Raftery, A.E.; Ševčíková, H.; Li, N.; Gu, D.; Spoorenberg, T.; Alkema, L.; Fosdick, B.K.; Chunn, J.; Lalic, N.; et al. World population stabilization unlikely this century. *Science* **2014**, *346*, 234–237. [CrossRef] [PubMed]

11. Alston, J.M.; Beddow, J.M.; Pardey, P.G. Agricultural Research, Productivity, and Food Prices in the Long Run. *Science* **2009**, *325*, 1209–1210. [CrossRef]

12. Singh, R.P.; Hodson, D.P.; Huerta-Espino, J.; Jin, Y.; Njau, P.; Wanyera, R.; Herrera-Foessel, S.A.; Ward, R.W. Will Stem Rust Destroy the World's Wheat Crop? *Adv. Agron.* **2018**, *98*, 271–309. [CrossRef]

13. Butler, E.E.; Huybers, P. Adaptation of US maize to temperature variations. *Nat. Clim. Chang.* **2013**, *3*, 68–72. [CrossRef]

14. Liang, X.Z.; Wu, Y.; Chambers, R.G.; Schmoldt, D.L.; Gao, W.; Liu, C.; Liu, Y.A.; Sun, C.; Kennedy, J.A. Determining climate effects on US total agricultural productivity. *Proc. Natl. Acad. Sci. USA* **2017**, *114*, E2285–E2292. [CrossRef] [PubMed]

15. Tebaldi, C.; Lobell, D. Estimated impacts of emission reductions on wheat and maize crops. *Clim. Chang.* **2015**, 1–13. [CrossRef]

16. Petersen, L.K. Real-Time Prediction of Crop Yields From MODIS Relative Vegetation Health: A Continent-Wide Analysis of Africa. *Remote Sens.* **2018**, *10*, 1726. [CrossRef]

17. Ray, D.K.; Gerber, J.S.; MacDonald, G.K.; West, P.C. Climate variation explains a third of global crop yield variability. *Nat. Commun.* **2015**, *6*, 5989. [CrossRef]

18. Ortiz-Bobea, A.; Just, R.E. Modeling the Structure of Adaptation in Climate Change Impact Assessment. *Am. J. Agric. Econ.* **2013**, *95*, 244–251. [CrossRef]

19. Lobell, D.B.; Schlenker, W.; Costa-Roberts, J. Climate Trends and Global Crop Production Since 1980. *Science* **2011**, *333*, 616–620. [CrossRef] [PubMed]

20. Kukal, M.S.; Irmak, S. Climate-Driven Crop Yield and Yield Variability and Climate Change Impacts on the U.S. Great Plains Agricultural Production. *Sci. Rep.* **2018**, *8*, 3450. [CrossRef]

21. Najafi, E.; Devineni, N.; Khanbilvardi, R.M.; Kogan, F. Understanding the Changes in Global Crop Yields Through Changes in Climate and Technology. *Earths Future* **2018**, *6*, 410–427. [CrossRef]

22. Peng, B.; Guan, K.; Chen, M.; Lawrence, D.M.; Pokhrel, Y.; Suyker, A.; Arkebauer, T.; Lu, Y. Improving maize growth processes in the community land model: Implementation and evaluation. *Agric. For. Meteorol.* **2018**, *250–251*, 64–89. [CrossRef]

23. Sheng, M.; Liu, J.; Zhu, A.X.; Rossiter, D.G.; Zhu, L.; Peng, G. Evaluation of CLM-Crop for maize growth simulation over Northeast China. *Ecol. Model.* **2018**, *377*, 26–34. [CrossRef]

24. Butler, E.E.; Huybers, P. Variations in the sensitivity of US maize yield to extreme temperatures by region and growth phase. *Environ. Res. Lett.* **2015**, *10*, 034009. [CrossRef]

25. Zhang, Q.; Zhang, J.; Guo, E.; Yan, D.; Sun, Z. The impacts of long-term and year-to-year temperature change on corn yield in China. *Theor. Appl. Climatol.* **2015**, *119*, 77–82. [CrossRef]

26. Lobell, D.B.; Tebaldi, C. Getting caught with our plants down: the risks of a global crop yield slowdown from climate trends in the next two decades. *Environ. Res. Lett.* **2014**, *9*, 074003. [CrossRef]

27. Tebaldi, C.; Lobell, D.B. Towards probabilistic projections of climate change impacts on global crop yields. *Geophys. Res. Lett.* **2008**, *35*, L08705. [CrossRef]

28. Gornott, C.; Wechsung, F. Statistical regression models for assessing climate impacts on crop yields: A validation study for winter wheat and silage maize in Germany. *Agric. For. Meteorol.* **2016**, *217*, 89–100. [CrossRef]

29. Tao, F.; Zhang, Z.; Zhang, S.; Ratter, R.P.; Shi, W.; Xiao, D.; Liu, Y.; Wang, M.; Liu, F.; Zhang, H. Historical data provide new insights into response and adaptation of maize production systems to climate change/variability in China. *Field Crops Res.* **2016**, *185*, 1–11. [CrossRef]

30. Ummenhofer, C.C.; Xu, H.; Twine, T.E.; Girvetz, E.H.; McCarthy, H.R.; Chhetri, N.; Nicholas, K.A. How Climate Change Affects Extremes in Maize and Wheat Yield in Two Cropping Regions. *J. Clim.* **2015**, *28*, 4653–4687. [CrossRef]

31. Wang, P.; Zhang, Z.; Song, X.; Chen, Y.; Wei, X.; Shi, P.; Tao, F. Temperature variations and rice yields in China: Historical contributions and future trends. *Clim. Chang.* **2014**, *124*, 777–789. [CrossRef]

32. Wang, P.; Zhang, Z.; Chen, Y.; Wei, X.; Feng, B.; Tao, F. How much yield loss has been caused by extreme temperature stress to the irrigated rice production in China? *Clim. Chang.* **2016**, *134*, 635–650. [CrossRef]

33. Anderson, C.J.; Babcock, B.A.; Peng, Y.; Gassman, P.W.; Campbell, T.D. Placing bounds on extreme temperature response of maize. *Environ. Res. Lett.* **2015**, *10*, 124001. [CrossRef]

34. Hamer, H.; Picanso, R.; Prusacki, J.J.; Rater, B.; Johnson, J.; Barnes, K.; Parsons, J.; Young, D.L. *National Agricultural Statistics Service*; United States Department of Agriculture: Washington, DC, USA, 2017.

35. Menne, M.J.; Durre, I.; Vose, R.S.; Gleason, B.E.; Houston, T.G. An Overview of the Global Historical Climatology Network-Daily Database. *J. Atmos. Ocean. Technol.* **2012**, *29*, 897–910. [CrossRef]

36. Hartmann, D.; Tank, A.K.; Rusticucci, M. *Climate Change 2013: The Physical Science Basis. Contribution of Working Group I to the Fifth Assessment Report of the Intergovernmental Panel on Climate Change*; Technical Report; IPCC: Geneva, Switzerland, 2013.

37. USDA. *Field Crops: Usual Planting and Harvesting Dates*; Technical Report; National Agricultural Statistics Service: Washington, DC, USA, 2010.

38. Crow, E.L.; Davis, F.A.; Maxfield, M.W. *Statistics Manual*; Dover Publications, Inc.: Mineola, NY, USA, 1960.

39. Gent, P.R.; Danabasoglu, G.; Donner, L.J.; Holland, M.M.; Hunke, E.C.; Jayne, S.R.; Lawrence, D.M.; Neale, R.B.; Rasch, P.J.; Vertenstein, M.; et al. The Community Climate System Model Version 4. *J. Clim.* **2011**, *24*, 4973–4991. [CrossRef]

40. Taylor, K.E.; Stouffer, R.J.; Meehl, G.A. An Overview of CMIP5 and the Experiment Design. *Bull. Am. Meteorol. Soc.* **2011**, *93*, 485–498. [CrossRef]

41. Abatzoglou, J.T.; Brown, T.J. A comparison of statistical downscaling methods suited for wildfire applications. *Int. J. Climatol.* **2012**, *32*, 772–780. [CrossRef]

42. Abatzoglou, J.T. Development of gridded surface meteorological data for ecological applications and modelling. *Int. J. Climatol.* **2011**, *33*, 121–131. [CrossRef]

43. Hayhoe, K.; Edmonds, J.; Kopp, R.; LeGrande, A.; Sanderson, B.; Wehner, M.; Wuebbles, D. Climate models, scenarios, and projections. **2017**, *1*, 133–160. [CrossRef]

44. Tebaldi, C.; Lobell, D. Differences, or lack thereof, in wheat and maize yields under three low-warming scenarios. *Environ. Res. Lett.* **2018**, *13*, 065001. [CrossRef]

45. Jones, J.W.; Hoogenboom, G.; Porter, C.H.; Boote, K.J.; Batchelor, W.D.; Hunt, L.A.; Wilkens, P.W.; Singh, U.; Gijsman, A.J.; Ritchie, J.T. The DSSAT cropping system model. *Eur. J. Agron.* **2003**, *18*, 235–265. [CrossRef]

46. USDA, NASS. *Agricultural Prices*; USDA, NASS: Washington, DC, USA, 2018.

47. USDA, NASS. *Crop Production 2016 Summary*; USDA, NASS: Washington, DC, USA, 2017.

48. Högy, P.; Wieser, H.; Köhler, P.; Schwadorf, K.; Breuer, J.; Franzaring, J.; Muntifering, R.; Fangmeier, A. Effects of elevated CO_2 on grain yield and quality of wheat: Results from a 3-year free-air CO_2 enrichment experiment. *Plant Biol.* **2009**, *11*, 60–69. [CrossRef] [PubMed]

49. Lobell, D.B.; Bänziger, M.; Magorokosho, C.; Vivek, B. Nonlinear heat effects on African maize as evidenced by historical yield trials. *Nat. Clim. Chang.* **2011**, *1*, 42–45. [CrossRef]

50. Jin, Z.; Ainsworth, E.A.; Leakey, A.D.B.; Lobell, D.B. Increasing drought and diminishing benefits of elevated carbon dioxide for soybean yields across the US Midwest. *Glob. Chang. Biol.* **2018**, *24*, e522–e533. [CrossRef]

51. Jin, Z.; Zhuang, Q.; Wang, J.; Archontoulis, S.V.; Zobel, Z.; Kotamarthi, V.R. The combined and separate impacts of climate extremes on the current and future US rainfed maize and soybean production under elevated CO_2. *Glob. Chang. Biol.* **2017**, *23*, 2687–2704. [CrossRef] [PubMed]

52. Obermeier, W.A.; Lehnert, L.W.; Kammann, C.I.; Müller, C.; Grünhage, L.; Luterbacher, J.; Erbs, M.; Moser, G.; Seibert, R.; Yuan, N.; et al. Reduced CO_2 fertilization effect in temperate C3 grasslands under more extreme weather conditions. *Nat. Clim. Chang.* **2016**, *7*, 137. [CrossRef]

Perspective

Climate-Smart Agriculture and Non-Agricultural Livelihood Transformation

Jon Hellin [1,*] **and Eleanor Fisher** [2]

1   Sustainable Impact Platform at the International Rice Research Institute (IRRI),
    Metro Manila 1301, Philippines
2   School of Agriculture, Policy and Development at the University of Reading, Reading RG6 6AH, UK;
    e.fisher@reading.ac.uk
*   Correspondence: j.hellin@irri.org

Received: 20 February 2019; Accepted: 27 March 2019; Published: 31 March 2019

Abstract: Agricultural researchers have developed a number of agricultural technologies and practices, known collectively as climate-smart agriculture (CSA), as part of climate change adaptation and mitigation efforts. Development practitioners invest in scaling these to have a wider impact. We use the example of the Western Highlands in Guatemala to illustrate how a focus on the number of farmers adopting CSA can foster a tendency to homogenize farmers, instead of recognizing differentiation within farming populations. Poverty is endemic in the Western Highlands, and inequitable land distribution means that farmers have, on average, access to 0.06 ha per person. For many farmers, agriculture per se does not represent a pathway out of poverty, and they are increasingly reliant on non-agricultural income sources. Ineffective targeting of CSA, hence, ignores small-scale farming households' different capacities for livelihood transformation, which are linked to the opportunities and constraints afforded by different livelihood pathways, agricultural and non-agricultural. Climate-smart interventions will often require a broader and more radical agenda that includes supporting farm households' ability to build non-agricultural-based livelihoods. Climate risk management options that include livelihood transformation of both agricultural and non-agricultural livelihoods will require concerted cross-disciplinary research and development that encompasses a broader set of disciplines than has tended to be the case to date within the context of CSA.

Keywords: climate-smart agriculture; livelihood transformation; Guatemala; climate change

---

1. Introduction

Climate change will have a detrimental impact on agricultural productivity in many parts of the developing world [1]. Farmers have long adapted to climate variability, but the severity of the predicted changes may be beyond many farmers' current ability to adapt and improve their livelihoods [2,3]. There is an urgent need to work with farmers to develop climate change adaptation, mitigation and transformation strategies. Sustainable development goal (SDG) 13 is on *Climate Action* and, hence, there is much interest in the promotion of climate-smart agricultural practices (CSA). These are practices that contribute to an increase in global food security (and other development goals), an enhancement of farmers' ability to adapt to a changing climate and the mitigation of emissions of greenhouse gases [3,4]. CSA, hence, not only contributes to the realization of SDG 13, but is also intrinsically linked to several other SDGs, for example, SDG 1: *No Poverty* and SDG 2: *Zero Hunger*.

Transformative approaches have gained traction in contemporary policy debates on climate impacts, stimulated, amongst other factors, by the United Nations Sustainable Development Goals (SDGs) and the Intergovernmental Panel on Climate Change (IPCC). CSA can be transformative in terms of its aims to ensure food security via a reorientation of agricultural development in the context of the realities of climate change. For example, recent research on the climate-smart village (CSV) approach [4] highlights the potential of scaling out so as to benefit larger number of farmers. However, there has been limited scaling of the CSV approach. One of the challenges is that the scaling of the CSV approach is premised largely on identifying a portfolio of CSA options and the financial or institutional mechanisms that enhance adoption by farmers, and targeting these at regions with similar agro-ecological conditions [4], with less attention being given to the local context [5,6]. The danger is the a priori belief that CSA is a pathway out of poverty. For many farmers, adaptation to climate change in ways that lead to an escape from poverty, and greater prosperity may not be via CSA [7].

Lipper et al. [3] stress that CSA results in higher resilience and lower risks to food security. While this may be the case, there are farmers for whom agricultural-based livelihoods are so precarious that even "climate-proofing" their agricultural systems represents a higher risk to food security and prosperity than non-agricultural livelihood options. The challenge, therefore, is that at the same time that international calls for transformative approaches are made, current and future rural livelihood conditions are so adverse that, for some, this is a matter of changing to grasp any livelihood opportunity, including adverse coping strategies, without any ability to improve agricultural practices in ways that could be considered synonymous with transformative change in a positive sense. Hence, "climate-smart" may actually mean the need for actions that focus on supporting people in building non-agricultural-based livelihoods [3,4]. If this livelihood transformation is to be positive, it will require concomitant policy and development support to provide enabling conditions for non-agricultural livelihoods to be built. Moreover, this needs to be performed in ways that improve household income and security, i.e. are prosperity-enhancing, thus avoiding recourse to adverse coping strategies. The Western Highlands of Guatemala illustrates this challenging development scenario.

2. Climate-Smart Agriculture in the Western Highlands, Guatemala

Scientific evidence points to negative impacts on agriculture in Guatemala, and other parts of Central America, due to changing temperature and rainfall patterns, e.g., [8]. Inequalities in land distribution have forced many resource-poor farmers to farm steep hillsides, areas that are very susceptible to soil and land degradation. The response has often been the promotion of CSA. Development practitioners are rediscovering technologies and practices that were promoted in the region in the 1980s and 1990s under the guise of soil and water conservation [9]. These included live barriers, stone terraces, cover crops, green manures and agro-forestry. Farmers' uptake of these technologies and practices was disappointing 20–30 years ago [10], largely because, as is the case worldwide, a technology-led approach tends to ignore the needs for institutional enabling factors, which are very important when it comes to farmers' uptake of agricultural technologies [6,11,12].

The promotion of CSA in Guatemala is particularly challenging. The country suffers from extreme rural poverty and food insecurity [13]. Guatemala is ethnically very diverse, and indigenous groups (who make up almost 40% of the total population) live mainly in the Western Highlands. The underpinnings of present-day poverty are rooted in conflict, linked to Guatemala's 36-year civil war, which ended in the mid-1990s, and during which tens of thousands of indigenous people died [14]. This has left a legacy of inequality and continued social tension.

Small-scale farmers practice largely subsistence and some market-oriented agriculture. The most important cultivated food crop is maize, which is intercropped with beans, chilies and squash [15]. Recent research has shown that land availability in the Western Highlands is 0.06 ha per person [13]. This contributes to considerable food insecurity: farm households produce enough maize (the main staple crop) for fewer than seven months of consumption per year, and for household consumption have to purchase maize to make up the deficit. As a consequence, the majority of farmers seek off-farm employment on a temporary basis, while a minority have managed to branch into the production of higher-value vegetable crops for the export market [16].

Donors have invested much in rural development projects [17,18]. One such rural development project in the Western Highlands was implemented from 2013–2018. The Buena Milpa project was supported by the United States Agency for International Development (USAID), through its Global Hunger and Food Security Initiative "Feed the Future". Its main objectives were to reduce poverty, food insecurity and malnutrition, while increasing the sustainability and resilience of maize-based farming systems (The ideas reported here stem from Hellin's involvement as a socio-economist in this project in the Western Highlands of Guatemala). More details are provided in [13,19]. A strong emphasis of the project was the promotion of CSA, and the project worked through a number of non-governmental organizations. During the course of the work, it became clear that more attention needed to be focused on farmers' different capacities to engage in climate risk management. The danger was that poor targeting of CSA would lead to weak farmer uptake, by implication excluding many poor farmers and/or including those farmers for whom farming (and improvements in farm productivity via the use of CSA) would do little to enable them to escape poverty.

That project brought to the fore the need to recognize more explicitly the heterogeneity of farm households and the need to broaden the portfolio of livelihood options available to them. A further challenge was to accommodate the understandable desires of the donor to see an impact on the ground. There was pressure to scale CSA, in terms of enhanced farmer uptake of technologies and practices. Implicitly, this served to dismiss emerging evidence that the role of non-agriculture-based livelihoods needed to be taken into account in decision-making regarding appropriate interventions; the promotion of livelihood improvement through CSA was inappropriate for some categories of farmer. There was a danger that the focus on numbers would distract from whether farmer uptake of CSA, while contributing to an improvement in food security, would still leave farmers trapped in poverty, not to mention the potential for other unanticipated impacts, such as when wealthier farmers are able to capture the benefits of CSA, with the consequence that their wealth grows at the expense of poorer farmers, leading, ultimately, to greater social inequality.

The project in Guatemala clearly demonstrates the importance of priority-setting and factoring in the varied possibilities and local conditions that farmers face when it comes to targeting project interventions. Thornton et al. [5] provide a useful framework for CSA priority-setting that is based on six elements, and is designed to help guide best-bet CSA intervention. There is widespread recognition of the trade-offs when implementing CSA among the three pillars of food security, adaptation and mitigation [5,6]. The example of the Western Highlands illustrates "higher-level" trade-offs between some of the SDGs. These include trade-offs between SDG 13: *Climate Action* and SDG 5: *Gender Equality* together with SDG 10: *Reduced Inequality*.

In short, the Guatemalan project illustrates that a focus on the number of farmers adopting CSA can divert attention from the far more important issue, which is to support farmers' adaptation to climate change, either through making their agriculture-based systems more climate-resilient and/or by expanding their envelope of prosperity-enhancing non-agricultural livelihoods. The latter has been less prevalent in CSA interventions, and this has been at the expense of potentially ensnaring poorer categories of small-scale farmers in an agricultural-based poverty trap.

3. Climate-Smart Agriculture and Poverty Reduction

Farm households can be distinguished based on their asset endowment, e.g., their amount of land, access to key agricultural inputs etc., coupled with characteristics that determine the livelihood strategies available to them. These livelihood strategies, in turn, influence the livelihood incomes that hopefully enable a household to maintain and strengthen its livelihood security. The livelihood pathways available to a farm household are determined by the household's characteristics (e.g., dependency ratio, availability of labour, etc.), along with the interaction between the available assets (financial, natural, social, human) and the enabling or disabling economic, institutional and policy environment. An understanding of these livelihood pathways informs decisions as to where to target CSA and where to develop enabling approaches that facilitate livelihood changes.

There is no doubt that CSA and agricultural interventions can contribute to poverty reduction and enhanced prosperity. Numerous examples abound, e.g., [20–22]. However, the agricultural future is bleak for some farmers struggling with few resources and the additional challenge of climate change. Harris and Orr [23] argue that for rain-fed agriculture, crop production could be a pathway from poverty where smallholders are able to increase farm size or where markets stimulate crop diversification, commercialization and increased farm profitability. The potential to improve productivity is also, of course, important. Nevertheless, as Cavanagh et al. [24] comment, *"the poor and less poor are […] more capable of diversifying into off-farm and non-farm activities compared to the very poor, whose small land holdings and poor access to capital constrain their ability to diversify away from on-farm income and seasonal off-farm wage labour"*. We certainly found this to be the case in the Western Highlands of Guatemala [13].

Agriculture is not a pathway out of poverty for all farm households. Hence, for certain categories of household, poverty reduction will come from farmers moving out of agriculture and into the non-farm economy. For poor households, non-agricultural livelihood transformation can, of course, represent nothing more than a negative coping strategy. The challenge is to ensure that non-agricultural livelihood options are positive, i.e., prosperity-enhancing. It is a challenge in all parts of the world due to profound rural changes. In many parts of the world, agricultural production will have to increase hugely, along with labour productivity; the latter will lead to fewer people engaging in agriculture [25]. This has already led, in parts of Asia, to what Li [26] refers to as a rural population that is "surplus" to the needs of capital, as many of those dispossessed from their land are also unable to find meaningful employment off-farm. It is also increasingly common in Latin America.

The idea of a "surplus" population mirrors the earlier thesis of "functional dualism", proposed by de Janvry et al. [27] and expanded on by Blaikie [28]. The authors suggest that farmers rely upon returns from market activities to complement their agricultural returns from farming plots of land that are too small to allow for self-sufficiency. Farmers are often obliged to work as part-time wage laborers due to their resulting food insecurity, thus needing to make up shortfalls of staples and cash requirements for household goods, as well as to pay for inputs for the production process itself on their farms. They are increasingly dependent on non-farm sources of income but are unable to find sufficient employment opportunities or capital to migrate (and abandon the agricultural sector) or to depend fully on wage earnings for their subsistence. Returns from subsistence-oriented agricultural activities provide a necessary complement to the low wages that farmers receive in the labour market. In addition, where opportunities for wage labour are primarily in the agricultural sector, poor returns from own-farm agricultural production are reinforced, given that peak demand for agricultural labour may coincide with labour demands on farmers' own land.

The situation in the Western Highlands of Guatemala, as described in the section above, is in keeping with the functional dualism thesis, and has major implications for identifying and targeting appropriate pathways leading to rural poverty reduction. As suggested, for many farmers in the Western Highlands, CSA may not be an attractive option because of labour and land shortages. In the case of many farmers in the Western Highlands of Guatemala, temporary migration in search of non-farm employment has been a traditional coping strategy, with farmers investing the earned off-farm income in their villages and/or diversifying into non-farm agricultural activities, such as setting up a local shop. For many farmers, labour, essential for investment in soil improvement or maintenance of conservation structures, is not available, because they are working off-farm (and households may also have high dependency ratios) [23]. Similarly, another refrain, when CSA practices such as conservation agriculture are promoted, is that farmers should not burn their fields to clear the vegetation prior to planting because of the adverse impact on soil quality, especially biological health. For farmers who have been working off-farm, and for whom labour is scarce, this can be an unattractive recommendation.

CSA is also very problematic when it comes to small-scale farmers with very small landholdings, as is the case in the Western Highlands. Firstly, in the case of cross-slope soil conservation technologies, such as live-vegetation barriers and stone walls, land is taken out of production. In the case of live barriers, however, this can be partly compensated for by using species that make a contribution to the farm household, e.g., edible products for humans and/or animals. Secondly, even if CSA were to lead to significant improvements in agricultural productivity, the increase (while a contribution to food security) would be unlikely to help the farmer escape poverty. *"For most smallholders, however, small farm size and limited access to markets mean that returns from improved technology are too small for crop production alone to lift them above the poverty line"* [23].

This raises the question of how best to support categories of farmers who are being targeted but whose small holdings, household structure and asset endowment may be inappropriate for the measures advocated under the guise of CSA. It may be the case that the CSV approach would be more successful, but in the absence of effective scaling of this approach and comprehensive impact studies, this remains an under-researched area.

4. Non-Agricultural Livelihood Transformation

In rural contexts where small-holder farmers are based, development involves decreasing livelihood vulnerability and increasing incomes, typically through changes in livelihood activities [29]. Ideally, CSA should enable farmers to pursue livelihood pathways that lead to greater prosperity, while also building resilience. Recent thinking has advocated addressing the need to support changes that can be transformative, in the face of climate-related impacts that imply dramatic changes to environmental conditions. There is much research on developing a framework for assessing and comparing different types of interventions that address the key elements of CSA [5,10]. This research is necessary and important; however, in the context of agricultural transformation, the focus needs to broaden to systematically factor in livelihood trajectories outside of agriculture. CSA, to enhance food security and meet the SDGs, will require a longer-term perspective and bolder action that comprehensively targets farmer livelihoods [5].

The reality is that positive, sustainable livelihood pathways within the agricultural sector may not be an option for all types of farmers, i.e., not all households face the agro-ecological and socio-economic conditions necessary to move from one asset threshold and livelihood pathway to another, enabling them to escape poverty, while still remaining in agriculture. Guatemala epitomizes this reality. A report produced for the United States Agency for International Development (USAID) noted that *"given the agricultural foundation and 'capital' that many Western Highland communities continue to hold, [there is a need to] re-assess the productive options available in agriculture or agriculture-based livelihoods; and engage youth (many of whom have written agriculture off as an option) in development of potential integrated economic/environmental/social development initiatives"* [18].

Incorporating non-agricultural livelihood transformation within CSA requires innovative and open thinking on the way forward for CSA. This has been acknowledged by proponents of CSA, e.g., [3,4], but it poses disciplinary challenges and has not led to the type of holistic and transformative changes are that needed. What is required is a broader and more comprehensive understanding of the realities faced by farmers and the changes needed to foster large-scale transformation in their livelihood trajectories [30]. This means that CSA thinking has to involve those from a plethora of disciplines from the natural and social sciences [31]. In the context of CSA, we have a practical example of how transformation also becomes a political issue [32]. The debate around adaptation to environmental change often avoids questioning the socio-economic and political reasons why farmers' livelihoods are so vulnerable [33].

In the context of Guatemala, serious discussion around climate change adaptation, mitigation and transformation will have to contend with politically divisive issues. In a small way, within a project, "politics" can mean challenging the premises that drive inappropriate scaling. Within the bigger picture, it also means taking into account the political economies within which CSA is advocated for small-scale farmers. In the Guatemalan context, this political economy relates to several decades of conflict and on-going socio-economic inequality that structurally disadvantage small-scale farmers in the Western Highlands. While there are, of course, specificities to the political economy of Guatemala, which have shaped its uptake of CSA, in any given context there will be political economy issues underpinning the implementation of CSA that cannot and should not be ignored. The climate change discourse has tended to focus on the adaptation and mitigation of greenhouse gas emissions, rather than *"problems of unevenly distributed power relations, networks of control and influence, and rampant injustices of the 'system"* [34].

Clearly, a disregard for issues of power and inequality is not tenable if CSA is to provide a viable mechanism for livelihood transformation and a contribution to the SDGs. Indeed, there is growing evidence of CSA proponents adopting a more "radical" agenda, factoring in more readily political and institutional issues and ensuring that CSA debate and implementation does not remain largely a discourse among "elite development and research agencies" [35]. Such recognition of the political realities of small-holder agricultural development are important if CSA is to have continued longevity and relevance within international agendas on climate change action and the SDGs.

5. Conclusions

Climate adaptation requires transformative change. The CSA approach needs to move even more squarely beyond a focus on resilience of food systems to encompass systematic thinking and action with respect to the resilience of farm households. This poses a real challenge, because CSA has tended to overlook targeting issues related to socio-economic differentiation within small-scale farming populations, although recent CSA initiatives have more readily included analyses of the institutional dimensions of CSA. Greater acknowledgement of institutional issues, and indeed the politics, of CSA interventions within rural planning are to be welcomed. CSA has nevertheless, in practice, tended to exclude systematic consideration of support for non-agricultural livelihood transformation that is positive for farm households in marginal contexts, such as the Western Highlands of Guatemala.

In some cases, CSA can lead to the triple win of increased productivity, adaptation and mitigation, but this is not the case for all types of farmers. We argue that more systematic attention be directed at climate risk management that moves beyond the more conventional adaptation and mitigation discourse, towards an approach that includes livelihood transformation from a broader perspective, i.e., one that does not just focus on rural–agricultural transformation, but also identifies (and embraces) where agriculture per se is not a pathway out of poverty and where support for positive non-agricultural livelihood trajectories are needed for small-scale farmers. This requires more disciplines working together, and, perhaps, meeting the challenge of addressing entrenched power balances, both within communities of scientists and in the small-scale farming populations that are the subject of CSA interventions.

Author Contributions: Conceptualisation, J.H. and E.F.; writing—original draft preparation, J.H. and E.F.; writing—review and editing, J.H. and E.F.

Funding: The research in Guatemala reported here was funded by the United States Agency for International Development (USAID) through its Global Hunger and Food Security Initiative, 'Feed the Future'. This work was also supported by the CGIAR Research Program (CRP) on Rice Agri-food Systems (RICE, 2017-2022) and the CRP on Climate Change, Agriculture and Food Security (CCAFS), which is carried out with support from CGIAR Fund Donors and through bilateral funding agreements (for details please visit https://ccafs.cgiar.org/donors#.WxqT_4onaUk). The views expressed in this document cannot be taken to reflect the official opinions of the aforementioned organizations.

Acknowledgments: The authors would also like to thank two anonymous reviewers who provided invaluable comments on earlier versions of this paper.

Conflicts of Interest: The authors declare no conflict of interest. The funders had no role in the design of the study; in the collection, analyses, or interpretation of data; in the writing of the manuscript, or in the decision to publish.

References

1. Vermeulen, S.J.; Challinor, A.J.; Thornton, P.K.; Campbell, B.M.; Eriyagama, N.; Vervoort, J.M.; Kinyangi, J.; Jarvis, A.; Läderach, P.; Ramirez-Villegas, J.; et al. Addressing uncertainty in adaptation planning for agriculture. *Proc. Natl. Acad. Sci. USA* **2013**, *110*, 8357–8362. [CrossRef]
2. Adger, W.N.; Huq, S.; Brown, K.; Conway, D.; Hulme, M. Adaptation to climate change in the developing world. *Prog. Dev. Stud.* **2003**, *3*, 179–195. [CrossRef]

3. Lipper, L.; Thornton, P.; Campbell, B.M.; Baedeker, T.; Braimoh, A.; Bwalya, M.; Caron, P.; Cattaneo, A.; Garrity, D.; Henry, K.; et al. Climate-smart agriculture for food security. *Nat. Clim. Chang.* **2014**, *4*, 1068–1072. [CrossRef]

4. Aggarwal, P.; Jarvis, A.; Campbell, B.; Zougmoré, R.; Khatri-chhetri, A.; Vermeulen, S.; Loboguerrero, A.M.; Sebastian, S.; Kinyangi, J.; Bonilla-Findji, O.; et al. The climate-smart village approach: Framework of an integrative strategy. *Ecol. Soc.* **2018**, *23*, 15. [CrossRef]

5. Thornton, P.K.; Friedmann, M.; Kilcline, K.; Keating, B.; Nangia, V.; West, P.C.; Howden, M.; Cairns, J.; Baethgen, W.; Claessens, L.; et al. A framework for priority-setting in climate smart agriculture research. *Agric. Syst.* **2018**, *167*, 161–175. [CrossRef]

6. Totin, E.; Segnon, A.C.; Schut, M.; Affognon, H.; Zougmoré, R.B.; Rosenstock, T.; Thornton, P.K. Institutional perspectives of climate-smart agriculture: A systematic literature review. *Sustainability* **2018**, *10*, 1990. [CrossRef]

7. Hellin, J.; Fisher, E. Building pathways out of poverty through climate smart agriculture and effective targeting. *Dev. Pract.* **2018**, *28*, 974–979. [CrossRef]

8. Lobell, D.B.; Burke, M.B.; Tebaldi, C.; Mastrandrea, M.D.; Falcon, W.P.; Naylor, R.L. Prioritizing Climate Change Adaptation Needs for Food Security in 2030. *Science* **2008**, *319*, 607–610. [CrossRef]

9. Partey, S.T.; Zougmoré, R.B.; Ouédraogo, M.; Campbell, B.M. Developing climate-smart agriculture to face climate variability in West Africa: Challenges and lessons learnt. *J. Clean. Prod.* **2018**, *187*, 285–295. [CrossRef]

10. Hellin, J.; Haigh, M.J. Better land husbandry in Honduras: Towards the new paradigm in conserving soil, water and productivity. *Land Degrad. Dev.* **2002**, *13*, 233–250. [CrossRef]

11. Hellin, J.; López Ridaura, S. Soil and water conservation on Central American hillsides: If more technologies is the answer, what is the question? *AIMS Agric. Food* **2016**, *1*, 194–207. [CrossRef]

12. Chambers, R.; Pacey, A.; Thrupp, L. (Eds.) *Farmer First: Farmer Innovation and Agricultural Research*; Intermediate Technology Publications: London, UK, 1989.

13. Hellin, J.; Cox, R.; López-Ridaura, S. Maize Diversity, Market Access, and Poverty Reduction in the Western Highlands of Guatemala. *Mt. Res. Dev.* **2017**, *37*, 188–197. [CrossRef]

14. Steinberg, M.; Taylor, M. Guatemala's Altos de Chiantla: Changes on the high frontier. *Mt. Res. Dev.* **2008**, *28*, 255–262. [CrossRef]

15. Isakson, S.R. The agrarian question, food sovereignty, and the on-farm conservation of agrobiodiversity in the Guatemalan highlands. *J. Peasant Stud.* **2009**, *36*, 725–759. [CrossRef]

16. Hamilton, S.; Fischer, E.F. Non-traditional agricultural exports in highland Guatemala: Understandings of Risk and Perceptions of Change. *Lat. Am. Res. Rev.* **2003**, *38*, 82–110. [CrossRef]

17. Copeland, N. 'Guatemala Will Never Change': Radical Pessimism and the Politics of Personal Interest in the Western Highlands. *J. Lat. Am. Stud.* **2011**, *43*, 485–515. [CrossRef]

18. Democracy International. *Legacies of Exclusion: Social Conflict and Violence in Communities and Homes in Guatemala's Western Highlands*; Democracy International: Bethesda, MD, USA, 2015; ISBN 3019611660.

19. Hellin, J.; Ratner, B.D.; Meinzen-Dick, R.; Lopez-Ridaura, S. Increasing social-ecological resilience within small-scale agriculture in conflict-affected Guatemala. *Ecol. Soc.* **2018**, *23*, 5. [CrossRef]

20. Kassie, M.; Shiferaw, B.; Muricho, G. Agricultural Technology, Crop Income, and Poverty Alleviation in Uganda. *World Dev.* **2011**, *39*, 1784–1795. [CrossRef]

21. Verkaart, S.; Munyua, B.G.; Mausch, K.; Michler, J.D. Welfare impacts of improved chickpea adoption: A pathway for rural development in Ethiopia? *Food Policy* **2017**, *66*, 50–61. [CrossRef]

22. Dinesh, D.; Frid-Nielsen, S.; Norman, J.; Mutamba, M.; Loboguerrero Rodriguez, A.; Campbell, B. *Is Climate-Smart Agricultre Effective? A Review of Case Studies*; CCAFS Working Paper; CCAFS: Wageningen, The Netherlands, 2015.

23. Harris, D.; Orr, A. Is rainfed agriculture really a pathway from poverty? *Agric. Syst.* **2014**, *123*, 84–96. [CrossRef]

24. Cavanagh, C.J.; Chemarum, A.K.; Vedeld, P.O.; Petursson, J.G. Old wine, new bottles? Investigating the differential adoption of 'climate-smart' agricultural practices in western Kenya. *J. Rural Stud.* **2017**, *56*, 114–123. [CrossRef]

25. Collier, P.; Dercon, S. African Agriculture in 50Years: Smallholders in a Rapidly Changing World? *World Dev.* **2014**, *63*, 92–101. [CrossRef]

26. Li, T.M. To Make Live or Let Die? Rural Dispossession and the Protection of Surplus Populations. *Antipode* **2009**, *41*, 66–93. [CrossRef]

27. de Janvry, A.; Sadoulet, E.; Young, L.W. Land and labour in Latin American agriculture from the 1950s to the 1980s. *J. Peasant Stud.* **1989**, *16*, 396–424. [CrossRef]

28. Blaikie, P. Explanation and policy in land degradation and rehabilitation for developing countries. *Land Degrad. Dev.* **1989**, *1*, 23–37. [CrossRef]

29. Dorward, A. Integrating Contested Aspirations, Processes and Policy: Development as Hanging In, Stepping Up and Stepping Out. *Dev. Policy Rev.* **2009**, *27*, 131–146. [CrossRef]

30. O'Brien, K. Responding to environmental change: A new age for human geography? *Prog. Hum. Geogr.* **2010**, *35*, 542–549. [CrossRef]

31. Reid, W.V.; Chen, D.; Goldfarb, L.; Hackmann, H.; Lee, Y.T.; Mokhele, K.; Ostrom, E.; Raivio, K.; Rockström, J.; Schellnhuber, H.J.; et al. Environment and development. Earth system science for global sustainability: Grand challenges. *Science* **2010**, *330*, 916–917. [CrossRef]

32. Castree, N. Geography and Global Change Science: Relationships Necessary, Absent, and Possible. *Geogr. Res.* **2015**, *53*, 1–15. [CrossRef]

33. O'Brien, K. From adaptation to deliberate transformation. *Prog. Hum. Geogr.* **2012**, *36*, 667–676. [CrossRef]

34. O'Brien, K. Global environmental change III: Closing the gap between knowledge and action. *Prog. Hum. Geogr.* **2013**, *37*, 587–596. [CrossRef]

35. Chandra, A.; McNamara, K.E.; Dargusch, P. Climate-smart agriculture: Perspectives and framings. *Clim. Policy* **2018**, *18*, 526–541. [CrossRef]

Article

Climate Change-Induced Impacts on Smallholder Farmers in Selected Districts of Sidama, Southern Ethiopia

Tafesse Matewos

Geography & Environmental Studies & PhD Fellow in Development Studies at Institute of Policy & Development Research (IPDR), Hawassa University, Ethiopia. Po Box 05 Hawassa, Ethiopia; tafessemk@gmail.com

Received: 20 April 2019; Accepted: 16 May 2019; Published: 22 May 2019

Abstract: Different factors control the types of adaptive strategies and likelihoods of experiencing climate change-induced impacts by smallholder farmers. By using a mixed research method, this study examines the types and determinants of climate change-induced impacts on smallholder rural farmers in drought-prone low lands of Sidama, Southern Ethiopia. Randomly selected (401) households were surveyed on climate change-induced impacts. Longitudinal climatic data were also collected from the Ethiopian National Meteorological Agency to assess the trend of rainfall (RF), temperature and drought incidents. The analyses of the data revealed that RF and temperature had shown decreasing and increasing trends, respectively, during the three decades under consideration (1983–2014). These changes in RF and temperature exposed farmers to climate-related epidemics, drought, harvest loss, and hunger. The logit model results revealed that different factors control the likelihood of exposure to climate change-induced impacts. The findings revealed that literacy level, involving women in family decisions and farmers' involvement in adaptation planning, reduces the likelihood of exposure to climate change-induced hunger. Therefore, there is a need to work on human capital of the farmers through expanding education, strengthening women's participation in family decision-making, and by improving public participation in climate change adaptation undertakings to minimize climate change-induced impacts.

Keywords: climate change-induced impacts; smallholder farmers; drought-prone low lands; rural Sidama; southern Ethiopia

---

1. Introduction

Climate change has emerged as one of the development challenges of the 21st century [1,2]. There is high confidence and agreement among the global scientific community that climate change poses a serious threat to current and future sustainable development. According to the forecasts, climate change-induced impacts will continue to affect people, even if anthropogenic greenhouse gasses emissions stop today [3]. Sub-Saharan Africa (SSA), where smallholder farmers dominate agriculture [4], is one of the global hotspots for climate change-induced impacts [5]. In SSA, agriculture directly employs about 175 million people who cultivate degraded lands where there is no reliable supply of water for irrigation [6]. These smallholder farmers in SSA are among the most vulnerable groups to climate change and variability-induced impacts due to dependency on RF, limited use of irrigation, and weak adaptive capacity. Furthermore, limited human and material capacity, poor infrastructure, fragile environments, political instability, and marginalization contributed to the vulnerability in this region [7–10]. The case is not different in Ethiopia where smallholder farmers, who produce 90% of the total agricultural yield and own 95% of the total cultivated land [11], suffer from different climate change induced-impacts. Most of these subsistent farmers practice rain-fed traditional farming, use little modern agricultural inputs, and have little surplus to sell in local markets. They are not resilient enough to cope with climate change-induced recurrent shocks and long-term impacts [12–14]. Although climate change-induced impacts have been recurrent in Ethiopia and in the study area, little is known about the factors that affect the likelihood of exposure to such impacts in Ethiopia and SSA. Thus, by using mixed research strategy, this study presents the types and determinants of climate change induced-impacts on smallholder rural farmers in drought-prone lowland context.

Climate change induced-impacts refers to effects on natural and human systems due to gradual changes in climate, variations in weather and climatic elements from the average, and that of the impacts of climatic extremes. As IPCC report, climate change-induced impacts refer to impacts on human lives, livelihoods, culture, economies, ecosystems, and material resources due to hazardous climatic events over a period of time [3]. Climate change-induced impacts also include consequences and outcomes of the direct impacts including droughts, hunger, famine, loss of life and property, and sea level rise which can be potential or residual (ibid). The former includes impacts that occur due to a projected change in climate without adaptation. The latter refers to the impacts with adaptation [15]. Therefore, in this research, climate change-induced impacts include the immediate effects as well as outcomes and consequences of the immediate effects on human and natural systems. As global warming increases the likelihood of experiencing severe, persistent, and irreversible impacts on natural and human systems will be stronger [3,16]. Greenhouse gas emissions have adverse effects on biodiversity, ecosystem services, and economic development, which are causing risks for livelihoods and human security [17,18]. The 4th IPCC report argues that … *by 2020, between 75 million and 250 million people are projected to be exposed to increased water stress due to climate change*[15] (p13). Agricultural crop yield in some African countries could be reduced by 50% by 2020 and this can endanger the existence of many smallholder rural farmers.

Climate change-induced impacts have hampered poverty reduction and sustainable development in the Global South [5,19,20]. This is because most climate-related impacts affect the poor, who are struggling to come out of poverty [21]. Furthermore, livelihood strategies and housing conditions of the poor are more vulnerable to climate change and variability-induced impacts [22,23]. The poor usually have houses made of mud, bamboo, straw, and other inexpensive materials that are the most vulnerable to extreme weather events. The poor also cannot buy climate insurance against climate-related risks. Besides the direct impacts on agriculture-related livelihoods, climate change is also indirectly affecting the health and well-being of the poor through its effect on human and livestock health [24,25]. For example, because of increasing surface temperature, malaria is expected to migrate to higher altitudes and cause further health problems, which affect the poor's income and productivity. According to the World Health Organization (WHO) estimates, global warming causes about 150,000 deaths per year [26]. RF and temperature variability also result in pests and diseases incidents that affect the quantity and the quality of the crop yield [27]. Besides impacts, social strains caused by increased resource scarcity may lead to greater conflict [5,28], with the poor again being the most likely victims.

Erratic RF and higher temperatures characterize the drought-prone lowlands of Sidama. The area is part of the Great East African Rift Valley (GEARV). Climate change-induced impacts such as drought, crop failure, livestock loss, flooding, and water-borne and related diseases (such as diarrhea and malaria) have been recurrent in this area. The problem worsened in 2016 despite various adaptive strategies of the households, communities, and other actors, including the government. This needed a large-scale emergency relief for about 100,000 people, which is highest in recent documented drought history of the area. Despite a few research works on climate change-induced impacts on smallholder farmers, there is a gap in the literature about factors that control the likelihood of exposure to the impacts. Therefore, the main goal of this study is to identify factors that determine the likelihood of exposure to climate-induced impacts on smallholder rural farmers. Identifying such determinants is important for policy-makers, practitioners, and smallholder farmers to identify leverage points to manage the imminent impacts of climate change in the country and beyond.

2. Climate Change-Induced Impacts in Ethiopia

While agriculture is the backbone of Ethiopia's economy, it is RF dependent and dominated by smallholder subsistent farmers. About 80% of the Ethiopian population lives in rural areas making drought-prone agriculture as the primary means of livelihood. This figure is much higher than the SSA average, which stood at 63% in the year 2014 [11,29]. Furthermore, agriculture constitutes 40% of the GDP, supports 80% of the total employment, and is a source of 90% of the export in Ethiopia [11,30,31]. Nonetheless, climate change-induced impacts are challenging the roles of agriculture for the overall growth of the country [12]. In the 20th century alone, twelve extreme droughts happened in Ethiopia that hampered the economic development of the country. The drought incidents claimed the lives of hundreds of thousands and affected the livelihoods of over 50 million people [32]. Drought incidents have been increasing in recent decades in the country. Eight major droughts had occurred in the last three decades in Ethiopia: 1984/85 1987/88 1991 1994 1997 2002/03 2010/11, and in 2015/16 [32,33].

Previous studies revealed that climate change and variability had exerted significant impacts on agriculture and the overall economic growth in Ethiopia [32,34,35]. Specifically, climate change-induced impacts have hampered the country's economic growth and effort to move out of poverty. For instance, it was reported that from 1991–2010, the growth estimates were reduced by 2–9% because of climate change-induced impacts [30]. Under the worst-case scenarios, the economic impacts of climate change may reduce Ethiopia's GDP by up to 8% [36]. On the other hand, another study on climate change impact on agriculture and related sectors revealed a 10% GDP drop from the target [37]. The same study revealed a widening income inequality that could reach up to 20% due to the effect of climate change. In general, with higher vulnerability, and lesser resilience to climate change-induced impacts, the country is projected to experience a 6% decline in total agricultural output in the coming years [38].

Besides the impacts on the national economy, climate change-induced impacts have been affecting health and livelihood status of smallholder farmers in Ethiopia. Some of these impacts include climate-related epidemics, crop failure, flooding, livestock mortality, shortage of water and pasture, migration, and food aid dependency [2,39,40]. Unsustainable adaptation strategies of the farmers, such as selling assets and charcoal production, have resulted in the degradation of assets, environment degradation, and biodiversity losses. Climate change-induced food and water shortages affect the health, educational, and economic status of the households [8]. For instance, during drought events, children are forced to drop out of the school to participate in such household activities as collecting water and fuel-wood from long distances. Malaria affects labor availability at the household level and increases the health expenses of the family.

3. Data and Methods

3.1. Study Area

Located in the Horn of Africa, Ethiopia had a population of 94.35 million in 2017, according to the Ethiopian Central Statistical agency. The country has nine regional states and two city administrations. The Southern Nations Nationalities and Peoples' Regional State (SNNPRS) is one of the nine administrative regions comprising about 20% of the country's total population and 10% of the land area. The Sidama Administration Zone is one of the 14 administrative zones in the SNNPRS. With a population of 3,677,370 in 2014 [41], it is the most populous zone in the region. The zone is located in the central-eastern part of the region bordered by Oromiya in the North, East, and Southeast, Gedeo Zone in the South, and Wolayita Zone in the west. The zone lies between 6°10′ to 7°05′ North latitude and 38°21′ to 39°11′ East longitude. The total area of the zone is 6981.9 km^2 [42].

Three drought-prone districts (Hawassa Zuria, Boricha, and Loka Abaya) were selected for this study (Figure 1). The districts are located in the heart of the East African Rift Valley and they were chosen based on their higher potential risk of exposure to adverse climate change and variability. Variable RF and higher temperatures characterize the districts compared to other districts of the zone. The population of the districts is 576,865 from 113,285 households. About 97% of the population in these districts lives in rural areas where drought-prone agriculture is their source of livelihood. Agro-pastoral communities dominate the southwestern parts of the study area, where climate-related animal diseases are prevalent.

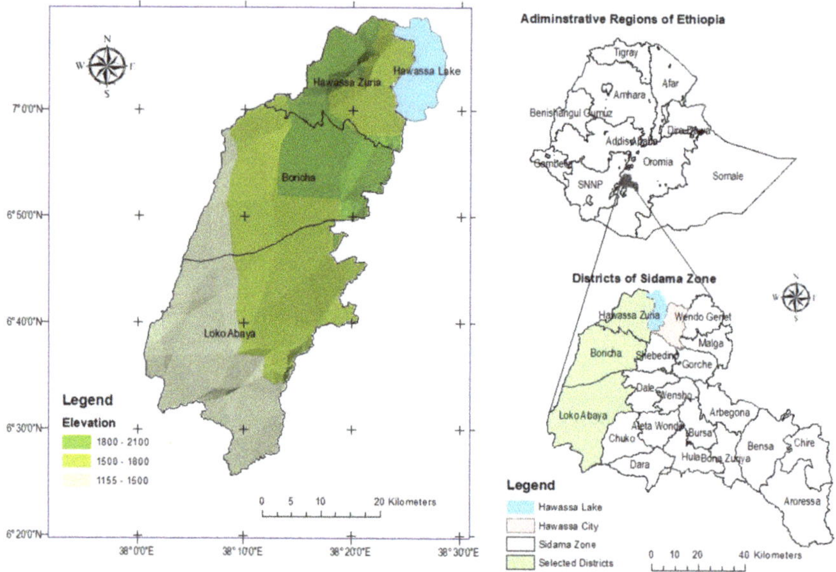

Figure 1. The geographic location of the study area.

Diverse climate characterizes the drought-prone districts of Sidama Zone. Altitude ranges from 560–2300 meters above sea level and the average annual RF ranges between 700 and 1200 mm. The RF pattern in the area is bimodal, which occurs in the summer and spring seasons. The summer (*kiremt*) rains last from June to September while the spring (*belg*) rains usually begin in February and end in late May. Both the *kiremt* and *belg* rains are irregular and unpredictable, resulting in frequent losses of harvest and cattle. The region has a bimodal precipitation distribution and high RF variability due to the seasonal progression of the inter-tropical convergence zone (ITCZ), an atmospheric circulation feature often modified by the El Niño–Southern Oscillation (ENSO). Also, there is severe livestock and human diseases (for example, malaria, cholera, and trypanosomiasis) in the area largely due to the inhospitable climate. The perennial Bilate River, which dissects the Wolayita and Sidama Zones, drains the western part of the area. The eastern part lacks perennial rivers, and households largely depend on artificial ponds and hand-dug shallow wells for water for domestic uses and watering livestock. The area experiences severe water shortages when the ponds run out of water during the December–February dry period.

3.2. Data

This study employed a cross-sectional household survey method. Randomly selected 401 rural household heads took part as respondents for the study. The following sampling procedure was applied. Based on their climatic conditions, accessibility, population size, and population density three kebeles from Boricha, two kebeles from Hawassa Zuria, and two kebeles from Loka Abaya districts (a total of seven kebeles) were selected. Accordingly, Hanja Cafa, Haldada Dela, and Korangoge from

Boricha district, Muticha Gorbe, and Sala Kore kebeles from Loka Abaya district, and Doyo Cala and Doyo Otilicha kebeles from Hawassa Zuria districts were selected for the study. The ideal sample size was calculated based on the Krejcie and Morgan formula; the maximum sample size at a 95% confidence interval and 5% margin of error for 750,000 people is 382 [43]. Considering attritions of the responses, a 5% contingency (19 HHs) gives 401 households, which were selected from the three districts. The 401 sample size was proportionally distributed based on the number of households in each district. So, 189 HHs from Boricha, 93 HHs from Loka Abaya, and 119 HHs from Hawassa Zuria districts were included for this study. The respondents were household heads who were selected by using systematic random sampling. Female-headed households were also included proportionately in all the districts (Table 1).

Table 1. Distribution of rural population and sampled households (HHs) in selected districts.

District Name	Total Rural Population	Total HHs	Kebele Name	Sampled HHs	Total
Boricha	299,175	58,823	Hanja Cafa	66	189
			Haldada Dela	66	
			Korangoge	57	
Loka Abaya	122,445	25,278	Muticha Gorbe	36	93
			Sala Kore	58	
Hawassa Zuriya	155,245	29,184	Doyo Chala	53	119
			Doyo Otilicho	66	
TOTAL	576,865	113,285		401(143)	401

( ) female headed HHs. Source: Own compilation from SNNPRS, BoFED 2015.

The field data collection process took place from April to May 2017 with trained data collectors supervised by the researcher. More data were gathered through focus group discussions, interviews with key informants, and field observation. The study also used RF data collected from the National Meteorological Agency of Ethiopia for the period 1983–2014 to support other data.

The RF data was analyzed by using trend analysis, Mann–Kendall's rank test, and standard RF anomalies (SRA). The SRA refers to standard RF anomaly and Pi and $P\mu$ represent RF of a given year and the mean RF, respectively. Accordingly, the result helps to classify temperature RF condition into different categories (Table 2).

$$\text{SRA} = \frac{P_i - P\mu}{\delta} \tag{1}$$

The collected quantitative data were encoded into STATA software (Version 14.2), and various descriptive and inferential techniques (mean, percent, standard deviation, and logit model) were used to analyze and interpret the results. The logit model analysis was carried out by using the STATA software to identify the determinants of climate change-induced effects on smallholder farmers. The qualitative data were analyzed via systematic thematization to augment the quantitative results.

3.3. Model Specifications

A logit model is a statistical approach used when the dependent variable (DV) is nominal with two or more categories commonly grouped as dummy variables (Table 3). On the other hand, the independent (predictor) variables (IVs) in a logit model can be qualitative (nominal or ordinal level) and interval/ratio level variables. This study used a logit model to identify factors that control the likelihood of households' exposure to climate change-induced impacts. Climate change-induced impacts considered in this study include the likelihoods to exposure to drought, harvest loss, flooding, and hunger. The exposure to hunger is examined vis-à-vis eighteen socio-economic and institutional s (Table 2). The eighteen dependent variables were selected among other socioeconomic and institutional variables after all the variables were tested to have a relationship with the four dependent variables considered for this study. The logit model showed that the selected eighteen variables have some explanatory power on the dependent variables considered.

Table 2. Variables description.

S.N	Variables	Variable Type	Description	Code
1	Sex	Explanatory	Sex of the HHs head	continuous
2	Family size	Explanatory	Number of HH members	continuous
3	Educational status	Explanatory	HHs head no. of years in school)	continuous
4	Total land size	Explanatory	Size of land in hectare	continuous
5	Land use certificate	Explanatory	I have land use right certificate	1) yes 0) no
6	PSNP beneficiary	Explanatory	Beneficiary of PSNP support	1) yes 0) no
7	Produce *enset*	Explanatory	Produce *enset* crop	1) yes 0) no
8	Membership in 1 to 5	Explanatory	Membership in 1 to 5	1) yes 0) no
9	Head of 1 to 5	Explanatory	Head of 1 to 5	1) yes 0) no
10	Communal grazing land	Explanatory	Communal grazing land	1) yes 0) no
11	Communal water pond	Explanatory	Communal water ponds	1) yes 0) no
12	Drought resistant variety	Explanatory	Drought resistant crop variety	1) yes 0) no
13	Fertilizer/urea/NPS	Explanatory	Fertilizer/urea/NPS	1) yes 0) no
14	Training on climate change	Explanatory	Get trained on climate change adaptation	1) yes 0) no
15	Early warning information	Explanatory	Get early climate change induced information	1) yes 0) no
16	Improved animal fodder	Explanatory	Use improved animal fodder	1) yes 0) no
17	HHs participation in adaptive decision	Explanatory	HHs participation in climate decision making	1) yes 0) no
18	Women participation in HHs decision	Explanatory	Women participation in HHs decision making	1) yes 0) no
19	Drought	Dependent	Exposure to the effects of drought	1) yes 0) no
20	Harvest loss	Dependent	Experienced harvest loss	1) yes 0) no
21	Flooding	Dependent	Affected by flooding	1) yes 0) no
22	Famine	Dependent	Experienced hunger	1) yes 0) no

1. Productive Safety Net Program is a government program being implemented in the study area to support poor and vulnerable people. 2. Are groups of 5 people organized by government and used for political, economic, and social purposes. 3. Drought resistant varieties are plant seeds that are supplied by the government and can grow in moisture stress conditions. 4. Urea and NPS are the two types of chemical fertilizers supplied to farmers by the government (usually in loan) NPS refers to Nitrogen-Phosphoric fertilizer containing Sulphur. 5. Improved animal fodder is a plant species (supplied by the government) that is used as a forage to feed animals during a shortage of pasture. 6. HHS are members of a family living in a house.

Table 3. Mann–Kendall's rank test result for inter-annual rainfall (RF).

Coefficient	n	Winter RF	Spring RF	Summer RF	Annual RF
Correlation Coefficient	1.000	0.117	0270 *	−0.020	−0.161
Sig. (2-tailed)		0.347	0.030	0.871	0.195
N	32	32	32	32	32

* Correlation is significant at the 0.05 level (2-tailed).

Logistic regression involves a linear function:

$$Z = b_0 + b_1 X_1 + b_2 X_2 + \ldots + b_n X_n \tag{2}$$

where Z is a dependent variable and $X_i \ldots X_n$ are explanatory variables.

The logistic regression function is thus, given by:

$$p = \frac{e^z}{1 + e^z} \tag{3}$$

where p is the probability that an event occurs; and the odds in favor of the occurrence are related according to $p = \frac{\text{Odds}}{1 - \text{odds}} = \frac{p}{1-p}$.

$$\log\left(\frac{p}{1-p}\right) = \beta_0 + \beta_1 X \tag{4}$$

Thus, odds is defined for an event with probability p and the logit is the log of the odds, i.e.;
Model - $p(x)$ = probability of the event occurring at $(Y=(x))$

$$p(y/x) \;=\; \frac{e^{\beta_0 + \beta_1 X_1 + \beta_2 X_2 + \ldots + \beta_k X_k}}{1 + e^{\beta_0 + \beta_1 X_1 + \beta_2 X_2 + \ldots + \beta_k X_k}} \tag{5}$$

A smaller sample size with a large number of predictors creates problems with the analysis, specifically when there are categorical predictors with limited cases in each category. Descriptive statistics were run for each of the predictors to solve the problem. The result showed that all categorical variables have a required number of cases set as a minimum standard [44]. A multicollinearity test was done to check for high intercorrelations among predictor (independent) variables. The test result suggested that the maximum r-value of the independent variables is 0.7, which is within an acceptable limit [44]. The presence of outliers or cases that would not well be explained by the model was also checked to make sure the model fits the data well. The model tried to find out factors that control the HHs likelihood of exposure to climate-induced impacts such as the exposure to the effects of drought, harvest loss, flooding, and hunger in the study area. R^2 and adjusted R^2 were also used to discuss the amount of variation explained by the independent variables. The former supposes that every independent variable in the model explains the variation in the dependent variable by explaining the variation in percent as if all independent variables in the model affect the dependent variable. On the other hand, the adjusted R^2 gives the percentage of variation explained by independent variables that in reality, affect the dependent variable. In this context, it is also called marginal effect after logit.

4. Results and Discussion

4.1. RF Condition and Drought Incidents

Studies on RF over Ethiopia revealed high variability and drought incidents [32,35,45]. Similarly, the analyses of RF data have shown declining RF trends and incidents of drought during the years under consideration (1983–2014). Annual and growing season (*belg* and *kiremt*) RF amounts had shown declining trends in the study area. The *belg* season (February, March, April, and May) RF accounts for 40% of the annual RF in the study area. It is an important season not only to grow short-growing season crops, but also it provides water for livestock and helps the growth of pasture. Though the current study does not check daily RF data of the growing seasons, the experts and focus group discussion participants reported that problems related to its onset, duration, and offset have serious consequence on food security in the study area. However, despite its huge role for livelihood security, it has shown a significant ($p \leq 0.05$) declining trend in the last three decades (Figure 2a, Table 3). The *kiremt* season (June, July, August, and September) is the main rain growing season in the study area which accounts for 41% of the total annual RF. *Kiremt* RF has also shown a declining trend over the years under consideration (Figure 2b), though the trend is not statistically significant (Table 3). The trend analysis of annual RF data has also revealed a declining trend, though the result is statistically insignificant (Table 3). The annual RF had varied from its average amount, which amounts to 1100 mm. The year 2009 was the driest year among the years under consideration with the annual amount of 796.37 mm. The years 1984, 1999, 2000, 2002, 2003, 2004, and 2009 were the drier with annual RF of 950.44 mm, 870.1 mm, 934.02 mm, 960.09 mm, 994.71 mm, and 925.08 mm, respectively.

The years 1983 and 1996 were the wettest years with annual RF amounts of 1420.42 mm and 1368.41 mm, respectively (Figure 3). On the other hand, the analysis of minimum, maximum, and mean monthly temperature had shown increasing trends, all of which are statistically significant

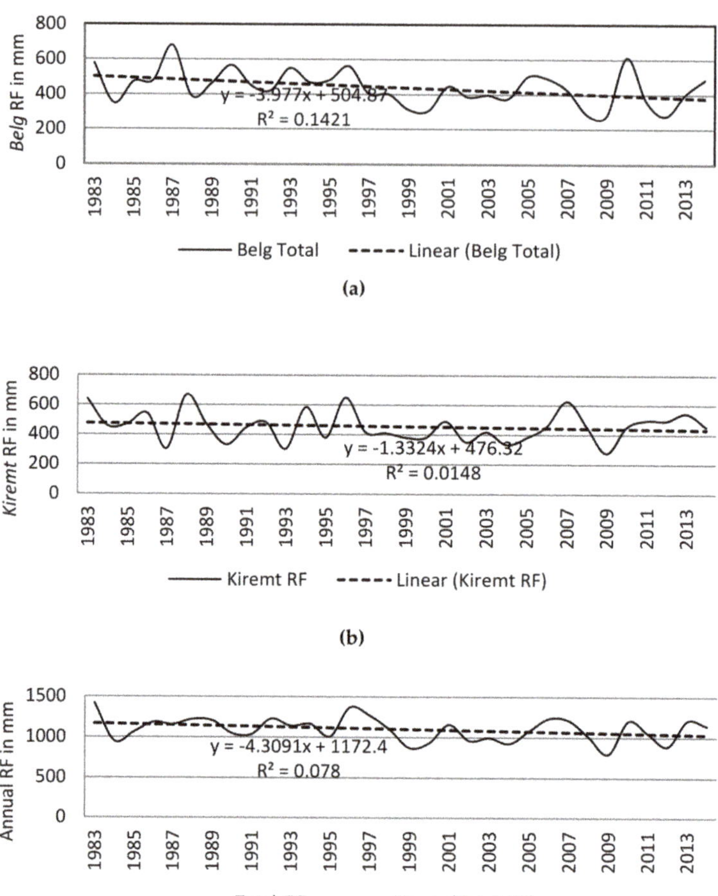

(a)

(b)

(c)

Figure 2. Rainfall trend in the study area (1983–2014). Source: Own computation from Ethiopian National Meteorological Agency data.

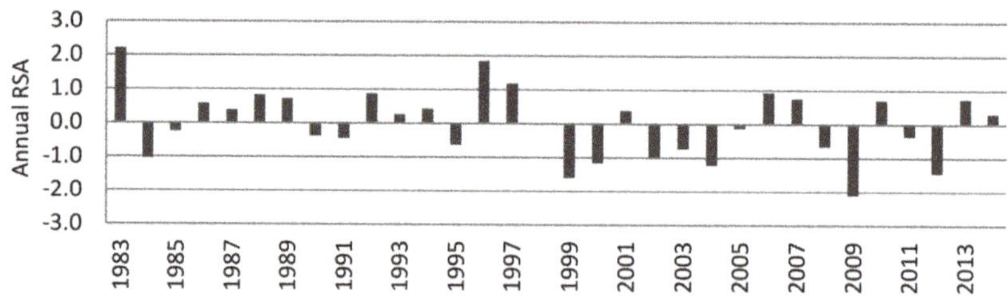

Source: Own computation from Ethiopian National Meteorological Agency data.

Figure 3. Annual RF standard anomalies in the study area (1983–2014).

The RF Standard Anomalies (RSA) helps to estimate the extent of drought based on the RF data [46] (Table 4).

Table 4. Drought severity index based on standard rainfall anomalies (SRA).

Anomaly	Temperature	RF
<−1.65	Extremely hot	Extreme drought
−1.65 to −1.28	Severe hot	Severe drought
−1.28 to −0.84	Moderate hot	Moderate drought
>−0.84	Cool	No drought

Source: Janowiak et al. 1986.

The analysis of annual RSA pointed out seven droughts incidents of different extent during the three decades under consideration. The year 2009 with annual SRA value of −2.11 was the driest year in the last three decades, which coincides with the 2009/10 extreme drought all over the country. The years 1999 and 2012 were also years of severe drought in the study area with SRA vales of −1.6 and −1.45, respectively. The other four moderate drought years were in 1984 (SRA = −1.04) 2000 (SRA = −1.16) 2002 (SRA = −0.98), and 2004 (SRA = −1.22) (Figure 3).

4.2. Types of Climate Change-Induced Impacts

Drought and flooding are the most common climate change variability-induced risks in Ethiopia [2,39,47]. The findings of this study nevertheless revealed other climate change-induced impacts that are recurrent in the study area. The common ones include climate-induced seasonal epidemics, drought, harvest loss, flooding, hunger, migration, and school dropout [48].

4.2.1. Climate Change and Variability Related Epidemics

Climate change-related epidemics globally claim over 150,000 lives per year [26]. Studies on the health impact of anthropogenic climate change show that climate change is affecting human health. Diseases such as cardiovascular mortality, respiratory illnesses, water-related contagious diseases, and malnutrition are directly or indirectly associated with climate change [24,26–50]. Malaria, cholera, and animal diseases were the major climate change and variability induced epidemics reported by the HHs in the study area. Of the total respondents asked about the incidence of malaria in their locality, 46% (n = 183) said that they had suffered from a malaria epidemic in their family. Malaria has been the most prevalent disease in the study area and its prevalence increases in the wet season from May to November. During wet seasons, the availability of moisture creates a conducive environment for mosquitos to breed and spread, such as those places left open to harvest rainwater.

On the other hand, thirty-five percent (n = 138) of the households reported cholera incidences. Water supply coverage in the study area has been marginal because of poor hydro-geological conditions and limited resources to improve water supply coverage. Water scarcity forces people to use untreated water for domestic consumptions. Key informants in Boricha district said that such incidences usually happen during water-scarce seasons where families resort to using unsafe water for domestic consumptions. Diarrheal diseases incidents are very common during the dry season (from December to March) affecting health and household labor availability in the area. Twenty-one percent (n = 85) of the study participants also reported the prevalence of animal diseases. The problem is more acute for agro-pastoral communities living in the western parts of the study area, especially in Loka Abaya district. Some HHS also reported the prevalence of animal diseases, such as trypanosomiasis. It affects animal farming in the study area. About 21% (n = 81) of respondents reported that they had experienced climate change-related livestock diseases. Trypanosomiasis is one of the tropical livestock diseases, which has widely spread all over the SSA [51,52]. The World Health Organization (WHO) has also pointed out that climate change induced animal diseases are affecting cattle header Maasai pastoralist communities living in Tanzania and Kenya [53].

4.2.2. Drought

Drought is a climatic condition, which occurs when there is a RF below its normal pattern for a long period of time [54]. Drought has happened in the study area every year since 2000, compelling large-scale relief intervention by government and international donors. A prolonged delay or a total absence of the rains during the two growing seasons (*belg* and *kiremt*) adversely affects crop and livestock productivity. About 84% (*n* = 336) of the households interviewed reported to have suffered from drought-related effects in the last ten years. During the dry season, pastoral household members travel eight to ten hours every other day to water their livestock at the Hawassa Lake or Bilate River. Such long-distance trekking undoubtedly hinders household heads or other family members from devoting enough time to other livelihoods and, therefore, incurring further economic losses. To support the farmers' drought incident experience with the RF data, the researchers carried out RSA analysis. The result pointed out decreasing annual, summer, and spring RF amounts for the years under consideration. The SRA results pointed out that seven drought events of different extent occurred in the study area of which one was extreme, two were severe, and four were moderate (Figure 3). The incidence of drought had increased in the study area because 88% of the drought happened in the last fifteen years of the three decades.

4.2.3. Harvest Loss

Climate change-induced harvest loss due to rainfall shortages in the growing season is one reason for food insecurity in developing countries [55]. There had been harvest losses because of climate change in the study area. This happened due to the shortage of RF during the growing season. Out of 401 HHs interviewed, eighty-two percent (*n* = 329) said that they have experienced harvest losses in the last ten years. The farmers in the study area practice a mixed livelihood: crop cultivation and animal husbandry. When drought occurs, the productivity of both suffers. The exception is for *enset* crop. Because of its biological characteristics, *enset* has strong drought resistance capacity and has a strong link with animal husbandry. The mixed livelihoods are dependent on RF since there are no small-scale irrigation projects in the study area. During the drought years when animal pasture/grass is scarce, farmers fed their cattle *enset* (false banana). This alternative is less feasible nowadays because of the scarcity of animal dung, which is a critical fertilizer to grow *enset*. As a result, as the number of cattle owned by households dwindles, *enset* production also declines. Thus, climate change is affecting the mixed (crop cultivation and animal husbandry) rural livelihood by degrading its natural connection.

4.2.4. Flooding

Climate change-induced flooding is one of the impacts of climate change, which has vast economic and social consequences. It has been occurring in Ethiopia for years inflicting heavy social and economic costs [56]. Flooding during rainy seasons is one of the primary climate-induced effects in the study area. Of the 402 households asked about their exposure to flooding, 74% (*n* = 297) said that flooding had affected them in the last ten years. Most parts of the study area, especially the western part, have suffered from devastating flooding due to the seasonal overflow of the Bilat River. On the other hand, since the area is located in lower elevations, there are run-offs from the surrounding highlands that destroy terraces built to control soil erosion. Also, several gullies occurred because of heavy rain that has affected the landscape as noted by the researcher during field visits. Flooding during rainy seasons causes property and harvest loss incurring a significant economic cost to the farmers. The social and economic effects mentioned by households include loss of human lives and livestock, property damages, destruction of crops, water-borne diseases, and soil erosion.

4.2.5. Climate Change-Induced Hunger

Famine is a condition where people lack food for a longer period whereas hunger is a seasonal lack of food because of natural or human-made causes. Hunger becomes famine when it persists and affects many people [57]. RF shortage is one of the main reasons for famine and hunger in SSA [4,5]. Of the 401 respondents included in this study, 77% ($n = 309$) did experience hunger in the last ten years. Data collected from the Sidama Zone Agriculture and Natural Resources Management Department show that climate change-induced hunger did happen in the three districts in the previous fifteen years. Many people received emergency food aid in 2000, 2004, 2009, and 2016. These years coincide with the years of lower SRA discussed in part 4.1. According to key informants, signs of hunger include a delayed start of rain during rainy seasons (*belg* or *kiremt*) or its absence during critical sowing seasons. Both affect crop and livestock productivity and food security.

4.3. Determinants of Climate Change-Induced Impacts on Smallholder Farmers

4.3.1. Determinants of Exposure to the Effects of Drought

The result of the logit model shows that several reasons decide the likelihood of exposure to the effects of drought in the study area. The model is statistically significant for the determinants of drought (N = 357, Chi^2 = 59.99, $p = 0.0000$). The pseudo R^2 and adjusted R^2 (after marginal effect) values for this category of explanatory variables are 18.56% and 88.83%, respectively (Table 5). Out of the explanatory variables considered in the model, educational status, growing enset, membership in 1 to 5 groups, and being head in 1 to 5 groups affect the exposure to the effects of drought. Besides, the use of chemical fertilizer and the participation of women in family decision-making are found to be the main determinants of exposure to the effects of climate change-induced drought. Specifically, the findings show that a one year increase in schooling decreases the likelihood of being affected by drought by 10% (at $p < 0.05$ level).

Growing enset is another determinant of climate change-induced drought. Households who grow enset are 111.3% less likely to experience the effects of drought than those without *enset* at $p < 0.05$ level (Table 5). *Enset* (false banana) is a perennial crop known for its drought-resistant nature and is wildly used as a staple food in the most southern part of Ethiopia [58,59]). *Enset* is important for the local economy because its products, locally called *waasa* and *bu'la*, are sold in markets and are a source of income. Its roots, leaves, and stem are also used to feed animals during drought years. The Peasant Association (Kebele Administration) is the lowest administrative unit in Ethiopia that is accountable to the district (district) administration. Locally, households are organized in small groups of five households known as the 1 to 5 group. The government designed it without having local farmers say in its formation. The government widely uses this group for economic, political, and social purposes. Of the total 402 households involved in this study, 60% ($n = 242$) were members in the 1 to 5 group organization and the remaining 40% ($n = 160$) were not. On the other hand, 11% ($n = 42$) were serving as group leaders. There is widespread debate and controversies on the merits and demerits of such grouping. However, the purpose here is as a social organization, to what extent being a member or a leader in 1 to 5 groups is related to the likelihood of facing climate change-induced impacts?

The model revealed that being a member or having a position in the 1 to 5 groups affects the likelihood of exposure to climate change-induced drought. Being a member of the group increases the likelihood of exposure to drought effects by 120.2% (at $p < 0.01$ level). On the other hand, having a position of leadership of a group decreases the exposure to climate change-induced drought by 149.7% at $p < 0.01$ level (Table 5). A further investigation on the socio-economic status of the 1 to 5 group heads pointed out that they are were individuals with a better resource base. Furthermore, the group heads have more decision making power on resources owned by the community. On the other hand, the members of these groups were poor farmers who joined the group to get government support. Despite the claims of the government to support the poor organized in the groups, the members were most vulnerable compared with others. Also, the model examined agricultural technology use and the likelihood of exposure to climate change-induced impacts. The results of the model show that except for improved animal fodder, other agricultural technologies did not decrease the likelihood of exposure to climate change-induced impacts. Using chemical fertilizers (Urea and NPS) increases the possibility of experiencing the effects of drought and harvest losses. It increases the likelihood of experiencing drought effects by 150.4% (at $p < 0.01$ level). This is contrary to the claims of government officials who argue using chemical fertilizers as rewarding for farmers. Interviews with farmers revealed that when RF onset is late, the application of chemical fertilizers is calamitous to farmers because the fertilizers wilt the crops in their early stages. Thus, applying chemical fertilizers during drought seasons increases their likelihood of exposure to the effects of drought. Moreover, those farmers who use chemical fertilizers are more likely to experience bankruptcy during drought years for they have to pay back the fertilizer loan, which is expensive according to key informants. Thus, the debt and risks associated with borrowing in precarious situations are found to be factors for the higher vulnerability of farmers.

Table 5. Marginal effects due to independent variables (the coefficients table).

Variables	Drought	Lost Harvest	Flooding	Hunger
Sex	−0.244	−0.370	−0.179	−0.559
	(0.360)	(0.355)	(0.278)	(0.327) *
Land certificate	0.468	0.100	−0.645	−0.273
	(0.424)	(0.448)	(0.379) *	(0.417)
Family size	0.026	0.063	0.009	0.021
	(0.080)	(0.087)	(0.063)	(0.073)
Educational status	−0.100	−0.030	0.069	−0.116
	(0.048) **	(0.051)	(0.042)	(0.045) ***
Total land size	−0.095	−0.202	−0.439	−0.571
	(0.160)	(0.172)	(0.181) **	(0.209) ***
PSNP beneficiary	0.003	−0.919	−0.923	−0.004
	(0.393)	(0.354) ***	(0.308) ***	(0.358)
Do you produce enset?	−1.113	−0.636	−0.209	0.094
	(0.497) **	(0.455)	(0.338)	(0.365)
Weather early warning information	0.176	0.436	−0.155	0.108
	(0.327)	(0.326)	(0.268)	(0.295)
Climate change adaptation training	0.202	0.127	−0.262	0.232
	(0.355)	(0.350)	(0.283)	(0.318)
Membership in 1 to 5	1.202	1.598	−0.852	1.475
	(0.380) ***	(0.387) ***	(0.301) ***	(0.343) ***
Head of 1 to 5	−1.497	−2.000	0.719	−1.427
	(0.510) ***	(0.515) ***	(0.486)	(0.482) ***
Grazing land	0.857	0.406	0.134	0.958
	(0.648)	(0.604)	(0.477)	(0.599)
Water pond	1.022	1.434	0.289	1.070
	(0.865)	(0.869) *	(0.542)	(0.870)
Improved animal fodder	−0.761	−0.816	−0.597	−0.887
	(0.521)	(0.480) *	(0.409)	(0.488) *
Fertilizer/urea/NPS	1.504	1.041	0.075	1.429
	(0.534) ***	(0.531) **	(0.506)	(0.506) ***
Drought resistant crop variety	0.437	−0.151	0.268	0.687
	(0.420)	(0.396)	(0.327)	(0.392) *
Women participation in family decision making	−1.089	−0.350	−0.020	−1.015
	(0.419) ***	(0.370)	(0.295)	(0.360) ***
Participation in adaptation decision	−0.135	−0.007	−0.529	−0.043
	(0.408)	(0.396)	(0.305) *	(0.360)
N	357	358	357	357
Chi^2	59.99	85.78	39.62	82.49
P	0.0000	0.0000	0.0024	0.0000
Pseudo R^2	0.1856	0.2444	0.0958	0.2118
Adjusted R^2 (Marginal effect after logit)	0.8883	0.8775	0.7570	0.8325

$* p < 0.1$; $** p < 0.05$; $*** p < 0.01$; (SE) = standard error.

4.3.2. Determinants of Facing Harvest Loss

The logit model results show that several factors affect the possibility of exposure to harvest loss in the study area. The model is statistically significant for the determinants of harvest loss (N = 358, Chi^2 = 85.78, $p = 0.0000$). The pseudo R^2 and adjusted R^2 values are 24.44% and 87.75%, respectively (Table 5). Among the explanatory variables, membership and headship in 1 to 5 groups, PSNP, and animal fodder control the likelihood of experiencing harvest loss. Also, the use of chemical fertilizer controls the likelihood of harvest loss. The PSNP is a social protection program that government instruments in drought-prone and food insecure areas of Ethiopia. Because of the widespread food insecurity, the government has been carrying out the PSNP in the study area. It aids vulnerable farmers in cash and kind in the study area. Of the sampled households, 29% ($n = 114$) were PSNP beneficiaries, whereas the remaining were not. The results of the model revealed that HHs supported by the program are less likely to experience the effects of harvest loss by 91.9% (at $p < 0.01$ level) than the non-users of the program. Like the case of drought exposure, membership in the 1 to 5 group increases the possibility of facing a harvest loss by 159.8% (at $p < 0.01$ level). On the contrary, bearing the leadership position in the 1 to 5 group reduces the likelihood of exposure to harvest loss by 200% (at $p < 0.01$ level). One of the challenges for farmers during the drought season is the lack of pasture and water for their livestock. To solve the problem, the government has introduced a plant species that farmers grow to feed their livestock. Farmers call it animal fodder in the study area. The logit model analysis shows that the use of this improved animal fodder reduces the likelihood of facing harvest loss. Farmers who use improved animal fodder are 81.6% less likely to experience harvest loss than the non-users (at $p < 0.1$ level). This is because, without the fodder, farmers use crops to feed livestock. Using chemical fertilizers in crop fields increases the chance of experiencing harvest loss. Farmers who apply chemical fertilizers in their crop field are 37% more likely to experience harvest loss during a drought season because crops wilt under inadequate rain conditions (Table 5).

4.3.3. Determinants of Exposure to the Effects of Flooding

Flooding is one of the climate change-induced impacts experienced by the HHs in the study area. The model is statistically significant for the determinants of experiencing flooding (N = 357, Chi^2 = 39.62.78, p = 0.0958). The pseudo R^2 and adjusted R^2 values of the model are 9.58% and 75.7%, respectively (Table 5). Among the variables considered, land use right certificate, being a beneficiary of PSNP, and membership in the 1 to 5 groups are the determinants of exposure to flooding. Also, taking part in climate change adaptation decisions affects exposure to climate change-induced flooding (Table 5). Land use right certificate affects the likelihood of experiencing climate change-induced impacts. The likelihood of facing flooding decreases by 64.5% for farmers who have land ownership certificate. This result is statistically significant (at $p < 0.1$ level). This is because farmers with a land use right certificate have a stronger landownership feeling than those without and this encourages farmers to carry out different soil and water conservation practices. Households with larger landholding size are less likely to experience flooding than HHs with a smaller landholding size. A unit increase in landholding size decreases the likelihood of experiencing flooding by 43.9% (at $p < 0.05$). Being a PSNP beneficiary also decreases the likelihood of facing climate change-induced flooding by 92.3% (at $p < 0.01$ level). On the other hand, membership in the 1 to 5 groups controls the likelihood of exposure to flooding impacts. Being a member in 1 to 5 group decreases the chance of experiencing flooding impacts by 85.2% (at $p < 0.01$ level). Participation in climate change adaptation decisions at the local level is also relevant. Farmers who take part in local level decision-making are 52% less likely to experience climate change induced flooding impacts than the non-participants (Table 5). The incident of flooding is common for low-lying areas, but the model did not consider topography and is a limitation of the model.

4.3.4. Determinants of Facing Hunger

Climate change and variability-induced hunger are one of the impacts experienced by the HHs in the study area, especially during drought periods. The model is statistically significant for the determinants of experiencing hunger by the HHs (N = 357, Chi^2 = 82.49, p = 0.0000). The pseudo R^2 and adjusted R^2 values are 21.18% % and 83.25%, respectively (Table 5). The logit model result identified sex, educational status, farmland size, membership, and headship in the 1 to 5 groups as determinants of exposure to hunger. Besides, the use of improved animal fodder, chemical fertilizer, and improved seed variety also control the chance of exposure to hunger. Furthermore, female participation in family decision making also decides the likelihood of exposure to change-induced hunger. The sex of the household head is associated with the likelihood of facing climate-induced impacts; being a male household head decreases the chance of exposure to hunger by 55.9% (at $p < 0.1$ level). This means female-headed HHs are more vulnerable to the exposure of hunger than the male counterpart. This is due to issues related to access to and control over resources owned by the family and the community.

On the other hand, a unit increase in education decreases the likelihood of facing hunger by 11.6% (at $p < 0.01$ level). This suggests the role of education in minimizing the impacts of climate change induced hunger. Household's total farm size also lowers the chance of facing hunger. A unit increase in farm size reduces the likelihood of facing hunger by 57.1% (at $p < 0.01$ level). This means as land holding size decreases, farmers become more vulnerable to climate change-induced hunger. Like the case of exposure to drought, membership in the 1 to 5 groups increases the likelihood of exposure to hunger. The likelihood of facing hunger increases by 147.5% for members in 1 to 5 group and decreases by 142.7% for the heads of the group (at $p < 0.01$ levels). Using improved animal fodder decreases the likelihood of experiencing hunger. Users of improved animal fodder are 88.7% less likely to face climate change-induced hunger than non-users (at $p < 0.1$ level). Women's involvement in family decision-making such as selling assets and renting out lands affects the likelihood of exposure to hunger. It reduces the chance of experiencing climate change-induced hunger by 101.5% (at $p < 0.01$ level). The use of chemical fertilizers and drought-resistant varieties also affects the possibility of experiencing climate change-induced hunger. Using these inputs increases the chance of experiencing hunger by 142.9% and 69.7%, respectively (Table 5). Key informants' interviews to justify the reasons behind these findings suggested the use of chemical fertilizer and improved seed varieties is disadvantageous for farmers. This is because farmers who use agricultural inputs usually get them in the form of a loan. The farmers have to sell their assets to pay back the debt since there is no surplus produce to sell in the market during drought years. This further degrades their asset base, exposing them to food security.

5. Conclusions

The analyses of data collected from different sources have shown that RF had been scarce, erratic and declining from time to time, exposing rained smallholder farmers to various climate change-induced impacts. Annual and growing season (*belg* and *kiremt*) RF had shown a declining trend in the study area. The analysis annual RSA has shown that the study area had experienced seven droughts of different extent during the three decades (1983–2014) under consideration. The monthly minimum, maximum, and average temperature had increased over the years under consideration, all of which are statistically significant. These changes in climatic variables resulted in climate change-induced epidemics, drought, harvest losses, flooding, and hunger.

The logit model results have shown that different several socio-economic and institutional factors control the likelihood of exposure to the impacts. The likelihood of exposure to the effects of drought is controlled by the educational status of the HH head, growing enset, and membership in the 1 to 5 groups. Furthermore, leadership in 1 to 5 groups, use of chemical fertilizers, and female involvement in family decision-making are other determinants. Furthermore, membership and headship in the 1 to 5 groups, being the beneficiary of the PSNP, the use of improved animal fodder, and the use of chemical fertilizer affect the likelihood of exposure to harvest loss. On the other hand, land use right certificate, participation in the PSNP, membership in the 1 to 5 groups, and participation in climate change adaptation decisions are determinants of climate change-induced flooding. Moreover, the logit model has identified that sex, educational status, farmland size, and membership and headship in the 1 to 5 groups affect the likelihood of experiencing climate change-induced hunger. Besides, the use of improved animal fodder, chemical fertilizers, and improved seed variety also control the likelihood of exposure to climate change-induced hunger. Participation of women in family decision-making also reduces climate change-induced hunger.

Climate change and variability-induced impacts are widely available in the study area despite various adaptive strategies of the government and the households. Government policies on education, female empowerment, land use right certification, and participation of the households in adaptation decisions can help to minimize climate change-induced impacts. The empirical literature also suggests that working on human capital is key to successful livelihood diversification [60–62]. In this regard, education (both formal and informal) and skills training need to be emphasized. On the contrary, organizing farmers into 1 to 5 groups contributed positively only in minimizing the incidence of flooding. Memberships in such organization did not help to reduce the possibility of facing the effects of drought, harvest loss, and hunger. Thus, the government needs to reform such organizations in a manner that they can contribute to lessening climate change impelled impacts. Besides, the supplies of agricultural inputs (improved seed varieties and chemical fertilizers) specifically during drought years increased the likelihood of exposure to climate change-induced impacts. Therefore, there is a need to build human capital through expanding education, strengthening female participation in family decision-making, and improve public participation in climate change adaptation undertakings to manage climate change-induced impacts sustainably.

Funding: This research received some funding from the NORHED-DEG Project of Norad in Hawassa University. The author is grateful for the financial support.

Acknowledgments: I would like to thank Tesfaye Semela and Sintayehu Hailu for commenting and shaping the statistical model used in the study. I also thank Girma Kebede for commenting the first draft of the article.

Conflicts of Interest: The author declares no conflicts of interest.

References

1. O'Brien, K.L.; Leichenko, R.M. Double exposure: Assessing the impacts of climate change within the context of economic globalization. *Glob. Environ. Chang.* **2000**, *10*, 221–232. [CrossRef]
2. Bewket, W.; Radeny, M.A.O.; Mungai, C. *Agricultural Adaptation and Institutional Responses to Climate Change Vulnerability in Ethiopia*; CGIAR Research Program on Climate Change, Agriculture and Food Security (CCAFS): Copenhagen, Denmark, 2015.
3. IPCC. *Climate Change 2014: Synthesis Report*; Contribution of Working Groups I, II and III to the Fifth Assessment Report of the Intergovernmental Panel on Climate Change; Pachauri, R.K., Meyer, L.A., Eds.; IPCC: Geneva, Switzerland, 2014; p. 151.
4. Kotir, J.H. Climate change and variability in Sub-Saharan Africa: A review of current and future trends and impacts on agriculture and food security. *Environ. Dev. Sustain.* **2011**, *13*, 587–605. [CrossRef]
5. Adeniyi, P.A. Climate change induced hunger and poverty in Africa. *J. Glob. Biosci.* **2016**, *5*, 3711–3724.
6. AGRA (Alliance for a Green Revolution in Africa Change). *Smallholder Agriculture in Sub-Saharan Africa*; AGRA: Nairobi, Kenya, 2014.
7. Adger, W.N.; Huq, S.; Brown, K.; Conway, D.; Hulme, M. Adaptation to climate change in the developing world. *Prog. Dev. Stud.* **2003**, *3*, 179–195. [CrossRef]
8. Morton, J.F. The impact of climate change on smallholder and subsistence agriculture. *Proc. Natl. Acad. Sci. USA* **2007**, *104*, 19680–19685. [CrossRef] [PubMed]
9. Schlenker, W.; Lobell, D.B. Robust negative impacts of climate change on African agriculture. *Environ. Res. Lett.* **2010**, *5*, 014010. [CrossRef]
10. FAO. Agriculture in Sub-Saharan Africa: Prospects and challenges for the next decade. In *Agricultural Outlook 2016–2025*; FAO: Rome, Italy, 2016.
11. MoA. *The Federal Democratic Republic of Ethiopia. Agriculture Sector Programme of Plan on Adaptation to Climate Change*; Technical Working Group: Addis Ababa, Ethiopia, 2011.
12. World Bank. *Measuring the Economic Impact of Climate Change on Ethiopian Agriculture: Ricardian Approach*; The World Bank: Washington, DC, USA, 2007.
13. Senbeta, A.F.; Olsson, J.A. Climate Change Impact on Livelihood, Vulnerability and Coping, Mechanisms: A Case Study of West-Arsi Zone, Ethiopia. Unpublished MA Thesis, Lund University, Lund, Sweden, 2009.
14. Tafesse, A.; Ayele, G.; Ketema, M.; Geta, E. Adaptation to climate change and variability in eastern Ethiopia. *J. Econ. Sustain. Dev.* **2013**, *4*, 91–103.

15. IPCC. *Climate Change 2007—The Physical Science Basis: Working Group I Contribution to the Fourth Assessment Report of the IPCC*; Solomon, S., Qin, D., Manning, M., Averyt, K., Marquis, M., Eds.; Cambridge University Press: Cambridge, UK, 2007.

16. IPCC. *Climate Change IPCC Third Assessment Report. Intergovernmental Panel on Climate Change*; IPCC Secretariat: Geneva, Switzerland, 2001.

17. Cooper, P.J.M.; Dimes, J.; Rao, K.P.C.; Shapiro, B.; Shiferaw, B.; Twomlow, S. Coping better with current climatic variability in the rain-fed farming systems of sub-Saharan Africa: An essential first step in adapting to future climate change? *Agric. Ecosyst. Environ.* **2008**, *126*, 24–35. [CrossRef]

18. Shiferaw, B.; Tesfaye, K.; Kassie, M.; Abate, T.; Prasanna, B.M.; Menkir, A. Managing vulnerability to drought and enhancing livelihood resilience in sub-Saharan Africa: Technological, institutional and policy options. *Weather Clim. Extrem.* **2014**, *3*, 67–79. [CrossRef]

19. Mertz, O.; Halsnæs, K.; Olesen, J.E.; Rasmussen, K. Adaptation to climate change in developing countries. *Environ. Manag.* **2009**, *43*, 743–752. [CrossRef] [PubMed]

20. Ravindranath, N.H.; Sathaye, J.A. Climate change and developing countries. In *Climate Change and Developing Countries*; Springer: Dordrecht, The Netherlands, 2002; pp. 247–265.

21. Mendelsohn, R.; Dinar, A.; Williams, L. The distributional impact of climate change on rich and poor countries. *Environ. Dev. Econ.* **2006**, *11*, 159–178. [CrossRef]

22. Carter, M.R.; Little, P.D.; Mogues, T.; Negatu, W. Poverty traps and natural disasters in Ethiopia and Honduras. *World Dev.* **2007**, *35*, 835–856. [CrossRef]

23. Ribot, J. Vulnerability does not fall from the sky: Toward multiscale, pro-poor climate policy. *Soc. Dimens. Clim. Chang. Equity Vulnerability A Warm. World* **2010**, *2*, 47–74.

24. McMichael, A.J. *Global Climate Change and Health: An Old Story Writ Large. Climate Change and Human Health: Risks and Responses*; World Health Organization: Geneva, Switzerland, 2003.

25. Kjellstrom, T. Climate change, direct heat exposure, health and well-being in low and middle-income countries. *Glob. Health Action* **2009**, *2*. [CrossRef]

26. Patz, J.A.; Campbell-Lendrum, D.; Holloway, T.; Foley, J.A. Impact of regional climate change on human health. *Nature* **2005**, *438*, 310. [CrossRef]

27. Rosenzweig, C.; Iglesias, A.; Yang, X.B.; Epstein, P.R.; Chivian, E. Climate change and extreme weather events; implications for food production, plant diseases, and pests. *Glob. Chang. Hum. Health* **2001**, *2*, 90–104. [CrossRef]

28. Nordás, R.; Gleditsch, N.P. Climate change and conflict. *Political Geogr.* **2007**, *26*, 627–638. [CrossRef]

29. UN-DESA. *World Urbanization Prospects, the 2014. Population Division, Department of Economic and Social Affairs 2014*; United Nations Secretariat: New York, NY, USA, 2014.

30. Asaminew, A.E. *Climate Change, Growth and Poverty in Ethiopia. The Robert S. Strauss Center for International Security and Law*; The University of Texas at Austin: Austin, TX, USA, 2013.

31. Evangelista, P.; Young, N.; Burnett, J. How will climate change spatially affect agriculture production in Ethiopia? Case studies of important cereal crops. *Clim. Chang.* **2013**, *119*, 855–873. [CrossRef]

32. Mersha, A.A.; van Laerhoven, F. The interplay between planned and autonomous adaptation in response to climate change: Insights from rural Ethiopia. *World Dev.* **2018**, *107*, 87–97. [CrossRef]

33. Seleshi, Y.; Zanke, U. Recent changes in RF and rainy days in Ethiopia. *Int. J. Climatol.* **2004**, *24*, 973–983. [CrossRef]

34. Adem, A.; Bewket, W. *A Climate Change Country Assessment Report for Ethiopia. Submitted to Forum for Environment*; Epsilon International R&D: Addis Ababa, Ethiopia, 2011.

35. Adnew, M.; Bewket, W. Variability and trends in RF amount and extreme event indices in the Omo-Ghibe River Basin, Ethiopia. *Reg. Environ. Chang.* **2014**, *14*, 799–810.

36. Yalew, A.W.; Hirte, G.; Lotze-Campen, H.; Tscharaktschiew, S. *Economic Effects of Climate Change in Developing Countries: Economy-Wide and Regional Analysis for Ethiopia*; Center of Public and International Economics: Dresdon, Germany, 2017.

37. Mideksa, T.K. Economic and distributional impacts of climate change: The case of Ethiopia. *Glob. Environ. Chang.* **2010**, *20*, 278–286. [CrossRef]

38. Kreft, S.; Eckstein, D.; Dorsch, L.; Fischer, L. *Global Climate Risk Index 2016: Who Suffers Most from Extreme Weather Events? Weather-related Loss Events in 2014 and 1995 to 2014*; Germanwatch: Bonn, Germany, 2015.

39. Bewket, W. Climate change perceptions and adaptive responses of smallholder farmers in central highlands of Ethiopia. *Int. J. Environ. Stud.* **2012**, *69*, 507–523. [CrossRef]
40. Atinkut, B.; Mebrat, A. Determinants of farmers choice of adaptation to climate variability in Dera district, south Gondar zone, Ethiopia. *Environ. Syst. Res.* **2016**, *5*, 6. [CrossRef]
41. SNNPRS-BoFED. *Annual Abstract SNNPRS Bureau of Finance and Economic Development Annual Statistical Abstract. 2003 E.C. Data Collection—Dissemination Core Process*; SNNPRS-BoFED: Hawassa, Ethiopia, 2015.
42. SZBoFED. *Sidama Zone Socio-Economic Profile. Sidama Zone Finance and Economic Development Department*; SZBoFED: Hawassa, Ethiopia, 2007.
43. Krejcie, R.V.; Morgan, D.W. Determining sample size for research activities. *Educ. Psychol. Meas.* **1970**, *30*, 607–610. [CrossRef]
44. Pallant, J. *SPSS Survival Manual: A Step by Step Guide to Data Analysis Using SPSS for Windows (Versions 10 and 11): SPSS Student Version 11.0 for Windows*; Open University Press: Milton Keynes, UK, 2001.
45. Bewket, W.; Conway, D. A note on the temporal and spatial variability of RF in the drought-prone Amhara region of Ethiopia. *Int. J. Climatol.* **2007**, *27*, 1467–1477. [CrossRef]
46. Janowiak, J.E.; Ropelewski, C.F.; Halpert, M.S. The precipitation anomaly classification: A method for monitoring regional precipitation deficiency and excess on a global scale. *J. Clim. Appl. Meteorol.* **1986**, *25*, 565–574. [CrossRef]
47. Arragaw, A.; Bewket, W. Smallholder farmers' coping and adaptation strategies to climate change and variability in the central highlands of Ethiopia. *Local Environ.* **2017**, *22*, 825–839.
48. Tafesse, M. An Assessment of Domestic Water Supply Status of Rural Household in Selecting Kebeles of Boricha District, Sidaama Zone, Southern Ethiopia. Unpublished Master's Thesis, Addis Ababa University, Addis Ababa, Ethiopia, 2009.
49. Haines, A.; Kovats, R.S.; Campbell-Lendrum, D.; Corvalán, C. Climate change and human health: Impacts, vulnerability and public health. *Public Health* **2006**, *120*, 585–596. [CrossRef]
50. McMichael, C.; Barnett, J.; McMichael, A.J. An ill wind? Climate change, migration, and health. *Environ. Health Perspect.* **2012**, *120*, 646–654. [CrossRef] [PubMed]
51. Allsopp, R. Options for vector control against trypanosomiasis in Africa. *Trends Parasitol.* **2001**, *17*, 15–19. [CrossRef]
52. Hotez, P.J.; Kamath, A. Neglected tropical diseases in sub-Saharan Africa: Review of their prevalence, distribution, and disease burden. *PLoS Negl. Trop. Dis.* **2009**, *3*, e412. [CrossRef]
53. Kissui, B.M. Livestock predation by lions, leopards, spotted hyenas, and their vulnerability to retaliatory killing in the Maasai steppe, Tanzania. *Anim. Conserv.* **2008**, *11*, 422–432. [CrossRef]
54. Philander, S.G. *Encyclopedia of Global Warming and Climate Change: AE*; Sage Publishers: Thousand Oaks, CA, USA, 2008.
55. Wheeler, T.; Von Braun, J. Climate change impacts on global food security. *Science* **2013**, *341*, 508–513. [CrossRef]
56. MoWR and NMA. *Climate Change National Adaptation Programme of Action (NAPA) of Ethiopia. National Meteorological Services Agency, Ministry of Water Resources*; Federal Democratic Republic of Ethiopia: Addis Ababa, Ethiopia, 2007.
57. Ayalew, M. *What Is Food Security and Famine and Hunger. Using Science against Famine: Food Security, Famine Early Warning, and El Nino*; National Center for Atmospheric Research: Boulder, CO, USA, 1997; pp. 1–8.
58. Tsegaye, A.; Struik, P.C. Analysis of enset (*Ensete ventricosum*) indigenous production methods and farm-based biodiversity in major enset-growing regions of southern Ethiopia. *Exp. Agric.* **2002**, *38*, 291–315. [CrossRef]
59. Tesfaye, B.; Lüdders, P. Diversity and distribution patterns of enset landraces in Sidama, Southern Ethiopia. *Genet. Resour. Crop Evol.* **2003**, *50*, 359–371. [CrossRef]
60. Ellis, F. *Rural Livelihood Diversity in Developing Countries: Evidence and Policy Implications*; Overseas Development Institute: London, UK, 1999.
61. Eneyew, A. Determinants of livelihood diversification in pastoral societies of southern Ethiopia. *J. Agric. Biodivers. Res.* **2012**, *1*, 43–52.
62. Adepoju, A.O.; Obayelu, O.A. Livelihood diversification and welfare of rural households in Ondo State, Nigeria. *J. Dev. Agric. Econ.* **2013**, *31*, 482–489.

Article

Warming Winters Reduce Chill Accumulation for Peach Production in the Southeastern United States

Lauren E. Parker [1,2,*] and John T. Abatzoglou [3]

[1]   USDA California Climate Hub, Davis, CA 95616, USA
[2]   John Muir Institute of the Environment, University of California, Davis, CA 95616, USA
[3]   Department of Geography, University of Idaho, Moscow, ID 83844, USA
*   Correspondence: leparker@ucdavis.edu

Received: 29 June 2019; Accepted: 25 July 2019; Published: 30 July 2019

Abstract: Insufficient winter chill accumulation can detrimentally impact agriculture. Understanding the changing risk of insufficient chill accumulation can guide orchard management and cultivar selection for long-lived perennial crops including peaches. This study quantifies the influence of modeled anthropogenic climate change on observed chill accumulation since 1981 and projected chill accumulation through the mid-21st century, with a focus on principal peach-growing regions in the southeastern United States, and commonly grown peach cultivars with low, moderate, and high chill accumulation requirements. Anthropogenic climate change has reduced winter chill accumulation, increased the probability of winters with low chill accumulation, and increased the likelihood of winters with insufficient chill for commonly grown peach cultivars in the southeastern United States. Climate projections show a continuation of reduced chill accumulation and increased probability of winters with insufficient chill accumulation for cultivars with high chill requirements, with approximately 40% of years by mid-century having insufficient chill in Georgia. The results highlight the importance of inter-annual variability in agro-climate risk assessments and suggest that adaptive measures may be necessary in order to maintain current peach production practices in the region in the coming decades.

Keywords: chill accumulation; climate change; peaches; perennial crops; Georgia; South Carolina

---

1. Introduction

The peach industry in the southeastern United States (SEUS) has been a part of the regional iconography since at least the mid-1920s, and was historically an important part of the agricultural economy [1]. While California's current peach production dwarfs that of Georgia and South Carolina [2], the industry in the SEUS continues to contribute millions to regional, state, and local economies [3], and peaches remain important to regional identity [1]. In 2017, approximately 80% of Georgia's peach crop and 90% of South Carolina's peach crop were damaged due to warm winter temperatures. The warm conditions resulted in insufficient winter chill accumulation in some areas, while other parts of the SEUS were impacted when an early bloom, due to unseasonably warm temperatures, was followed by a mid-March freeze. In Georgia, an estimated 70% of the total 2017 peach losses were attributed to inadequate chill and 10–15% of the losses, the result of a spring freeze [4]. The combined impacts of anomalously low chill accumulation and spring freeze yielded substantial economic damage across the region [5]. Given the role of the peach industry in both the economy and culture of the SEUS, the 2017 crop failure garnered much public interest including whether such warm winters and impacts to perennial agriculture may become more commonplace in the coming decades.

Like other fruit trees, peaches undergo a series of physiological changes during the fall that allow for the onset of dormancy, when growth and development are slowed or stopped and the plant is better able to tolerate cold temperatures. Many perennial crops must be exposed to a certain amount of cold temperatures, or chill, during this period of dormancy to continue their development in the spring [6]. Peach cultivation is governed by a number of climatic factors such as cold hardiness, frost tolerance, and sufficient heat accumulation. Peach cultivars are frequently selected based on climatological chill accumulation [7] as insufficient chill accumulation can reduce flower quality, inhibit pollination and fruit development, and lower fruit quality and yield [6,8,9], with subsequent economic impacts to both growers and consumers [10].

Observational studies have shown warming in both the mean and extreme cold winter temperatures over the past half century across the US [11–13], much of which is consistent with anthropogenic forcing [14] and is expected to continue under climate change [15,16]. The exceptions of observed warming trends are primarily found in the warming hole across parts of the SEUS where winter temperatures cooled and spring onset trended later over the latter half of the 20th century [17,18]. The warming hole is likely a consequence of internal variability of the climate system that has buffered the influence of anthropogenic forcing to date, but is not expected to persist into the coming decades [17]. While it is acknowledged that chill accumulation is only one of many thermal-metrics that might directly impact crop suitability in a changing climate [19], declines in chill accumulation have been observed in some regions [20] and are projected to decline further [21]. Likewise, increases in winter temperatures are projected to reduce chill accumulation below the thresholds needed for peach cultivars in many peach-growing portions of the US [20,22].

In view of recent crop impacts due to warm winters, we examine chill accumulation across the SEUS in the context of ongoing climate change with a focus on implications for peach cultivation. First, a first-order estimate is provided of the contribution of anthropogenic climate change to observed low chill accumulation winters in the SEUS and years with insufficient chill in prime peach-growing areas in Georgia and South Carolina during 1981-2017. Secondly, using a suite of downscaled climate projections, changes in chill accumulation, the frequency of low-chill winters, and changes in the risk of winters with insufficient chill for common peach cultivars in the coming decades were investigated. Comprehensively, this study presents methodologies that may be applied to agro-climate metrics for conducting climate change risk and impact analyses for perennial crop systems globally, and provides a risk assessment of insufficient chill for peaches—and general chill accumulation for other perennials—in the SEUS, presenting information useful for climate-informed decision making.

2. Materials and Methods

Two primary datasets were used in this study (available at https://data.nkn.uidaho.edu/). First, the observed daily maximum and minimum temperature (T_{max}, T_{min}) at a ~4-km spatial resolution for the period 1981–2017 for the SEUS [25°–35.2° N, 78.5°–88.5° W (see Figure 1a)] were acquired from the gridded surface meteorological dataset (gridMET) of [23]. Previous validation of gridMET showed high correlation and low bias of temperature when compared to meteorological station observations across the US [23], and comparisons in chill accumulation between gridMET and data from 50 SEUS meteorological stations from 1980–2017 showed strong spatial correlation ($r = 0.99$), with a mean absolute error of 50 chill hours and a median bias of -17 chill hours (analysis not shown). Second, the projections of daily T_{max} and T_{min} from 20 global climate models (GCMs) that participated in the fifth phase of the Climate Model Intercomparison Project (CMIP5) were statistically downscaled using the multivariate adaptive constructed analogs (MACA) method [24]. The MACA used gridMET as training data, thereby ensuring compatibility in contemporary climate statistics between the downscaled GCM experiments and gridded observations. The analysis of climate projections was constrained to simulations for the early (2010–2039) and mid- (2040–2069) 21st century periods given the limited ability for developing meaningful management strategies relevant to the end-of-century projections. Further, we focused on future experiments run under the Representative Concentration Pathway 4.5 (RCP 4.5) to provide a conservative estimate of projected changes in chill accumulation. The projections using RCP 8.5 would likely show similar qualitative changes, but with larger magnitudes, particularly for the mid-21st century where multi-model mean changes in winter mean temperatures show an additional

0.6 °C warming above RCP 4.5, although the variability among models exceeds the difference between RCP 4.5 and RCP 8.5 for the time horizons highlighted herein.

A first-order estimate is provided on the influence of anthropogenic climate change on observed 1981–2017 chill accumulation using a large ensemble of CMIP5 simulations and a pattern scaling approach that allows for comparisons between rates of local and global change [25]. The differences in monthly T_{max} and T_{min} as simulated by 23 different GCMs at their native spatial resolution were taken between two 30-year periods, 1850–1879 and 2070–2099. The pattern scaling approach allows the expression of modeled rates of regional change for an individual variable and month to modeled rates of change in the global mean annual temperature. This approach assumes a linear relationship between the variables, which is reasonable for climate change timescales [25]. The pattern scaling was calculated separately for each model, as well as for the 23-model median. For each model, the anthropogenic climate change signal was defined for monthly T_{max} and T_{min} by multiplying the monthly varying pattern scaling function by an 11-year moving average of the change in the modeled global mean annual temperature relative to each model's 1850–1879 baseline. It is acknowledged that this is one of several first-order approaches for approximating the modeled influence of anthropogenic climate change over the historical record [26,27].

Following [26], a time series of daily T_{max} and T_{min} for 1981–2017 for the SEUS was created that preserves the observed interannual climate variability, but removes the influence of modeled anthropogenic climate change by subtracting the estimated difference in modeled monthly temperature anomalies (relative to the 1850–1879 baseline) using pattern scaling from the observed temperatures. These counterfactual scenarios do not make an effort to discern the sources of change in the observed data. Rather, they provide an approach for estimating the proximal effects of modeled anthropogenic climate change in the context of real-world observations.

The peach location data were obtained from the 2016 United States Department of Agriculture—National Agricultural Statistics Service Cropland Data Layer (CDL, available at https://www.nass.usda.gov/Research_and_Science/Cropland/Release/index.php) for the SEUS states of Alabama, Georgia, South Carolina, and Florida [28]. Approximately 94-km^2 were classified as peach in the 2016 Southern CDL with nearly all of the orchards located in Georgia (~34.5-km^2) and South Carolina (~58-km^2) (Figure 1a). The 30-m resolution CDL data was aggregated to the common 4-km resolution of the climate data for analyzing chill accumulation over peach-growing locations, summing the number of 30-m peach cells within each 4-km grid cell. The peach-growing locations were classified as those 4-km grid cells with >0.01% peach density. Finally, in order to provide locally-relevant results in addition to the regional analysis, our peach cultivar-specific analysis focused on peach locations within a 4-county area of central Georgia and a 3-county area in the Piedmont region of South Carolina that are responsible for ~75% and ~50% of each state's peach production, respectively.

Estimates of chill accumulation derived from chilling models are used for selecting appropriate crop species and cultivars, and to track plant phenology for farm management practices [29,30]. While there are multiple modeling approaches for calculating chill, the Weinberger Chilling Hours Model [31] was utilized as chill requirements for SEUS peaches are most commonly reported in chilling hours. The chill thresholds for peach cultivars examined in this study were quantified using the Weinberger model in central Georgia. Further, this model is commonly used to track winter chill accumulation across the SEUS as part of the online tools available through regional university consortiums and university extension programs (e.g., http://agroclimate.org/; http://weather.uga.edu/), and as such, using this chill model allows for the most direct translation of this work to end users.

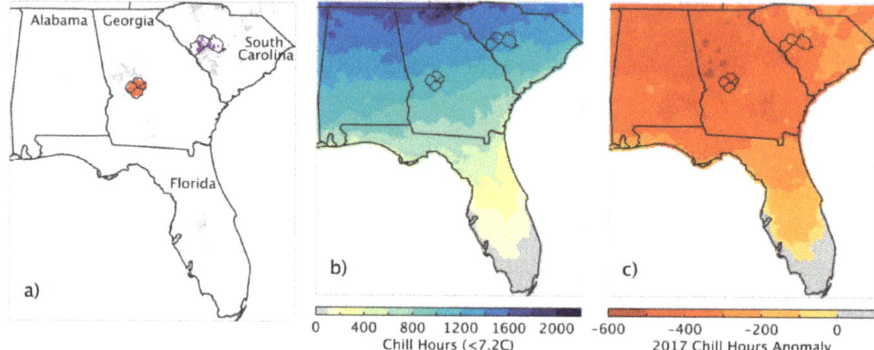

Figure 1. (**a**) The southeastern US study area. 4-km cells with >0.01% peach density are highlighted in grey. Georgia and South Carolina peach-growing counties examined explicitly in this study are outlined in grey and those cells with >0.01% peach density within these counties are highlighted in red (Georgia) and purple (South Carolina). (**b**) The average annual number of chill hours for the 1981–2017 observed period. Areas with <100 chill hours are masked in grey. (**c**) The winter chill accumulation anomaly in 2017 compared to the 1981–2017 average. Areas masked in grey as in (**b**).

The Chilling Hours Model sums the number of hours per day with temperatures <7.2 °C; hourly data were temporally disaggregated from daily T_{max} and T_{min} using a modified sine curve model [32]. Annual chill accumulation was considered from 1 October to 15 February, as is standard in the SEUS peach industry [33]. Peach chill requirements were obtained from the University of Georgia [34] for three cultivars grown in the SEUS. Gulfprince and Juneprince peaches require 400 and 650 chill hours, respectively, and are hereafter referred to as low- and moderate-chill cultivars. The Elberta peach cultivar (hereafter referred to as high-chill) requires 850 chill hours and is a cultivar standard to which the phenology of other peach cultivars is compared [34]. It is noted that not all of these cultivars are grown across all peach-growing locations of Georgia and South Carolina. Gulfprince is a cultivar grown primarily in southern Georgia, while central Georgia principally grows peaches with chill requirements ≥600 chill hours (Dario Chavez, University of Georgia Extension Specialist, personal communication). However, these three cultivars have been included as exemplary of the range of chill requirements across SEUS-grown peaches. By including the low-chill cultivar in our analyses of selected South Carolina and Georgia peach-growing counties, we show the capacity for these counties to continue to produce peaches under future climate conditions should future chill accumulation limit the productivity of the currently-grown moderate- and high-chill cultivars.

Chill accumulation was calculated over the 1981–2017 period with the observational data, and for the counterfactual scenarios using the observed data from 1981–2017 after removing the influence of anthropogenic climate change. The 1981–2017 data were further used to quantify changes in the frequency of low-chill winters, defined as the bottom decile (10th percentile). This provides both additional context for the peach-focused analysis herein and may be of broader interest to the SEUS fruit and nut industry reliant on understanding exposure of low-chill winters as it pertains to the economics of orchard operations [29]. The observed and counterfactual scenarios for 1981–2017 were used to quantify the degree to which modeled climate change influenced the average chill accumulation, the probability of experiencing a low-chill winter, and the risk of insufficient chill accumulation for the three peach cultivars across the key Georgia and South Carolina peach-growing regions. Chill accumulation was also calculated for the 2010–2069 period for each of the 20 downscaled climate datasets. A similar set of tests were applied to projections including changes in average chill accumulation and the probability of experiencing a low-chill winter across the SEUS. Finally, the probability of insufficient chill was estimated for the early and mid-21st century conditions for the key peach cultivars and regions in order to highlight the potential risk to peach cultivation. Given our focus on the changes to chill accumulation with respect to perennial fruit cultivation, areas with <100 chill hours over the 1981–2017 observed period were masked out.

3. Results

The average chill accumulation for the observed period 1981–2017 across the SEUS ranged from less than 100 h in southern Florida, to more than 2000 h in the Blue Ridge mountains of northeastern Georgia (Figure 1b). The majority (>65%) of the region—from northern Florida to northern Alabama, Georgia, and South Carolina—averaged 500–1500 chill hours, including approximately 1100 h in the central Georgia peach-growing region and 1350 h in the Piedmont peach-growing region of South Carolina. With the exception of southern Florida, the 2017 chill accumulation was substantially lower than the 1981–2010 normal. The accumulated chill in 2017 showed an SEUS average anomaly of approximately 330 h below normal. The Georgia peach regions showed an anomaly of ~430 h below normal, and South Carolina peach regions showed an anomaly of ~360 h below normal (Figure 1c).

The observed average chill accumulation over 1981–2017 was less than that modeled in the absence of anthropogenic climate change, consistent with the expectations from modeled warming (Figure 2a). A distinct geographic pattern of the reduced chill hours due to climate change was evident across the SEUS, with nominal differences in southern Florida and reductions of more than 120 h in northern Alabama, Georgia and South Carolina. The peach-growing regions showed average reductions of ~115–120 chill hours in Georgia and South Carolina. Notably, these reductions are averages over the 37-year period as the modeled estimate was larger in more recent years. Complementary to average reductions in chill hours, the percent of years experiencing low winter chill was substantially higher across the SEUS over the 1981–2017 period than it would have been in the absence of climate change (Figure 2b). These trends were found across models. The 23-model range for declines in chill was ~68–140 h, while the range for the probability of low-chill winters was 1.6–4.6% of years (from a reference of 10% of years).

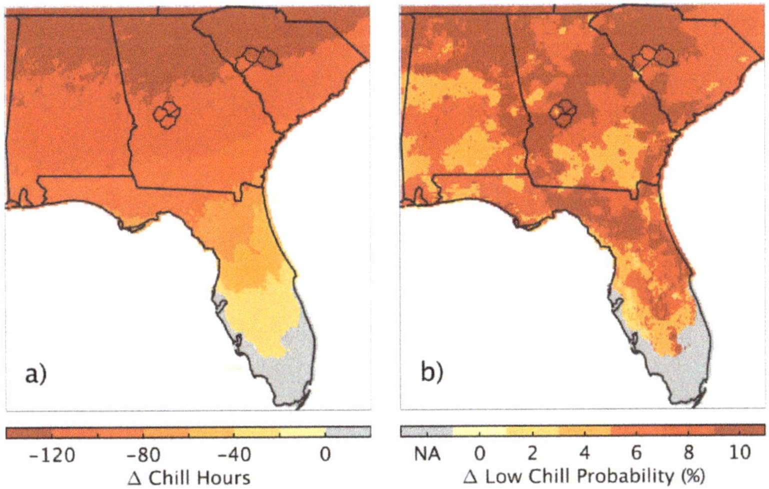

Figure 2. (**a**) The average change in 1981–2017 observed winter chill hours due to the influence of anthropogenic climate change (23-model median). (**b**) The change in the probability of a low-chill winter as a result of climate change, shown as 1981–2017 observed minus 1981–2017 counterfactual (23-model median). For both panels, the areas with <100 chill hours for the 1981–2017 observed climatology are masked in grey.

The reductions in chill accumulation and increases in the occurrence of low-chill winters may be inconsequential for agriculture unless there are direct impacts to plant physiology or indirect crop impacts (e.g., pathogens, pests). For the three peach cultivars, we show that the Georgia peach-growing region had five winters from 1981–2017 that did not accumulate sufficient chill for the high-chill cultivar (Figure 3a). No winters in the South Carolina peach-growing regions had insufficient chill for the cultivars considered from 1981–2017 (Figure 3b). By contrast, the counterfactual scenarios all showed greater chill accumulation and reduced occurrence of winters with insufficient chill for high-chill cultivars in Georgia. Notably, we show that the chill accumulation in 2017 would have been the lowest in the 37 year period in Georgia without climate change, suggesting that it was primarily driven by natural variability. However, the estimated 2017 chill accumulation excluding the modeled first-order influence of climate change for the peach-growing area of Georgia ranged from ~760 to ~920 chill hours across 23 models, with a median of ~825 h, well above the threshold of 650 chill hours required for moderate-chill cultivar and the ~660 chill hours observed that winter.

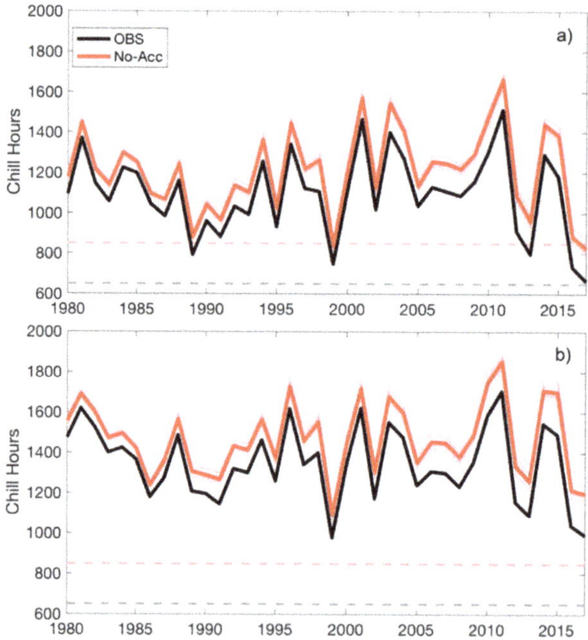

Figure 3. Time series of 1981–2017 chill accumulation for (**a**) the Georgia peach-growing region, and (**b**) the South Carolina peach-growing region. The observed data (OBS) are shown in black, while modeled chill accumulation estimates excluding the influence of anthropogenic climate change (No-Acc) are shown in red, with lighter red lines indicating individual models and the heavy red line indicating the 23-model median. The light pink dashed line indicates the chill requirement for a high-chill peach cultivar and the dashed grey line indicates the chill requirement for a moderate-chill peach cultivar.

The reduced chill accumulation across the SEUS was modeled relative to contemporary 1981–2017 averages for the early and mid-21st centuries, with multi-model mean SEUS declines of ~100 h, and ~185 h, respectively (Figure 4a,b). The geographic patterns of reductions in chill hours were similar to those shown for the influence of modeled climate change for the 1981–2017 period. Over Georgia (South Carolina) peach-growing regions, the average declines in chill were calculated as ~110 (~135) hours by the early 21st century and ~210 (~250) hours by the mid-21st century. In addition to declines in the average chill accumulation, the probability of experiencing a year with low winter chill accumulation increased. Averaged across the SEUS and across all models, approximately 20% of years by the early 21st century and 40% of years by the mid-21st century experienced low winter chill, with the greatest increases across western and northern Alabama, northern and central Georgia, and northern and central South Carolina (Figure 4c,d). By the early and mid- 21st century, Georgia (South Carolina) peach regions saw ~15% (30%) and 32% (52%) of years having low winter chill, respectively.

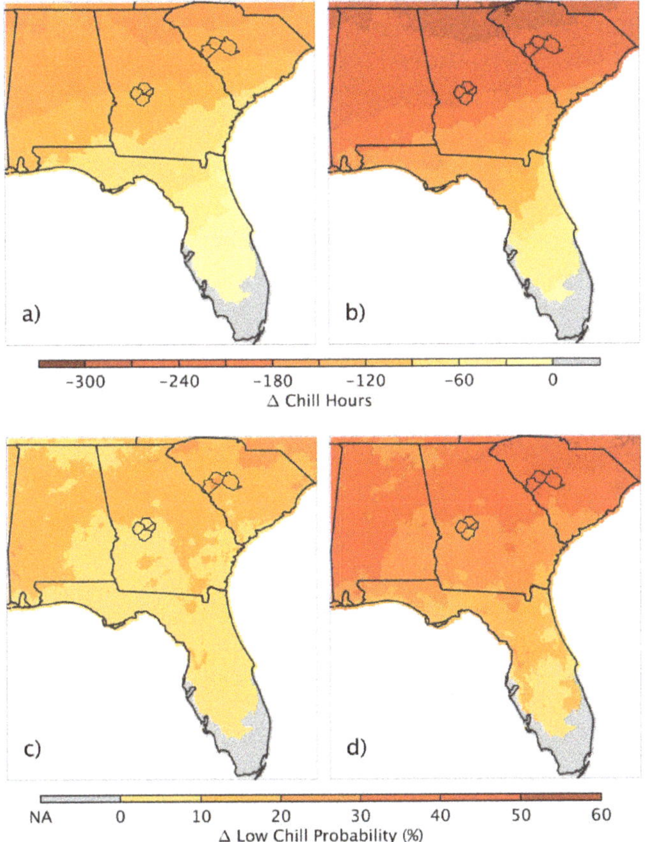

Figure 4. The difference in climatological chill hours for (**a**) the early 21st century (2010–2039) and (**b**) mid-21st century (2040–2069), relative to the observed 1981–2017 period. Panels (**c**) and (**d**) show differences in the probability of a low-chill winter for 2010–2039, and 2040–2069, respectively, relative to the observed 1981–2017 period. For all panels, the areas with <100 chill hours for the 1981–2017 observed climatology are masked in grey.

With respect to peach cultivar-specific chill requirements, 23% (4%) percent of years showed insufficient chill for the high- (moderate-) chill cultivar in prime peach-growing counties in Georgia by the early 21st century, rising to 43% (11%) percent of years by the mid-21st century (Figure 5a). The peach-growing regions in South Carolina, which did not see chill accumulations below established thresholds from 1980–2017, had 5% (0.25%) percent of years with insufficient chill for the high-chill (moderate-chill) cultivar by the early 21st century, and 12% (1.5%) percent of years by the mid-21st century (Figure 5b). Notably, there was substantial inter-model variability in the risk of winters with insufficient chill. For example, the percent of winters with insufficient chill for the moderate-chill cultivar in Georgia ranged from 0–16% for the early 21st century, and 0–30% for the mid-21st century.

By contrast, chill accumulation was sufficient for the low-chill peach cultivar under both future time periods in both states' peach-growing regions.

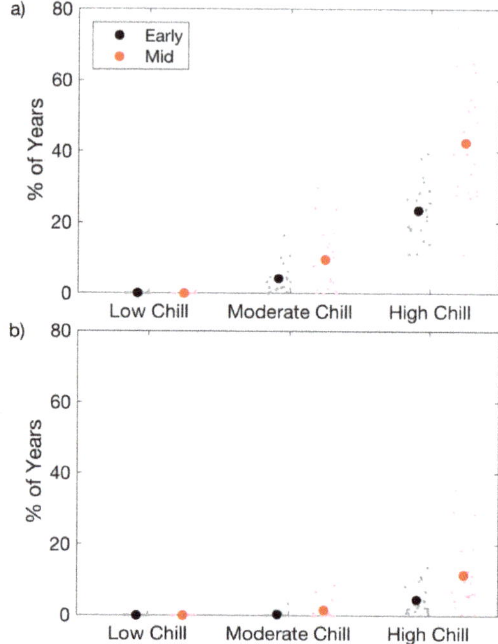

Figure 5. (**a**) For the Georgia peach-growing region, the percent of years with insufficient chill for a low-, moderate-, and high-chill peach cultivar under early 21st (black/grey) and mid-21st (red/pink) century conditions. Small grey and pink dots indicate the percent years for individual models, while the larger black and red dots indicate the 20-model average (**b**) As in (**a**) but for the South Carolina peach-growing region.

4. Discussion

Recent studies have shown that extreme events around the globe would not have been possible without the influence of human-induced warming [35–37]. Temperatures in the SEUS during the October-February chill accumulation periods of 2015–2016 and 2016–2017 were the 2nd and 3rd warmest since 1895, with the 1931–1932 winter being the warmest [38], suggesting that such warm winters are possible within the bounds of natural variability and can occur without significant contributions from anthropogenic climate change. While we do not undertake a detailed attribution analysis, our modeling exercise provides support that recent insufficient chill accumulation in the SEUS peach regions, such as in 2017, would not likely occur under the same synoptic conditions in the absence of climate change. Further, our results showing an increased probability of low-chill winters due to climate change add to the growing body of literature defining the contribution of anthropogenic climate change to observed adverse climate impacts [27,39,40].

Although insufficient chill accumulation is not a principle cause of loss for federally-insured crops in the SEUS [41,42], previous work has postulated that projected declines in chill may reduce suitability for perennial crop production [19,43]. Similarly, the projected future declines in chill accumulation across the SEUS complement previous work showing increases in the average and coldest winter minimum temperatures [16], and declines in chill accumulation in regions around the globe [21]. While this warming may offer range expansion for cold-intolerant crops, the related reduction in the winter chill accumulation in subtropical climates like the SEUS is projected to have negative impacts on warm-region fruit and nut crops, particularly those with moderate and high chill requirements [20,21]. However, the degree to which these declines may impact crop yield is unclear as uncertainties remain regarding the chill requirements that are physiologically needed for production, and the overall effect of marginal chill accumulation on crop yield and quality [44,45]. For example, while a common commercial peach cultivar grown in central Georgia has a stated chill requirement of 850 h, Georgia peach specialists have suggested that only 800 h are needed for a suitable crop [4]. Consequently, we underscore that this work is not predictive of yield impacts related to reduced chill accumulation.

Compounding the problem of crop chill requirements is the questionable accuracy of the chilling model. While this model has been widely used for quantifying crop chill requirements, it may be overly sensitive to warming, potentially overestimating the impact of climate change [43]. However, while it is acknowledged that previous studies have shown that the Dynamic Model may provide a more accurate representation of chill accumulation [21], the 20-model mean changes in the average chill accumulation show an agreement of declines across the SEUS and other warm-winter regions, regardless of the chilling model (Figure 6). Further, we recognize that familiarity with chill portions (the units of the Dynamic Model) may be lacking among extension agents and fruit industry professionals (Pamela Knox, University of Georgia Agricultural Climatologist, personal communication), and that regionally-defined chill portion thresholds do not yet exist for SEUS peach cultivars (Dario Chavez, personal communication). Finally, we acknowledge the limitations of using temporally disaggregated daily data [46], and that the microclimates of orchard sites and orchard management practices may augment or abate the projected changes and impacts.

Despite research suggesting that declines in crop suitability due to climate change may not be as severe as shown in our results [45], it is worth noting that we examined changes in chill accumulation under a conservative, moderate warming scenario. Provided that some degree of reductions in suitability are anticipated for peach crops across the SEUS—as well as for other crops with similar chill requirements—adaptive measures may be warranted to maintain production. These measures may include altering orchard management practices and selective planting. For existing orchards, the application of chemicals such as hydrogen cyanamide may effectively break dormancy in insufficiently-chilled peach crops [47], overhead irrigation to encourage evaporative cooling may aid chill accumulation, and orchard management practices such as controlling tree vigor may help to lower the chill needed for successful bud break [48]. For future orchards, site selection with preferential planting in sites with cooler microclimates, such as low-lying cool-air sinks, may provide an opportunity to increase exposure to chilling temperatures. Orchard managers may also consider specific scion and rootstock combinations that may help mitigate the negative impacts of low chill [49]. Moreover, a transition to crop cultivars with lower chill requirements (e.g., Gulfcrest or other varieties developed for warmer climates) may reduce or eliminate the negative impacts of declining chill accumulation under climate change, as evidenced by the minimal impact of future warming to the low-chill cultivar examined in this study. However, it is noted that orchards planted in cool-air microclimates may be at increased risk of frost damage, and lower-chill cultivars may be more susceptible to early bloom and subsequent frost damage. While quantifying the complex relationships between chill accumulation, bloom, and the relative risks of insufficient chill and spring frost damage are beyond the scope of this work, the interactions between these physiological and climatic conditions highlight the need to consider a broader suite of environmental and economic considerations in planning for future orchard management.

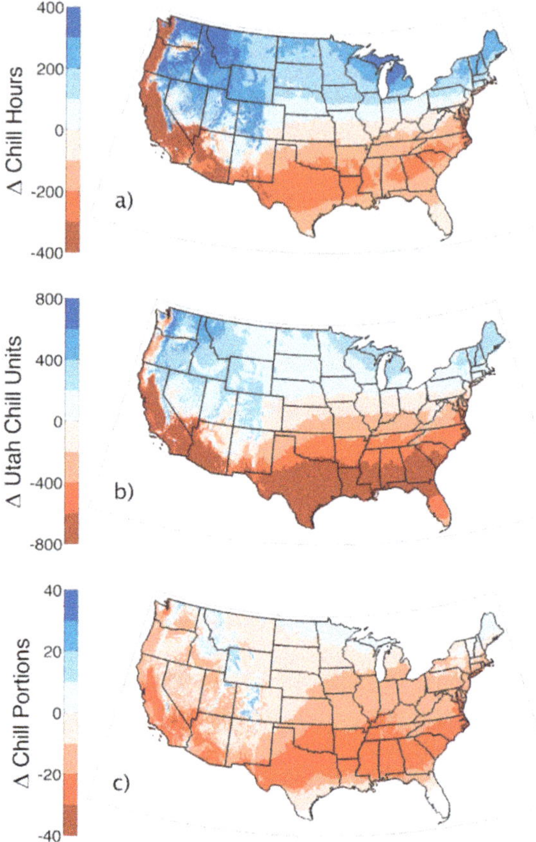

Figure 6. The 20-model average difference in annual accumulated chill between the modeled historical period (1971–2000) and the mid-century (2040–2069) period under RCP 4.5, where chill was accumulation was calculated over the October 1—April 30 cool season using (**a**) the Modified Chill Hour Model as chill hours 32–45 °C, (**b**) the Utah Model as chill units, and (**c**) the Dynamic Model as chill portions. The red shades indicate a reduction in chill accumulation under RCP 4.5, while the blue shades indicate an increase in chill accumulation. The white regions in (**c**) indicate areas with no chill accumulation under historical conditions. These data can be viewed and downloaded from the Climate Toolbox (https://climatetoolbox.org/) at (**a**) https://bit.ly/2QjbT2l (**b**) https://bit.ly/2AtEpZI and (**c**) https://bit.ly/2SBDcqm.

As has been suggested for perennial crop adaptation in other regions [19,50], the translocation of crops to cooler climates may also provide an adaptive measure for maintaining peach cultivation in the SEUS, particularly for those cultivars with higher chilling requirements. Historically, peach cultivation in Georgia extended into the northern portion of the state, but the favorability of that region declined over time due to frequent freeze damage [51]. If climate change reduces the freeze risk in northern Georgia, the area may provide a refuge within the state for continued cultivation of high-chill peach cultivars and other similarly at-risk perennials. However, any future translocation would require significant capital and be contingent upon economic viability, which is likely to be predicated on factors such as topography and soils, the costs associated with the purchase of farmland and the packing or processing facilities, competing land use, and market forces.

5. Conclusions

Quantifying the potential consequences of warming winters on chill accumulation may have implications for long-term orchard management and land use planning and may provide insights useful for climate-informed decision making for a variety of perennial crops that require winter chill. Our results show that anthropogenic climate change has negatively affected chill accumulation in the SEUS over the observed 1981–2017 period, and that ongoing climate change is likely to continue to reduce chill accumulation, with notable impacts on high- and moderate-chill peach cultivars in Georgia. We also highlight the importance of examining interannual variability when assessing climate change risks to agriculture, be that impacts to crop climatic niche or crop yield [19,52]. The adaptation measures (e.g., investments in lower-chill varieties) may be necessary in order for the SEUS, particularly Georgia, to continue to cultivate the crop that has historically been central to its cultural identity. Further, given the relationship between mild winter temperatures, early bloom, and damages due to a false spring—as also seen in 2017—we recommend future work consider the interaction between multiple agro-climatic variables to provide a more complete assessment of future crop suitability and identify the most appropriate adaptive efforts. Finally, as our study employs a methodology that is applicable across other geographic locations, perennial crop cultivars, and agro-climatic metrics, we recommend that similar work be undertaken across agricultural systems and regions to help identify potential crop-specific risks and adaptation opportunities.

Author Contributions: Conceptualization, L.E.P. and J.T.A.; Methodology, L.E.P. and J.T.A.; Formal Analysis, L.E.P. and J.T.A.; Writing—Original Draft Preparation, L.E.P. and J.T.A.; Writing—Review & Editing, L.E.P. and J.T.A.; Visualization, L.E.P. and J.T.A.; Supervision, J.T.A.; Project Administration, J.T.A.; Funding Acquisition, L.E.P. and J.T.A.

Funding: This research was partially supported by the USDA's Northwest Climate Hub under award 15-JV-11261944-093.

Acknowledgments: The authors wish to acknowledge the feedback and local context provided by University of Georgia Assistant Professor and Extension Specialist Dario Chavez, and University of Georgia Agricultural Climatologist Pamela Knox. The authors also wish to acknowledge Katherine Hegewisch for her efforts in incorporating chill accumulation data into the ClimateToolbox. Finally, the authors wish to thank four anonymous reviewers and the journal editor for their comments, which improved the quality of the manuscript.

References

1. Okie, T. Under the Trees: The Georgia Peach and the Quest for Labor in the Twentieth Century. *Agric. Hist.* **2011**, *85*, 72–101. [CrossRef] [PubMed]
2. USDA-NASS. *2016 State Agricultural Overview: California, Georgia, South Carolina*; USDA-NASS: Washington, DC, USA, 2016. Available online: https://www.nass.usda.gov/Statistics_by_State/Ag_Overview/ (accessed on 29 June 2017).
3. Wolfe, K.; Stubbs, K. *Georgia Farm Gate Value Report*; CAES: Athens, GA, USA, 2016; Available online: https://www.caes.uga.edu/content/caes-subsite/caed/publications/farm-gate.html (accessed on 29 June 2017).
4. Thompson, C. Georgia's peach farmers hoping for colder winter this year. Available online: https://newswire.caes.uga.edu/story.html?storyid=6416 (accessed on 29 June 2017).
5. SERCC. *Annual 2017 Climate Report for the Southeast Region*; SERCC: Chapel Hill, NC, USA, 2017.
6. Saure, M.C. Dormancy release in deciduous fruit trees. *Hortic. Rev.* **1985**, *7*, 239–300.
7. Janick, J.; Paull, R.E. *The Encyclopedia of Fruit and Nuts*; CABI: Oxfordshire, UK, 2008; ISBN 0851996388.
8. Atkinson, C.J.; Brennan, R.M.; Jones, H.G. Declining chilling and its impact on temperate perennial crops. *Environ. Exp. Bot.* **2013**, *91*, 48–62. [CrossRef]
9. Weinberger, J.H. Effects of high temperatures during the breaking of the rest of Sullivan Elberta peach buds. *Proc. Am. Soc. Hortic. Sci.* **1954**, *63*, 157–162.

10. Medellín-Azuara, J.; Howitt, R.E.; MacEwan, D.J.; Lund, J.R. Economic impacts of climate-related changes to California agriculture. *Clim. Chang.* **2011**, *109*, 387–405. [CrossRef]

11. Peterson, T.C.; Heim, R.R., Jr.; Hirsch, R.; Kaiser, D.P.; Brooks, H.; Diffenbaugh, N.S.; Dole, R.M.; Giovannettone, J.P.; Guirguis, K.; Karl, T.R. Monitoring and understanding changes in heat waves, cold waves, floods, and droughts in the United States: State of knowledge. *Bull. Am. Meteorol. Soc.* **2013**, *94*, 821–834. [CrossRef]

12. Abatzoglou, J.T.; Barbero, R. Observed and projected changes in absolute temperature records across the contiguous United States. *Geophys. Res. Lett.* **2014**, *41*, 6501–6508. [CrossRef]

13. Walsh, J.; Wuebbles, D.; Hayhoe, K.; Kossin, J.; Kunkel, K.; Stephens, G.; Thorne, P.; Vose, R.; Wehner, M.; Willis, J. Our changing climate. In *Climate Change Impacts in the United States: The Third National Climate Assessment*; U.S. Government Printing Office: Washington, DC, USA, 2014; pp. 19–67.

14. Deser, C.; Terray, L.; Phillips, A.S. Forced and internal components of winter air temperature trends over North America during the past 50 years: Mechanisms and implications. *J. Clim.* **2016**, *29*, 2237–2258. [CrossRef]

15. Sillmann, J.; Kharin, V.V.; Zwiers, F.W.; Zhang, X.; Bronaugh, D. Climate extremes indices in the CMIP5 multimodel ensemble: Part 2. Future climate projections. *J. Geophys. Res. Atmos.* **2013**, *118*, 2473–2493. [CrossRef]

16. Parker, L.E.; Abatzoglou, J.T. Projected changes in cold hardiness zones and suitable overwinter ranges of perennial crops over the United States. *Environ. Res. Lett.* **2016**, *11*, 34001. [CrossRef]

17. Meehl, G.A.; Arblaster, J.M.; Branstator, G. Mechanisms contributing to the warming hole and the consequent US east–west differential of heat extremes. *J. Clim.* **2012**, *25*, 6394–6408. [CrossRef]

18. Schwartz, M.D.; Ault, T.R.; Betancourt, J.L. Spring onset variations and trends in the continental United States: Past and regional assessment using temperature-based indices. *Int. J. Climatol.* **2013**, *33*, 2917–2922. [CrossRef]

19. Parker, L.E.; Abatzoglou, J.T. Shifts in the thermal niche of almond under climate change. *Clim. Chang.* **2018**, *147*, 211–224. [CrossRef]

20. Baldocchi, D.; Wong, S. Accumulated winter chill is decreasing in the fruit growing regions of California. *Clim. Chang.* **2008**, *87*, 153–166. [CrossRef]

21. Luedeling, E. Climate change impacts on winter chill for temperate fruit and nut production: A review. *Sci. Hortic.* **2012**, *144*, 218–229. [CrossRef]

22. Carbone, G.J.; Schwartz, M.D. Potential impact of winter temperature increases on South Carolina peach production. *Clim. Res.* **1993**, *2*, 225–233. [CrossRef]

23. Abatzoglou, J.T. Development of gridded surface meteorological data for ecological applications and modelling. *Int. J. Climatol.* **2013**, *33*, 121–131. [CrossRef]

24. Abatzoglou, J.T.; Brown, T.J. A comparison of statistical downscaling methods suited for wildfire applications. *Int. J. Climatol.* **2012**, *32*, 772–780. [CrossRef]

25. Mitchell, T.D. Pattern scaling: An examination of the accuracy of the technique for describing future climates. *Clim. Chang.* **2003**, *60*, 217–242. [CrossRef]

26. Abatzoglou, J.T.; Williams, A.P. Impact of anthropogenic climate change on wildfire across western US forests. *Proc. Natl. Acad. Sci. USA* **2016**, *113*, 11770–11775. [CrossRef]

27. Williams, A.P.; Seager, R.; Abatzoglou, J.T.; Cook, B.I.; Smerdon, J.E.; Cook, E.R. Contribution of anthropogenic warming to California drought during 2012–2014. *Geophys. Res. Lett.* **2015**, *42*, 6819–6828. [CrossRef]

28. Boryan, C.; Yang, Z.; Mueller, R.; Craig, M. Monitoring US agriculture: The US department of agriculture, national agricultural statistics service, cropland data layer program. *Geocarto Int.* **2011**, *26*, 341–358. [CrossRef]

29. Luedeling, E.; Zhang, M.; Girvetz, E.H. Climatic changes lead to declining winter chill for fruit and nut trees in California during 1950–2099. *PLoS ONE* **2009**, *4*, e6166. [CrossRef] [PubMed]

30. Parker, L.E.; Abatzoglou, J.T. Comparing mechanistic and empirical approaches to modeling the thermal niche of almond. *Int. J. Biometeorol.* **2017**, *61*, 1593–1606. [CrossRef] [PubMed]

31. Weinberger, J.H. Chilling requirements of peach varieties. *Proc. Am. Soc. Hortic. Sci.* **1950**, *56*, 122–128.

32. Linvill, D.E. Calculating chilling hours and chill units from daily maximum and minimum temperature observations. *HortScience* **1990**, *25*, 14–16.

33. Okie, W.R.; Blackburn, B. Increasing chilling reduces heat requirement for floral budbreak in peach. *HortScience* **2011**, *46*, 245–252. [CrossRef]

34. CAES. College of Agriculture and Environmental Sciences (CAES). Available online: http://www.caes.uga.edu/extension-outreach/commodities/peaches/cultivars.html (accessed on 21 March 2017).
35. Knutson, T.R.; Kam, J.; Zeng, F.; Wittenberg, A.T. CMIP5 model-based assessment of anthropogenic influence on record global warmth during 2016. *Bull. Am. Meteorol. Soc.* **2018**, *99*, S11–S15. [CrossRef]
36. Imada, Y.; Shiogama, H.; Takahashi, C.; Watanabe, M.; Mori, M.; Kamae, Y.; Maeda, S. Climate change increased the likelihood of the 2016 heat extremes in Asia. *Bull. Am. Meteorol. Soc.* **2018**, *99*, S97–S101. [CrossRef]
37. Walsh, J.E.; Thoman, R.L.; Bhatt, U.S.; Bieniek, P.A.; Brettschneider, B.; Brubaker, M.; Danielson, S.; Lader, R.; Fetterer, F.; Holderied, K.; et al. The high latitude marine heat wave of 2016 and its impacts on Alaska. *Bull. Am. Meteorol. Soc.* **2018**, *99*, S39–S43. [CrossRef]
38. NOAA. *Climate at a Glance: Regional Time Series*; NOAA: Silver Spring, MD, USA, 2018. Available online: https://www.ncdc.noaa.gov/cag (accessed on 29 June 2017).
39. Fischer, E.M.; Knutti, R. Anthropogenic contribution to global occurrence of heavy-precipitation and high-temperature extremes. *Nat. Clim. Chang.* **2015**, *5*, 560–564. [CrossRef]
40. Stott, P.A.; Christidis, N.; Otto, F.E.L.; Sun, Y.; Vanderlinden, J.; Van Oldenborgh, G.J.; Vautard, R.; Von Storch, H.; Walton, P.; Yiou, P.; et al. Attribution of extreme weather and climate-related events. *Wiley Interdiscip. Rev. Clim. Chang.* **2016**, *7*, 23–41. [CrossRef]
41. Reyes, J.; Elias, E. Spatio-temporal variation of crop loss in the United States from 2001 to 2016. *Environ. Res. Lett.* **2019**, *14*, 074017. [CrossRef]
42. AgRisk Viewer. Agricultural Risk in a Changing Climate: A Geographical and Historical View of Crop Insurance. Available online: https://swclimatehub.info/rma (accessed on 31 March 2019).
43. Luedeling, E.; Brown, P.H. A global analysis of the comparability of winter chill models for fruit and nut trees. *Int. J. Biometeorol.* **2011**, *55*, 411–421. [CrossRef] [PubMed]
44. Campoy, J.A.; Darbyshire, R.; Dirlewanger, E.; Quero-García, J.; Wenden, B. Yield potential definition of the chilling requirement reveals likely underestimation of the risk of climate change on winter chill accumulation. *Int. J. Biometeorol.* **2019**, *63*, 183–192. [CrossRef] [PubMed]
45. Pope, K.S.; Dose, V.; Da Silva, D.; Brown, P.H.; DeJong, T.M. Nut crop yield records show that budbreak-based chilling requirements may not reflect yield decline chill thresholds. *Int. J. Biometeorol.* **2015**, *59*, 707–715. [CrossRef]
46. Luedeling, E. Interpolating hourly temperatures for computing agroclimatic metrics. *Int. J. Biometeorol.* **2018**, *62*, 1799–1807. [CrossRef] [PubMed]
47. Dozier, W.A.; Powell, A.A.; Caylor, A.W.; McDaniel, N.R.; Carden, E.L.; McGuire, J.A. Hydrogen cyanamide induces budbreak of peaches and nectarines following inadequate chilling. *HortScience* **1990**, *25*, 1573–1575.
48. Erez, A. Means to compensate for insufficient chilling to improve bloom and leafing. *Acta Hortic.* **1995**, *395*, 81–96. [CrossRef]
49. Ghrab, M.; Mimoun, M.B.; Masmoudi, M.M.; Mechlia, N.B. Chilling trends in a warm production area and their impact on flowering and fruiting of peach trees. *Sci. Hortic.* **2014**, *178*, 87–94. [CrossRef]
50. Lobell, D.B.; Field, C.B.; Cahill, K.N.; Bonfils, C. Impacts of future climate change on California perennial crop yields: Model projections with climate and crop uncertainties. *Agric. For. Meteorol.* **2006**, *141*, 208–218. [CrossRef]
51. Taylor, K. Peaches. Available online: http://www.georgiaencyclopedia.org/articles/arts-culture/peaches (accessed on 16 May 2018).
52. Ray, D.K.; Gerber, J.S.; MacDonald, G.K.; West, P.C. Climate variation explains a third of global crop yield variability. *Nat. Commun.* **2015**, *6*, 5989. [CrossRef] [PubMed]

 *climate*

Article

Defining Crop–Climate Departure in West Africa: Improved Understanding of the Timing of Future Changes in Crop Suitability

Temitope S. Egbebiyi *, Olivier Crespo and Chris Lennard

Climate System Analysis Group (CSAG), Department of Environmental and Geographical Science, University of Cape Town, Private Bag X3, Rondebosch, Cape Town 7701, South Africa
* Correspondence: EGBTEM001@myuct.ac.za. or temitope@csag.uct.ac.za

Received: 20 May 2019; Accepted: 16 August 2019; Published: 21 August 2019

Abstract: The future climate is projected to change rapidly with potentially severe consequences for global food security. This study aims to improve the understanding of future changes in the suitability of crop growth conditions. It proposes a definition of crop realization, of the climate departure from recent historical variability, or crop–climate departure. Four statistically downscaled and bias-corrected Global Climate Models (GCMs): CCCMA, CNRM5, NOAA-GFDL, and MIROC5 performed simulations for the period 1960–2100 under the Representative Concentration Pathway RCP8.5 scenario to compute 20 year moving averages at 5-year increments. These were used to drive a crop suitability model, Ecocrop, for eight different crops across the three Food and Agriculture Organizations (FAO) AgroEcological Zones (AEZs) of West Africa (Guinea, Sahel, and Savanna). Simulations using historical climate data found that all crops except maize had a suitability index value (SIV) ≥ 0.50 outside the Sahel region, equivalent to conditions being suitable or strongly suitable. Simulations of future climate reveal that warming is projected to constrain crop growth suitability for cassava and pineapple in the Guinea zone. A potential for the northward expansion of maize is projected by the end of the century, suggesting a future opportunity for its growth in the southern Sahel zone. Crop growth conditions for mango and pearl millet remain suitable across all three AEZs. In general, crops in the Savanna AEZ are the most sensitive to the projected changes in climate. The changes in the crop–climate relationship suggests a future constraint in crop suitability, which could be detrimental to future food security in West Africa. Further studies to explore associated short- and long-term adaptation options are recommended.

Keywords: climate-departure; crop–climate departure; crop suitability; Ecocrop; food security; West Africa

1. Introduction

The livelihood and economies of most Sub Saharan African (SSA) countries are driven by rainfed agriculture [1–3]. About 96% of agricultural lands in SSA are rainfed [1,4]. Agriculture employs over 65% of the active labour force of the region, the majority of whom are practicing subsistence rainfed farming [5]. The agricultural sector is also responsible for 75% of SSA domestic trade [6,7]. It adds significantly to the economy of the region by contributing up to 15–20% to the Gross Domestic Product (GDP) [1,4,8,9]. In 2000, about 80% of the cereals consumed in SSA were domestically produced locally [4]. However, West Africa has been identified as one of the most vulnerable regions of the world owing to its low adaptive capacity and a fast-growing population, with many citizens whom are faced with malnourishment [1,10–12]. An adverse change in the climate over West Africa, both spatially and temporally, coupled with inadequate institutional and economic capacity to cope or adapt to its impact could become a determinant threat to agricultural production. Thus, food security and socio-economic activities across the region may be affected [13–17].

Climate strongly affects rainfed agriculture with direct consequences on food security [18–20]. This has resulted in different studies focusing on the response of crops and agriculture to the impact of increased greenhouse gas emissions across different regions of SSA owing to malnutrition and the need to improve food security [1,11,20–22]. Extreme changes in climate are projected to increase [23], translating into increased occurrence of both droughts and floods, which already account for 70% of economic losses through soil erosion and drought in West Africa [24,25]. The fifth Intergovernmental Panel on Climate Change (IPCC) assessment reported a projected warming across the different seasons over SSA to be larger than the global annual mean temperature increase [23]. The projected warming (1.5–4 °C by 2100) is likely to affect the agricultural sector by a reduction of up to 50% in crop yield and 90% in revenue across the region by the end of the century [7,23]. However, this may be further aggravated in regions like West Africa where the climate is warming faster and may lead to a radical departure from the regions' historical variability [26].

The definition of departure varies across disciplines. Broadly, a departure refers to a deviation or variation from a norm, standard rule, or behaviour. It can also mean starting out on a new course of action. In climate science, climate departure can be defined as a shift in the climate pattern of a region outside the range of historical variability and may be described in terms of mean local temperature exceeding historical highs [27–29]. Mora et al. [29] described climate departure as the year in which the average temperature of the coldest year after 2005 was warmer than the historic hottest year at a given location. Here we have defined climate departure as a deviation from the historical mean and/or variance of the local climate of an area or region induced by global warming [30].

The projected global warming level and the timing over the continent may intensify the impact of climate change on crop suitability. Severe temperature fluctuations and other extreme weather conditions such as droughts and floods may also threaten crop suitability thresholds. They vary spatially, resulting in potential yield declines where crop growth conditions are currently suitable and possible yield increases in other areas [31,32]. Challinor et al. [33], for instance, projected a future decline in crop yield of up to 5% for every degree of warming above the historical level in Africa.

Given the current state of climate departure research and the direct impact of climate on crop production systems (particularly rainfed), we are interested in the climate change induced crop realizations when climate departs from historical variability, which we term crop–climate departure. This study explores and proposes the information value of a comprehensive definition of crop–climate departure as "a departure from historical crop suitability threshold, whether in terms of variability, mean or both, due to warming of the climate over a location both in space and time resulting from climate change whether of radical climatic nature or not". This is in the context of recent climate historical variability and future climate projections using three West African weather stations, within three Food and Agriculture Organizations (FAO) Agro-Ecological Zones (AEZs). Mora et al. [29] suggests that West Africa will experience a climate departure with a mean temperature about two decades (2029) earlier than the global mean temperature (2047). Thus, we use the region as our proof of concept and to examine any likely large-scale crop suitability consequent changes the region may already be experiencing. Section 2 describes the data and methods used. Results from the study are outlined in Section 3. The discussion of the results and concluding remarks and recommendations for future are in Sections 4 and 5, respectively.

2. Data and Methods

2.1. Study Area

The demonstration area for this work is West Africa (Figure 1), which has rainfed agriculture as its mainstay economy. It is located at latitude 4–20° N and 16° W–20° E. The region comprises of 15 countries namely Benin, Burkina Faso, Gambia, Ghana, Guinea Bissau, Guinea, Ivory Coast, Liberia, Mali, Mauritania, Niger, Nigeria, Senegal, Sierra Leone, and Togo. It is divided into three FAO AEZs: Guinea (4–8° N), Savanna (8–12° N), and the Sahel (12–20° N) [25,34]. The temperature increases to the north of the region, while precipitation increases towards the south [25,34,35]. The Sahel zone is the warmest and driest, while the Guinea zone is the coolest and wettest of the three AEZs in West Africa. The climate of the region is mainly controlled by the West African Monsoon (WAM) which accounts for about 70% of the annual rainfall [20,34]. The WAM is an important and dynamic characteristic of the West African climate during the summer period [36]. It is produced from the reversal of the land and ocean differential heating and dictates the seasonal pattern of rainfall over West Africa between latitudes 9° and 20° N. The WAM is characterized by winds that blow south–westerly during the warmer months (June–September) and north-easterly during the cooler months (January–March) of the year [25,36]. It is the major system that influences the onset, variability, and pattern of rainfall over West Africa [3,37]. This affects rainfall producing systems with an impact on rainfed agriculture, which influences crop growth suitability and consequently food production in the region.

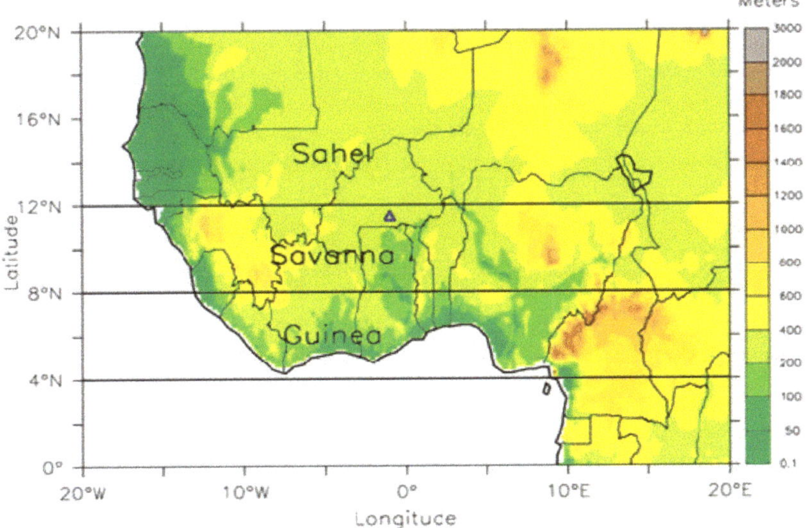

Figure 1. West African topography and the three Food and Agriculture Organizations (FAO)-Agro-ecological zones (AEZs): the Guinea, Savanna, and Sahel zones [25,34].

Different crops are grown in various parts of West Africa. Some of the major crops grown in the region are cassava, groundnut, millet, maize, sorghum, yam, plantain, cocoa, rice, and cowpea [20,38–40]. Millet and sorghum accounted for 64% of cereal production within the region in the year 2000, making them among the more important staple crops in West Africa [20,41]. Cassava is also an important staple food crop in terms of production in West Africa owing to its high resilience to drought [20,40,42]. This also applies to yam production, which accounts for about 91% of the world's production [20,41,43]. Maize provides about 20% of the calorie intake in West Africa and is adjudged the most important staple food overall in SSA [20,41]. Other crops such as cocoa and plantain, to mention a few, contribute significantly to the economy of the region.

2.2. Data

This study used three dataset types: observational weather station data, climate modelled data (statistically downscaled at the weather station level), and crop suitability data. The observed weather station data validated the mean monthly temperature and total monthly rainfall across the three AEZs. The crop suitability time series were simulated based on output from the Ecocrop suitability model [43] and the modelled climate data.

2.2.1. Climatic Variables

Temperatures (minimum and mean) and rainfall are important climate variables used in determining the impacts of climate change at subcontinental to global scales [44,45]. These two climate variables also have a significant effect on crop yield [46]. While rainfall affects the crop production in relation to its photosynthesis activities and leaf area, temperature affects the length of the crop growing season [47,48]. For this study, we used mean monthly minimum temperature (t-min) and mean monthly temperature (t-mean) and total monthly precipitation (prec.) of weather station data from Tabou, Ivory Coast; Sokode, Togo; Magaria, Niger. These weather stations each lie in three AEZs, Guinea, Savanna, and Sahel, respectively over West Africa. For the study, we used four statistically downscaled and bias corrected Global Climate Models (GCMs) in our analysis (CCCMA, CNRM5, GFDL, and MIROC) under a high-end climate change emission scenario (no adaptation), RCP8.5 (See Table 1 below for a description of the model). The GCMs were statistically downscaled using the Conditional Interpolation method as described in Hewitson and Crane [49]. The Conditional Interpolation downscaling method calculates the local phase relationships (PMI) for each weather station and each synoptic state combination. The bias relationship (BSI) between the weather station and its surroundings is then calculated. The method estimates the spatial extent of precipitation accurately and derives spatially referenced values representative of the area average. Overall, the interpolation conditioned by the synoptic state appears to better estimate realistic gridded values appropriate for use with model simulation output. For the temperature variable, the conditional interpolation employs the information content of the source data coupled with additional assumptions that may be physically justified (such as lapse-rate effects). The climate data were sourced from the Climate Information Portal (CIP) of the Climate System Analysis Group (CSAG), University of Cape Town (http://www.csag.uct.ac.za/climate-services/cip/). Data from this portal are at the station scale and weather stations in each AEZ are representative of that area.

Table 1. List of statistically downscaled and bias-corrected GCMs used in the study.

Modelling Institution	Institute ID	Model Name	Resolution
Canadian centre for climate modelling and analysis	CCCMA	CanESM2	$2.8° \times 2.8°$
Centre National de Recherches Meteorolo-Giques/Centre Europeen de Recherche et Formation Avanceesencalcul scientifique	CNRMCERFACS	CNRM-CM5	$1.4° \times 1.4°$
National Oceanic and Atmospheric Administration Geophysical Fluid Dynamic Laboratory	NOAAGDFL	GFDL_ESM2M	$2.5° \times 2.0°$
Japan agency for Marine-Earth Science and Technology	MIROC	MIROC5	$1.4° \times 1.4°$

2.2.2. Crop Thresholds to Suitability

The results of field experiments apply globally. A database of crop thresholds that translate into climate suitability has been collected and used in many locations to describe the suitability range of many plant and crop species using prec., t-min, t-mean, and the length of the growing season [43,50]. The climate threshold hosted by FAO dataset was obtained from the "dismo" package of the cran R software [51] (https://cran.r-project.org/web/packages/dismo/index.html). It was used in computing the climate suitability of each crop evaluated. It is acknowledged that thresholds will vary depending on finer resolution of the species (e.g., different varieties) or location (e.g., different soil, different rain distribution). However, the concept of crop suitability and the general validation of the thresholds makes this a useful tool to assess the impact of climate change and the emergence of novel regional climates on crop suitability over large areas examining the concept of crop–climate departure. The Ecocrop suitability model assessed four broad crop types and eight crops in total: cereals (pearl millet and maize); horticultural crops (tomato and pineapple); root and tuber crops (plantain and cassava) and fruit crops (mango and orange), using Ecocrop. The crop thresholds are listed in Table 2.

Table 2. Crop growth thresholds for eight crops as generated by the Ecocrop model.

Crop Name	Growing Duration (Days)	Temperature (°C)				Rainfall (mm)			
		Tmin	Topmin	Topmax	Tmax	Rmin	Ropmin	Ropmax	Rmax
Pearl millet	60–120	12	25	35	40	200	400	900	1700
Maize	65–365	10	18	33	47	400	600	1200	1800
Cassava	180–365	10	20	29	35	500	1000	1500	5000
Plantain	365	16	23	28	38	1000	1300	2000	5000
Pineapple	330–365	10	21	30	36	550	800	2500	3500
Tomato	70–150	7	20	27	35	400	600	1300	1800
Orange	180–365	13	20	38	38	450	1200	2000	2700
Mango	150–365	8	24	30	48	300	600	1500	2600

Where Tmin, Topmin, Topmax, and Tmax represents monthly minimum temperature, minimum optimum temperature, maximum optimum temperature, and maximum temperature, respectively; Rmin, Ropmin, Ropmax, and Rmax represents total monthly minimum rainfall, minimum optimum total monthly rainfall, maximum optimum total monthly rainfall, and maximum total monthly rainfall, respectively; optimum values represent the most suitable period for crop planting.

2.2.3. Model Description

The Ecocrop model is a crop suitability model. It uses a crop growth suitability threshold dataset hosted by the FAO [43]. It is an empirical model originally developed by Hijmans et al. [43] and based on the FAO-Ecocrop database [50] (Figure 2). The computation of optimal, suboptimal, and non-optimal conditions based on these datasets allows for the simulation of the suitability of crops in response to 12-month climate via t-min, t-mean, and prec. [43] (Figure 2). The Ecocrop model evaluates the relative suitability of crops in response to a range of climates including rainfall, temperature, and the growing season for optimal crop growth. A suitability index is generated as follows: 0 < 0.25 (not suitable), 0.25 < 0.5 (marginally suitable), 0.5 < 0.75 (suitable), and 0.75 < (highly suitable) [50,52]. The default Ecocrop parameters were assumed. Although those thresholds may vary with different geographical and/or climatic conditions, previous studies report a close correlation between the Ecocrop model and the climate change impact projections from other crop models [16,27,33,50]. A paucity of data over regions of interest like SSA limits the validation of these processes [53]. Nevertheless, the method contributes to the demand for regional scale assessment of crop response to future climate projections. The 12 coloured lines observed for the Ecocrop climate suitability simulations in Figure 2a,b below represents 12 months. Each describes the most suitable conditions for the crop under consideration in any given month. A highly seasonal crop (e.g., maize) has suitable growth conditions for a limited number of months. The conditions for non-seasonal crops are suitable throughout the year. Crops with a growing cycle longer than a year (e.g., pineapple and plantain) are represented by a single 12-month period.

2.3. Method

Four GCMs for the period 1960–2100 under the RCP8.5 scenario computed a 20-year moving average at 5 year time increments. It generated one mean 12-month value per 20-year window period for t-min, tmean, and prec. The mean 12-month climate value informed the crop suitability model, Ecocrop, for each GCM based on the methodologies described in Ramirez-Villegas, Jarvis, and Läderbach [50] for eight crops across the three AEZs of West Africa. The Ecocrop model simulated crop suitability indices characterized the crop–climate relationship and the impact global warming has on this relationship for each AEZ both spatially and temporally for each climate window. The suitability index scores were calculated for a range of climate variables for the period 1980–2000 using observed weather station data. This was used as a baseline to evaluate the downscaled GCM results spanning 1960–2100 at the three West African weather stations. It assessed the crop growth suitability in the zone for past climate conditions in reference to the published literature. Present day climate data was used as the preference for this zone owing to the constraint and paucity of weather station data and this data being the best available data to overlap with the Ecocrop model for the zone in the given study period.

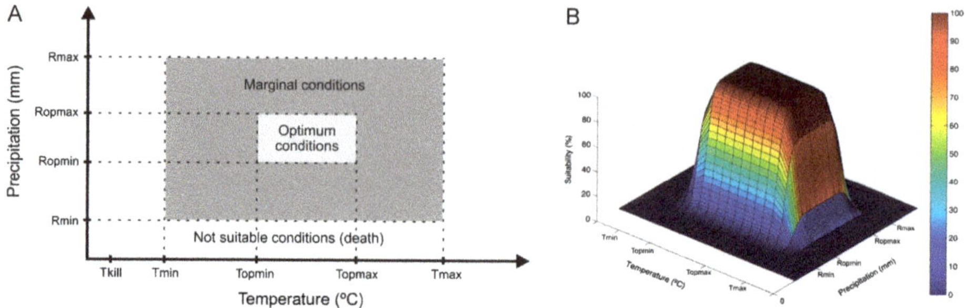

Figure 2. (**a**) two dimensional and (**b**) three dimensional diagram describing climate thresholds and its translation into crop suitability (Adapted from [50]).

3. Results

3.1. Evaluation of the GCMs in Simulating Rainfall and Temperature over West Africa

The downscaled climate data was firstly validated with the observed weather station data. Where missing records occurred, the corresponding month in the model's data were removed before computing the relationship between the datasets. Each GCM was correlated with the observed data for prec. and t-mean over the three weather stations, despite some discrepancy in precipitation over the Guinea zone (Figures 3 and 4). For temperature, the four models were correlated (r ≥0.6) with the observed t-mean across the three AEZs of West Africa with the highest correlation (r = 0.9) over Magaria. The models were also correlated (r ≥0.6) with the observed prec. in the Savanna and Sahel AEZs. A moderate correlation (r ≥0.3) with observed weather station data was evident in the Guinea AEZ. This weak correlation may be due to the low resolution of the GCMs in capturing the total monthly rainfall in the Guinea zone. The validated GCM data was then input into the Ecocrop model.

3.2. GCMs Representation of AEZs, Seasons, and Suitability over West Africa AEZs

A correlation exists between the Ecocrop suitability model simulated with climate inputs from four GCMs, CCMA, CNRM, GFDL, and MIROC (hereafter Eco-GCMs), although with minor variations in amplitude and time. However, it is worth stating that the variation in simulated suitability by the four GCMs may be attributed to the inter-annual variability of the GCMs or the GCMs parametrization scheme. Nevertheless, Eco-GCMs simulated crop suitability is similar across the three AEZs over West Africa for the eight crops considered in the study. For example, cassava shows a similar suitability pattern across the AEZs (Figure 5). It is unsuitable (Eco-CCMA and CNRM) to marginally suitable (Eco-GFDL and MIROC) for cassava crop growth in the Sahel AEZ. In the Savanna AEZ, it is currently highly suitable for cassava, but this is predicted to decline in the future to become marginally unsuitable. The Guinea AEZ suitability for cassava does not change. The variability in crop growth suitability curves may be attributed to the variation in yield and production of cassava across the region due to the impact of climate change, corroborating previous studies [8,38].

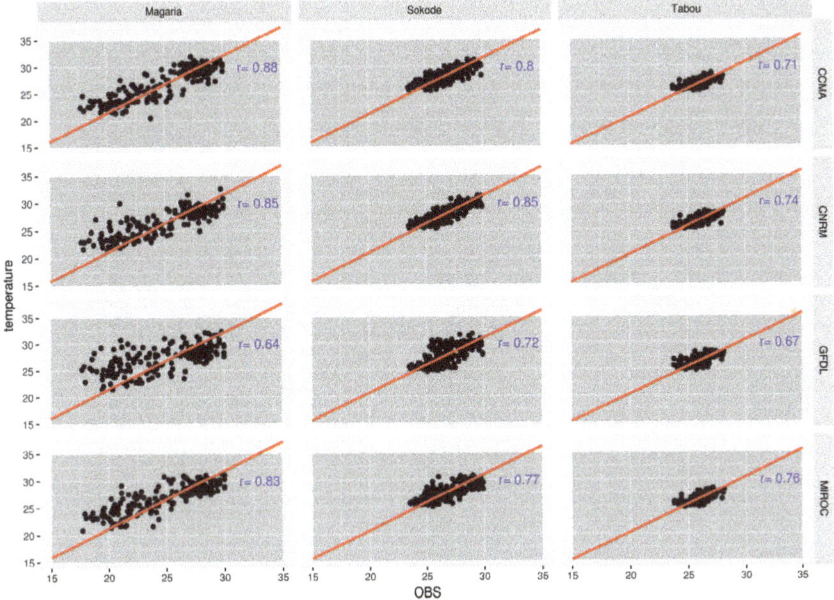

Figure 3. Mean monthly temperature (°C) as depicted by station observations and statistically downscaled CMIP5 GCMs (CCMA, CNRM5, GFDL, and MIROC) across the three agro ecological zones, Tabou, Sokode, and Magaria, of West Africa for the period 1980–2000. The top right corner r-values in each panel represent the correlations between the simulated and observed mean monthly temperature.

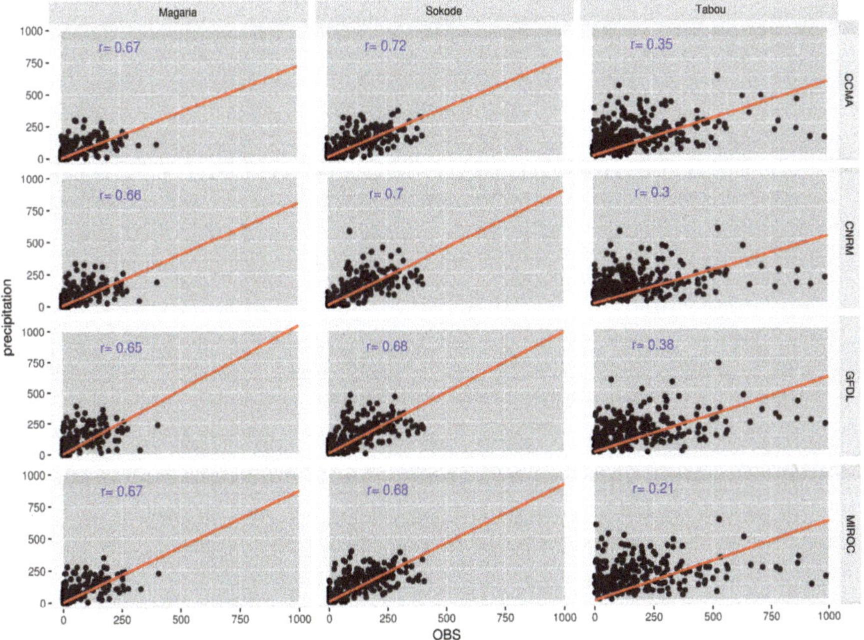

Figure 4. Total monthly rainfall (mm/month) as depicted by station observations and statistically downscaled CMIP5 GCMs (CCMA, CNRM5, GFDL, and MIROC) across the three agro ecological

zones, Tabou, Sokode, and Magaria, of West Africa for the period 1980–2000. The top right corner r-values in each panel represent the correlations between the simulated and observed mean monthly temperature.

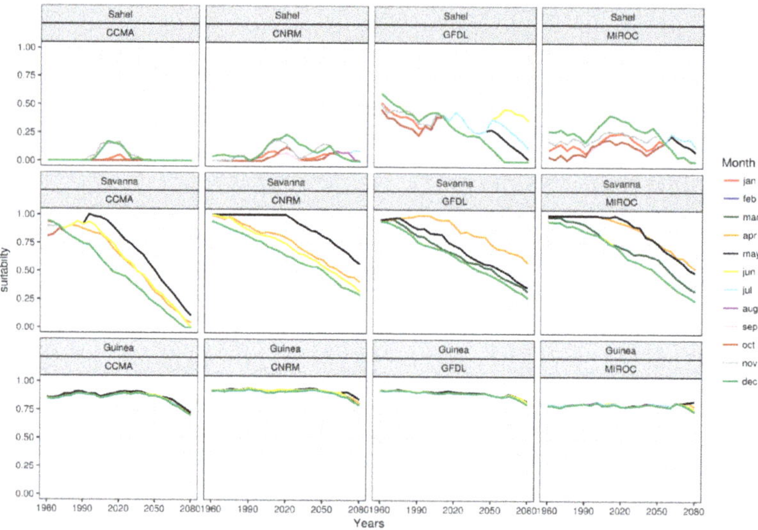

Figure 5. Cassava planting month suitability plots in the Guinea, Sahel, and Savanna as simulated by the four GCMs (CCMA, CNRM, GFDL, and MIROC) used as climate inputs into the Ecocrop suitability model.

Variability in the suitability of the month of planting for cassava crops in response to both AEZ and time increment is observed across the GCMs (Figure 5). The Guinea AEZ is currently the most suitable AEZ in which to grow cassava and is predicted to remain so. Suitable planting months in the Savanna AEZ as identified by all four models includes April–June and December, however a notable decline is observed, consistent for all four GCMs. Suitability declines from just below 1.0 to below 0.5 by 2050 in most cases. Conditions in the Sahel are presently and remain of low suitability. The simulation of the cassava crop growing season and period of planting across the three AEZs: Guinea (January–July and September–December), Savanna (April–November), and the Sahel (May/June–November) corroborates with previous findings with respect to the planting period and growing season in West Africa [54–56].

The concept of crop–climate departure allows for a consolidation of climate outputs from an ensemble of four GCMs into simulated crop suitability indices. Despite the marginal scale differences, the four GCMs consistently represent the unsuitability of cassava in the Sahel AEZ, its fast-declining suitability in the Savanna AEZ, as well as the high suitability of the Guinea AEZ. The cassava growing season is 12 months. Due to the predicted decline in the suitability of growth conditions, it is expected that it will become seasonal i.e., the suitability remains subject to appropriate seasonal planting, but conditions will then become unsuitable in the Savanna AEZ. Farmers will be required to adapt their practices in this AEZ. There is no reason at this stage to prefer one over the other GCMs, thus we use GCMs ensemble data as future climate scenario to simulate crop suitability in the subsequent sections of the paper plots of crop suitability from the Eco-GCMs. As seen from Figure 5, the ensemble suitability plots give a good representation of the Eco-GCMs model simulated suitability across the AEZs over West Africa. It shows the non-suitability of cassava in the Sahel, the fast-declining suitability and the observed seasonality in the Savanna AEZ, and high suitability in the Guinea zone. Thus, the crops ensemble suitability simulations are used in the results and discussion in the subsequent sections of this paper. A summary of crop suitability index values are given in Table 3.

Table 3. Ecocrop simulated crop Suitability Index Value (SIV) for the eight (8) different crops across the three Agro-ecological zones (AEZs) of West Africa.

Years	1960			1990			2020			2050			2080		
Crops/AEZs	GUI	SAV	SAH	GUI	SAV	SAH	GUI	SAV	SAH	GUI	SAV	SAH	GUI	SAV	SAH
Cassava	>0.75	>0.75	<0.25	>0.75	>0.75	<0.25	>0.75	>0.50	<0.25	>0.75	≤0.50	<0.25	>0.75	<0.50	<0.25
Maize	≤0.50	≥0.50	<0.75	≤0.50	≥0.50	0.50–0.75	<0.50	>0.50	>0.50	<0.50	>0.50	>0.50	<0.50	>0.75	>0.75
Mango	0.50–0.75	1.00	≤0.75	0.50–0.75	1.00	≤0.75	0.50–0.75	>0.75	>0.75	0.50–0.75	>0.75	0.75	0.50–0.75	>0.75	0.75
Orange	>0.75	>0.75	<0.25	>0.75	>0.75	<0.25	>0.75	>0.75	>0.25	>0.75	0.50–0.75	<0.25	>0.75	0.50–0.75	<0.25
Pearl millet	>0.75	>0.50	>0.50	>0.75	>0.50	>0.50	>0.75	>0.50	>0.50	>0.75	>0.50	>0.50	>0.75	>0.50	>0.50
Pineapple	>0.75	>0.75	<0.5	>0.75	<0.50	<0.50	>0.75	>0.75	0.50	>0.75	0.50–0.75	<0.50	>0.75	<0.50	<0.50
Plantain	1.00	>0.75	0.00	1.00	>0.75	0.00	>0.75	<0.75	0.00	>0.75	<0.75	0.00	>0.75	<0.50	0.00
Tomato	>0.75	>0.25	<0.5	>0.75	>0.50	<0.5	>0.75	>0.50	<0.5	<0.75	<0.50	<0.25	<0.75	<0.50	<0.25

GUI—Guinea AEZ, SAV—Savanna AEZ, SAH—Sahel AEZ. Ecocrop suitability Index: 0.0–0.25—Unsuitable/No suitability, 0.25–0.50—Marginally suitable, 0.50–0.75—Suitable, 0.75–1.00—Highly suitable.

3.3. Crop Suitability Response with Past Climate

Past climatic conditions (1960–2010) indicate that a crop growth suitability gradient existed from south to north across the three AEZs for each crop type considered. A Suitability Index Value (SIV) (0.75–1.00) is observed for each crop type in the Guinea and Savanna zones throughout the year (Figures 6 and 7). An exception is the cereal crop maize. Maize is marginally suitable (0.25–0.50) in the Guinea AEZ, suitable (0.50–0.75) for planting in May, September, and December in the Savanna AEZ, and January–February in the Sahel AEZ. The suitability increases (0.75–1.00) in 2050 (Figure 6). In the Sahel AEZ, other cereals and mango are suitable (above 0.50). Crop growth suitability increased for pineapple crops in this AEZ.

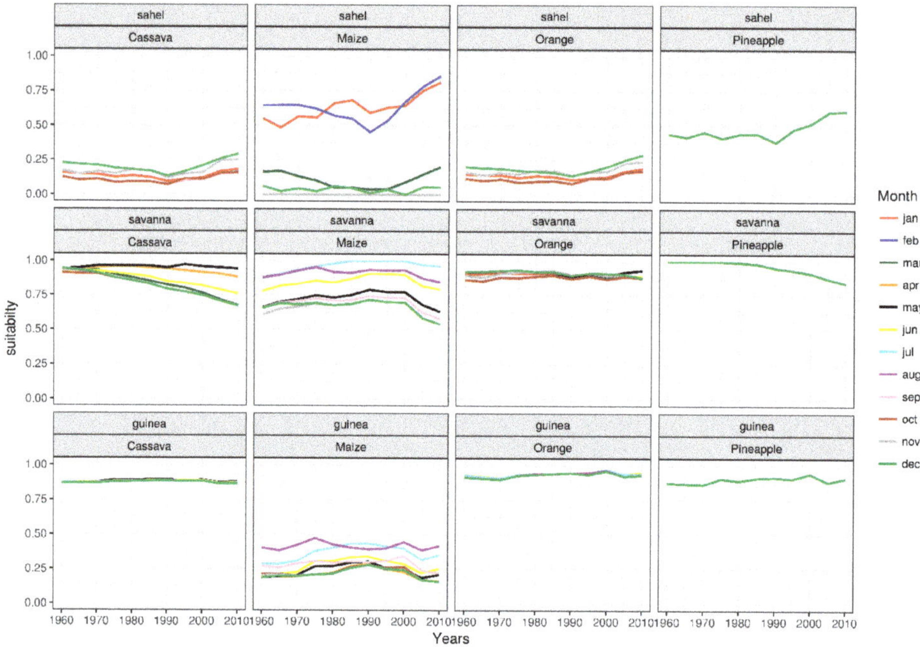

Figure 6. Ensemble crop suitability plots in the past climate (1960–2010) for cassava, maize, orange, and pineapple.

The Ecocrop simulations of crop growth suitability for the period 1960–2010 and the crop types evaluated corroborate previous findings with respect to the type of crops actually grown in the region. Cereals and root and tuber crops were the principal agricultural commodities in the region during this period [57,58]. Cassava and maize crops demonstrated modest yield increases of 6.3 to 10.3 and 1.1 to 1.8 tons ha^{-1}, respectively in the last 40 years [58,59]. The historical suitability of growth conditions for maize across the region (although marginal in the Guinea zone) highlights its importance as a staple crop here, accounting for almost 20% of the calorie intake for the population of West Africa [20,41]. The large area grown and high yield of pearl millet between 1960 and 2010 can be linked to the high suitability indices across the AEZs of West Africa over this period, contributing considerably to the livelihoods and economies of the countries in this region [60,61]. Increased productivity has also been witnessed in crops such as orange, mango, pineapple, and tomatoes in the last 40 years, again correlating with the high crop growth suitability indices identified for these crops [62,63]. Given the importance of these crops in the regional economy, a key question is, how is the projected change in climate predicted to impact on the crop growth suitability of these key crops in West Africa?

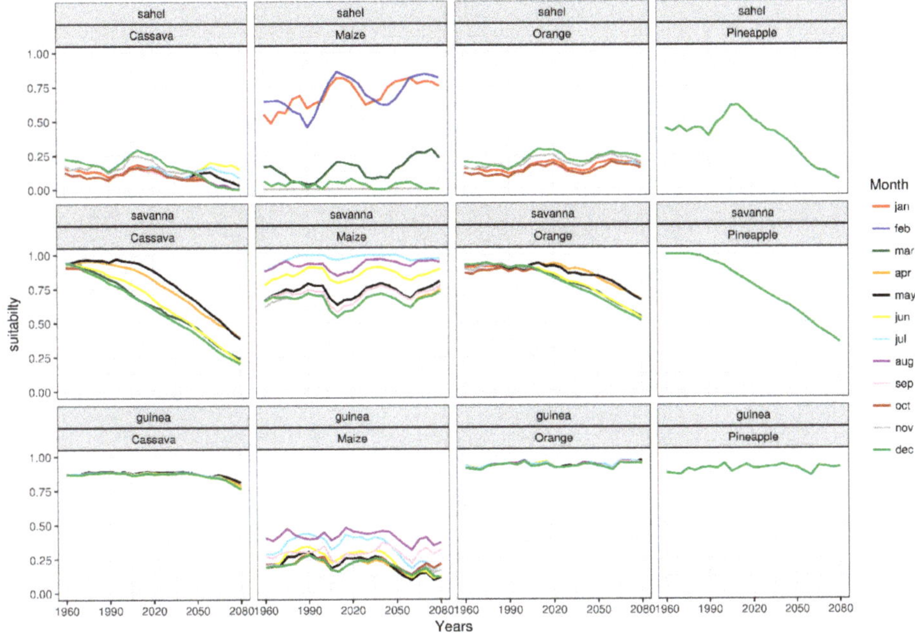

Figure 7. Projected model ensemble suitability over West Africa between 1960–2100 for cassava, maize, orange, and pineapple.

3.4. Projected Changes in Crop Suitability and Time of Planting over West Africa

The projected increase in global temperature is predicted to have a varied impact on crop growth suitability in West Africa (Figures 8 and 9). The Guinea AEZ remains largely unchanged with respect to crop growth suitability, evidently a more resilient area. Drastic declines are predicted for multiple crops in the Savanna AEZ including for cassava, orange, and pineapple. The main staple crop, maize, remains stable with an SIV of 0.5–1.0. It is interesting to note that the SIV for maize in the Sahel AEZ is projected to increase, shifting from suitable in 2020 to highly suitable by 2050 (Figure 7). Conditions for pearl millet will remain highly suitable (Figure 9), although the SIV for mango will decline post 2020.

The impact of future warming will affect crop seasonality, i.e the suitability of the time of planting. For root and tuber crops and cassava, in the months of April and May, they will become marginally suitable for cultivation by mid-century in the Savanna AEZ. Conditions will be unsuitable if planted in March, June, or December, which are currently optimal seasons. No change is predicted for cereal crops mango or orange.

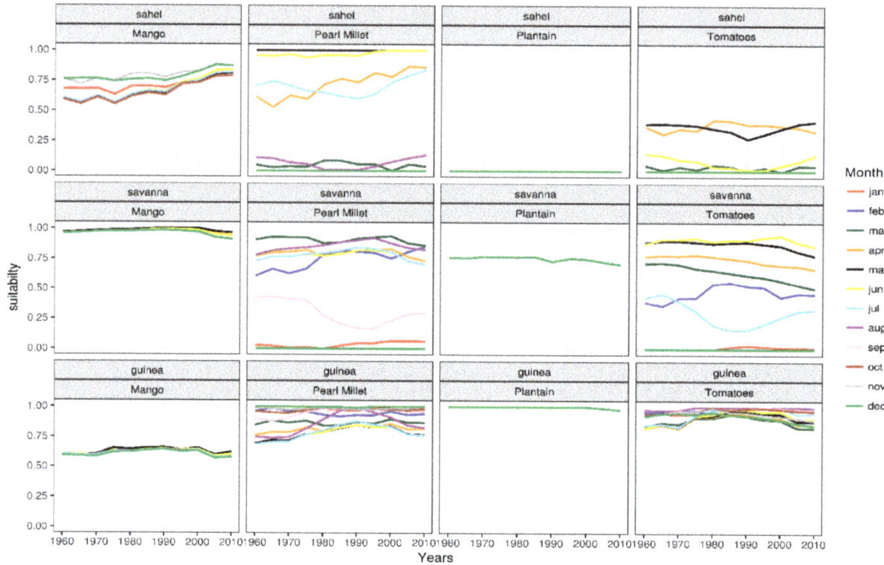

Figure 8. Ensemble crop suitability plots in the past climate (1960–2010) for mango, pearl millet, plantain, and tomatoes.

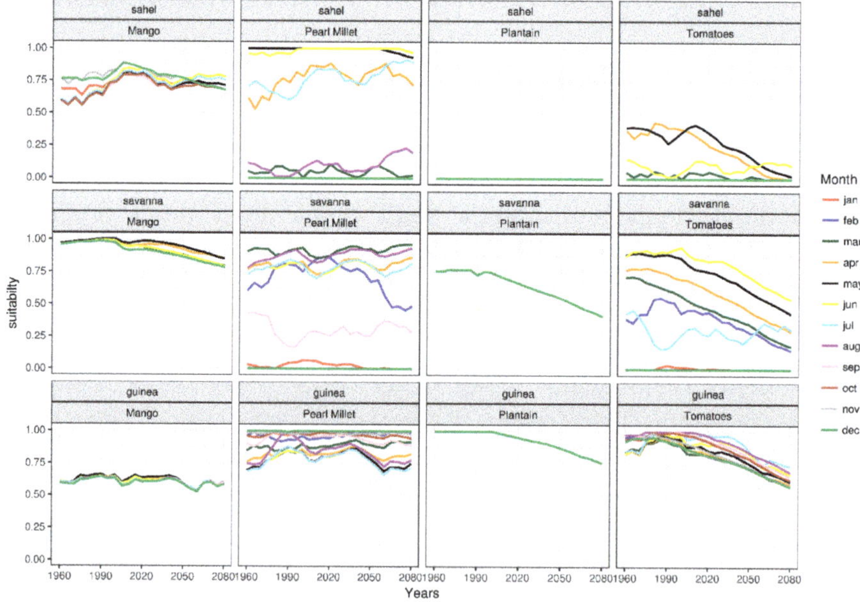

Figure 9. Projected model ensemble suitability over West Africa between 1960–2100 for mango, pearl millet, plantain, and tomatoes.

4. Discussion

4.1. From Climate Departure to Crop–climate Departure

The impact of global warming on the crop–climate relationship in the three AEZs of West Africa varies depending on the crop grown. The combination of a changing climate and crop growth suitability thresholds results in a projected deviation for the SIV from historical data. This may be a predicted increase in SIV, as observed for maize, or a decline as noted for crops such as cassava. A further variable is the AEZ itself. The predicted increase in SIV for maize in the Sahel zone by 2100 results from the projected increase in temperature and precipitation in this location. The decline in SIV forecast for cassava and pineapple in the Savanna AEZ decreases the suitability of the SSA region for these crops, spatially constraining suitable areas to the Guinea AEZ by 2100. The warming climate also influences the suitability of the month of planting of many crops. A potential cause for concern is the timing of crop–climate departure in the Savanna AEZ, which is already evident in root and tuber, cassava, and pineapple crops, and projected for orange crops by 2100. The diversity of crops grown in this AEZ may be diminished in the future.

The projected shifts in suitability and the variation of suitability between different crops and AEZ due to global warming highlights the importance of local climatic conditions in determining the extent of crop growth, development, and yield in response to climate suitability thresholds [31,64]. The above characteristics observed from the crop–climate relationship of different crops in the three AEZs of West Africa supports our proposed crop–climate departure definition of "a departure/shift from a historical crop suitability threshold, whether in terms of variability, mean or both, resulting from climate change whether of radical climatic nature or not due to warming of the climate over a location both in space and time". This definition can be used to inform adaptation responses to impacts resulting from climate change that will influence crop suitability, especially in a vulnerable region with low adaptive capacity like West Africa. Cultivating crops with high suitability as projected by the models such as cassava in the Guinea and maize in the southern Sahel may be the best solution. However, with improved/hybrid seedlings of the crop projected to decline that can withstand the variability in climate, this improved suitability may be considered another option.

4.2. Crop–Climate Departure and the Spatio-Temporal Variability of Crop-Suitability in West Africa

The temporal and spatial targeting of adaptation measures under an increasing global temperature will be crucial in maintaining and improving food security in the future. Identifying the timing and location of changes in crop growth suitability due to climate change can play a potentially key role in addressing the challenge of food production [16,33]. This is particularly relevant for the crops of vital importance in the West African region assessed here. The projected decrease in growth suitability conditions for cassava crops in the Savanna AEZ between 2020 and 2050, for example, depends on the time of planting or season. This projected departure is critical for the Savanna region as it may impact negatively on both the economy and livelihoods within the region. The improved understanding of crop–climate departure timing may permit timely adaptation plans such as modification to crop management regimes that account for this change in crop seasonality. If this were to be included in combination with increased use of key varietal traits, e.g., drought resistance, it will greatly assist in improving the adaptive capacity of such crops and mitigating the future impacts of a warmer climate to increase the resilience of current cropping systems within the region [16,65]. Improvements to the underlying knowledge base can therefore potentially improve both crop yield and crop quality.

In AEZs where the continued growing of given crop types is no longer possible, a further adaptation strategy is a shift to other more resilient crop species [66], or the substitution of existing crops with crops not previously grown within a given AEZ. Maize, for example, is projected to increase in yield by up to 7% in comparison to non-adapted crops under future climate change scenarios in SSA [33]. An increase in the planting area of crops such as maize further northwards into the southern Sahel AEZ due to the projected rainfall increase in this AEZ [31], or a shift of cassava or pineapple cropping into the Guinea AEZ due to reduced rainfall and the crops ability to withstand drought [16,31,65,67] represent further opportunities. A spatio-temporal projection of potential crop growth suitability can help provide information on future opportunities and constraints that will arise from shifts in the location of suitable crop lands within each AEZ [65]. Adaptation measures may then be prioritized for individual countries in response to the predicted changes [31,32], resulting in the maximum utilization of suitable areas for specific crop types, which will greatly assist in mitigating the future impacts from a warmer climate.

5. Summary and Conclusion

In order to improve the understanding of climatic impacts on agriculture, we conceptualized and explored the notion of crop–climate departure from historical variability in West Africa. We used four downscaled CMIP5 GCMs (CCMA CNRM5, GFDL, and MIROC) for the period 1960–2100 under RCP8.5 emission scenario and a crop suitability model, Ecocrop, across three weather stations representative of the Guinea, Savanna, and Sahel AEZs. In summary, all four GCMs correlate with observed weather station data in their simulation of monthly mean temperature and total monthly rainfall in the Savanna and Sahel zones, but moderately over the Guinea zone. It is recommended that future simulations acquire data from additional weather stations and utilize additional CMIP5 GCMs such as CSIRO, ICHEC, HADGEM, IPSL, MPI etc. In terms of crop rotations, the current climate is suitable for maize in the Savanna and Sahel AEZs, while future projections predict a potential for the expansion of maize further into the Sahel zone. The Guinea zone remains less suitable for maize but provides the correct climate both currently and in the future for crops such as cassava and pineapple. The predicted range for pearl millet and mango will remain stable in all three AEZs. Importantly, the Savanna AEZ, given its current cropping regime, is the most sensitive to climate change and shows the least resilience of the three AEZs considered. Climate change adaptation strategies will require prioritization in this zone. The climate-departure concept has been used to characterize crop–climate relationships i.e., crop–climate departure with increased warming and how it can, with appropriately planned adaptation and mitigation strategies, increase food security in the future.

Author Contributions: T.S.E. was responsible for developing the initial content of the manuscript, including literature search and data analysis. O.C. and C.L. were the supervisors for the research and provided guidance in terms of the article structure, data analysis, and finalization of the manuscript.

Funding: This research was supported with funding from National Research Foundation (NRF, South Africa), Alliance Centre for Climate and Earth Systems Science (ACCESS, South Africa), JW Jagger Centenary Scholarship, and Sari Johnson scholarship from the Postgraduate Funding Office, University of Cape Town, South Africa. Interpretation of the findings and the conclusion drawn from the study were the responsibilities of the authors and not on any part of NRF, ACCESS, JW Jagger Centenary Scholarship, and the Sari Johnson scholarship.

Acknowledgments: This study was supported with bursaries from the National Research Foundation (NRF, South Africa), Alliance Centre for Climate and Earth Systems Science (ACCESS, South Africa), and the JW Jagger Centenary Scholarship and Sari Johnson scholarship from the Postgraduate Funding Office, University of Cape Town, South Africa. We also acknowledge the anonymous reviewers and editor for their constructive comments which helps improve the quality of the study.

Conflicts of Interest: The authors declare that they have no financial or personal relationships that may have inappropriately influenced them in writing this article.

Abbreviations

ACCESS	Alliance for Collaboration for Climate and Earth System Sciences
AEZs	AgroEcological Zones
CCCMA	Canadian Centre for Climate Modelling and Analysis
CIP	Climate Information Portal
CMIP5	Coupled Model Intercomparison Project (Phase 5)
CNRM5	Centre National de Recherches Meteorolo-Giques
CSAG	Climate System Analysis Group
FAO	Food and Agriculture Organisation

GCMs	Global Climate Models
GFDL	Geophysical Fluid Dynamic Laboratory
MIROC	Japan agency for Marine-Earth Science and Technology Model
NRF	National Research Foundation
SSA	sub Saharan Africa

Declarations

Ethics approval and consent to participate	Not applicable
Consent for publication	Not applicable
Availability of data and materials	Not applicable

References

1. Roudier, P.; Sultan, B.; Quirion, P.; Berg, A. The impact of future climate change on West African crop yields: What does the recent literature say? *Glob. Environ. Chang.* **2011**, *21*, 1073–1083. [CrossRef]
2. Diasso, U.; Abiodun, B.J. Drought modes in West Africa and how well CORDEX RCMs simulate them. *Theor. Appl. Climatol.* **2017**, *128*, 223–240. [CrossRef]
3. Omotosho, J.B.; Abiodun, B.J. A numerical study of moisture build-up and rainfall over West Africa. *Meteorol. Appl.* **2007**, *14*, 209–225. [CrossRef]
4. World Bank. Making Development Climate Resilient. A World Bank Strategy for Sub-Saharan Africa. Report No. 46947-AFR. 2009, p. 144. Available online: http://siteresources.worldbank.org/INTAFRICA/Resources/ClimateChange-StrategyReport2010-Full_vNoImages.pdf (accessed on 21 August 2019).
5. Blein, R.; Soulé, B.G.; Dupaigre, B.F.; Yérima, B. *Agricultural Potential of West Africa.* Available online: http://www.fondation-farm.org/IMG/pdf/potentialites_rapport_ang_mp.pdf (accessed on 21 August 2019).
6. McCarthy, J.; Canziani, O.; Leary, N.; Dokken, D.; White, C. Climate Change 2001: Impacts, Adaptation, and Vulnerability. Available online: https://library.harvard.edu/collections/ipcc/docs/27_WGIITAR_FINAL.pdf (accessed on 21 August 2019).
7. World Bank. *World Development Report 2013: Jobs*; License: Creative Commons Attribution CC BY 3.0; World Bank: Washington, DC, USA, 2012. [CrossRef]
8. Benhin, J.K.A. South African crop farming and climate change: An economic assessment of impacts. *Glob. Environ. Chang.* **2008**, *18*, 666–678. [CrossRef]
9. Schlenker, W.; Lobell, D.B. Robust negative impacts of climate change on African agriculture. *Environ. Res. Lett.* **2010**, *5*, 014010. [CrossRef]
10. Slingo, J.M.; Challinor, A.J.; Hoskins, B.J.; Wheeler, T.R. Introduction: Food crops in a changing climate. *Philos. Trans. R. Soc. B Biol. Sci.* **2005**, *360*, 1983–1989. [CrossRef] [PubMed]
11. Knox, J.; Hess, T.; Daccache, A.; Wheeler, T. Climate change impacts on crop productivity in Africa and South Asia. *Environ. Res. Lett.* **2012**, *7*, 034032. [CrossRef]
12. Abatan, A.A. West African Extreme Daily Precipitation in Observations and Stretched-grid Simulations by CAM-EULAG. Available online: https://www.researchgate.net/publication/241809235_West_African_extreme_daily_precipitation_in_observations_and_stretched-grid_simulations_by_CAM-EULAG (accessed on 21 August 2019).
13. Grolle, J. Heavy rainfall, famine, and cultural response in the West African Sahel: The 'Muda' of 1953–54. *GeoJournal* **2014**, *43*, 205–214. [CrossRef]
14. Tarhule, A.; Woo, M.K. Towards an interpretation of historical droughts in northern Nigeria. *Clim. Chang.* **1997**, *37*, 601–616. [CrossRef]
15. Challinor, A.; Wheeler, T.; Garforth, C.; Craufurd, P.; Kassam, A. Assessing the vulnerability of food crop systems in Africa to climate change. *Clim. Chang.* **2007**, *83*, 381–399. [CrossRef]
16. Rippke, U.; Ramirez-Villegas, J.; Jarvis, A.; Vermeulen, S.J.; Parker, L.; Mer, F.; Diekkrüger, B.; Challinor, A.J.; Howden, M. Timescales of transformational climate change adaptation in sub-Saharan African agriculture. *Nat. Clim. Chang.* **2016**, *6*, 605–609. [CrossRef]

17. Paeth, H.; Hall, N.M.; Gaertner, M.A.; Alonso, M.D.; Moumouni, S.; Polcher, J.; Ruti, P.M.; Fink, A.H.; Gosset, M.; Lebel, T.; et al. Progress in regional downscaling of west African precipitation. *Atmos. Sci. Lett.* **2011**, *12*, 75–82. [CrossRef]

18. Hansen, J.; Sato, M.; Ruedy, R. Perception of climate change. *Proc. Natl. Acad. Sci. USA* **2012**, *109*, E2415–E2423. [CrossRef] [PubMed]

19. Pretty, J.N.; Morison, J.I.L.; Hine, R.E. Reducing food poverty by increasing agricultural sustainability in developing countries. *Agric. Ecosyst. Environ.* **2003**, *95*, 217–234. [CrossRef]

20. Sultan, B.; Gaetani, M. Agriculture in West Africa in the Twenty-First Century: Climate Change and Impacts Scenarios, and Potential for Adaptation. *Front. Plant Sci.* **2016**, *7*, 1–20. [CrossRef] [PubMed]

21. Lobell, D.B.; Burke, M.B.; Tebaldi, C.; Mastrandrea, M.D.; Falcon, W.P.; Naylor, R.L. Prioritizing Climate Change Adaptation Needs for Food Security in 2030 Region. *Science* **2008**, *319*, 607–610. [CrossRef]

22. Zinyengere, N.; Crespo, O.; Hachigonta, S. Crop response to climate change in southern Africa: A comprehensive review. *Glob. Planet. Chang.* **2013**, *111*, 118–126. [CrossRef]

23. IPCC. Summary for Policymakers. In *Climate Change 2013: The Physical Science Basis. Contribution of Working Group I to the Fifth Assessment Report of the Intergovernmental Panel on Climate Change*; Stocker, T.F., Qin, D., Plattner, G.K., Tignor, M., Allen, S.K., Boschung, J., Nauels, A., Xia, Y., Bex, V., Midgley, P.M., Eds.; Cambridge University Press: Cambridge, UK, 2013.

24. World Bank. *World Development Report 2010: Development and Climate Change*; World Bank: Washington, DC, USA, 2010; Available online: https://openknowledge.worldbank.org/handle/10986/4387 (accessed on 30 November 2018).

25. Egbebiyi, T.S. Future Changes in Extreme Rainfall Events and African Easterly Waves over West Africa. Master's Thesis, University of Cape Town, Cape Town, South Africa, 2016.

26. Sui, Y.; Lang, X.; Jiang, D. Time of emergence of climate signals over China under the RCP4.5 scenario. *Nat. Clim. Chang.* **2013**, *39*, 1–5.

27. Vermeulen, S.J.; Challinor, A.J.; Thornton, P.K.; Campbell, B.M.; Eriyagama, N.; Vervoort, J.M.; Kinyangi, J.; Jarvis, A.; Läderach, P.; Ramirez-Villegas, J.; et al. Addressing uncertainty in adaptation planning for agriculture. *Proc. Natl. Acad. Sci. USA* **2013**, *110*, 8357–8362. [CrossRef]

28. Hawkins, E.; Sutton, R. Time of emergence of climate signals. *Geophys. Res. Lett.* **2012**, *39*, 1–6. [CrossRef]

29. Mora, C.; Frazier, A.G.; Longman, R.J.; Dacks, R.S.; Walton, M.M.; Tong, E.J.; Sanchez, J.J.; Kaiser, L.R.; Stender, Y.O.; Anderson, J.M. The projected timing of climate departure from recent variability. *Nature* **2013**, *502*, 183–187. [CrossRef]

30. Thornes, J.E. IPCC, 2001: Climate change 2001: Impacts, adaptation and vulnerability, Contribution of Working Group II to the Third Assessment Report of the Intergovernmental Panel on Climate Change, edited by J. J. McCarthy, O.F.; Canziani, N.A.; Leary, D.J., Dokken, A. *Int. J. Climatol.* **2002**, *22*, 1285–1286. [CrossRef]

31. Porter, J.R.; Semenov, M.A. Crop responses to climatic variation. *Philos. Trans. R. Soc. B Biol. Sci.* **2005**, *360*, 2021–2035. [CrossRef]

32. Lobell, D.B.; Schlenker, W.; Costa-Roberts, J. Climate trends and global crop production since 1980—Supporting Online Material. *Science* **2011**, *333*, 616–620. [CrossRef]

33. Challinor, A.J.; Watson, J.; Lobell, D.B.; Howden, S.M.; Smith, D.R.; Chhetri, N. A meta-analysis of crop yield under climate change and adaptation. *Nat. Clim. Chang.* **2014**, *4*, 287–291. [CrossRef]

34. Abiodun, B.J.; Adeyewa, Z.D.; Oguntunde, P.G.; Salami, A.T.; Ajayi, V.O. Modeling the impacts of reforestation on future climate in West Africa. *Theor. Appl. Climatol.* **2012**, *110*, 77–96. [CrossRef]

35. Klutse, N.A.; Ajayi, V.O.; Gbobaniyi, E.O.; Egbebiyi, T.S.; Kouadio, K.; Nkrumah, F.; Quagraine, K.A.; Olusegun, C.; Diasso, U.; Abiodun, B.J. Potential impact of 1.5 °C and 2 °C global warming on consecutive dry and wet days over West Africa. *Environ. Res. Lett.* **2018**, *13*, 055013. [CrossRef]

36. Janicot, S.; Caniaux, G.; Chauvin, F.; De Coëtlogon, G.; Fontaine, B.; Hall, N.; Kiladis, G.; Lafore, J.P.; Lavaysse, C.; Lavender, S.; et al. Intraseasonal variability of the West African monsoon. *Atmos. Sci. Lett.* **2011**, *12*, 58–66. [CrossRef]

37. Nicholson, S.E. The West African Sahel: A Review of Recent Studies on the Rainfall Regime and Its Interannual Variability. *ISRN Meteorol.* **2013**, *2013*, 1–32. [CrossRef]

38. Paeth, H.; Capo-Chichi, A.; Endlicher, W. Climate Change and Food Security in Tropical West Africa—A Dynamic-Statistical Modelling Approach. *Erdkunde* **2008**, *2*, 101–115. [CrossRef]

39. Nelson, G.C.; Van Der Mensbrugghe, D.; Ahammad, H.; Blanc, E.; Calvin, K.; Hasegawa, T.; Havlik, P.; Heyhoe, E.; Kyle, P.; Lotze-Campen, H.; et al. Agriculture and climate change in global scenarios: Why don't the models agree. *Agric. Econ.* **2014**, *45*, 85–101. [CrossRef]

40. Jarvis, A.; Ramirez-Villegas, J.; Campo, B.V.H.; Navarro-Racines, C. Is Cassava the Answer to African Climate Change Adaptation? *Trop. Plant Biol.* **2012**, *5*, 9–29. [CrossRef]

41. FAOSTAT. *FAO Statistical Yearbook 2014*. 2014. Available online: http://www.fao.org/3/a-i3590e.pdf (accessed on 21 August 2019).

42. Srivastava, A.K.; Gaiser, T.; Ewert, F. Climate change impact and potential adaptation strategies under alternate climate scenarios for yam production in the sub-humid savannah zone of West Africa. *Mitig. Adapt. Strateg. Glob. Chang.* **2016**, *21*, 955–968. [CrossRef]

43. Hijmans, R.J.; Guarino, L.; Cruz, M.; Rojas, E. Computer tools for spatial analysis of plant genetic resources data: 1. DIVA-GIS. *Plant Genet. Resour. Newsletter* **2001**, *127*, 15–19.

44. Cong, R.G.; Brady, M. The interdependence between rainfall and temperature: Copula analyses. *Sci. World J.* **2012**, *12*. [CrossRef]

45. IPCC. IPCC Expert Meeting on Climate Change, Food, and Agriculture. Available online: https://www.ipcc.ch/site/assets/uploads/2018/05/Food-EM_MeetingReport_FINAL-1.pdf (accessed on 21 August 2019).

46. Medori, M.; Michelini, L.; Nogues, I.; Loreto, F.; Calfapietra, C. The impact of root temperature on photosynthesis and isoprene emission in three different plant species. *Sci. World J.* **2012**, *2012*. [CrossRef]

47. Olesen, J.E.; Bindi, M. Consequences of climate change for European agricultural productivity, land use and policy. *Eur. J. Agron.* **2002**, *16*, 239–262. [CrossRef]

48. Cantelaube, P.; Terres, J.M. Seasonal weather forecasts for crop yield modelling in Europe. *Tellus Ser. A Dyn. Meteorol. Oceanogr.* **2005**, *57*, 476–487. [CrossRef]

49. Hewitson, B.C.; Crane, R.G. Gridded area-averaged daily precipitation via conditional interpolation. *J. Clim.* **2005**, *18*, 41–57. [CrossRef]

50. Ramirez-Villegas, J.; Jarvis, A.; Läderach, P. Empirical approaches for assessing impacts of climate change on agriculture: The EcoCrop model and a case study with grain sorghum. *Agric. For. Meteorol.* **2013**, *170*, 67–78. [CrossRef]

51. Hijmans, A.R.J.; Phillips, S.; Leathwick, J.; Elith, J.; Hijmans, M.R.J.; Hijmans, R.J. Species Distribution Modeling. Available online: https://cran.r-project.org/web/packages/dismo/dismo.pdf (accessed on 21 August 2019).

52. Hunter, R.; Crespo, O. *Large Scale Crop Suitability Assessment Under Future Climate Using the Ecocrop Model: The Case of Six Provinces in Angola's Planalto Region*; Springer: Cham, Switzerland, 2018.

53. White, J.W.; Hoogenboom, G.; Kimball, B.A.; Wall, G.W. Methodologies for simulating impacts of climate change on crop production. *F. Crop. Res.* **2011**, *124*, 357–368. [CrossRef]

54. Vrieling, A.; de Leeuw, J.; Said, M.Y. Length of growing period over africa: Variability and trends from 30 years of NDVI time series. *Remote Sens.* **2013**, *5*, 982–1000. [CrossRef]

55. Butt, B.; Turner, M.D.; Singh, A.; Brottem, L. Use of MODIS NDVI to evaluate changing latitudinal gradients of rangeland phenology in Sudano-Sahelian West Africa. *Remote Sens. Environ.* **2011**, *115*, 3367–3376. [CrossRef]

56. Brown, M.E.; de Beurs, K.M. Evaluation of multi-sensor semi-arid crop season parameters based on NDVI and rainfall. *Remote Sens. Environ.* **2008**, *112*, 2261–2271. [CrossRef]

57. Dixon, J.A.; Gibbon, D.P.; Gulliver, A. *Farming Systems and Poverty: Improving Farmers' Livelihoods in a Changing World*; FAO & World Bank: Rome, Italy; Washington, DC, USA, 2001.

58. Alliance for Green Revolution in Africa (AGRA). The Africa Agriculture Status Report 2014: Climate Change and Smallholder Agriculture in Sub-Saharan Africa. Available online: https://ccafs.cgiar.org/publications/africa-agriculture-status-report-2014-climate-change-and-smallholder-agriculture-sub#.XVy0U0G-lPY (accessed on 21 August 2019).

59. Bationo, A.; Hartemink, A.E.; Lungo, O.; Naimi, M.; Okoth, P.; Smaling, E.M.A.; Thiombiano, L. African Soils: Their Productivity and Profitability of Fertilizer Use. Available online: https://library.wur.nl/WebQuery/wurpubs/fulltext/26759 (accessed on 21 August 2019).

60. Mason, S.C.; Maman, N.; Palé, S. Pearl millet production practices in semi-arid West Africa: A review. *Exp. Agric.* **2015**, *51*, 501–521. [CrossRef]

61. Singh, P.; Boote, K.J.; Kadiyala, M.D.; Nedumaran, S.; Gupta, S.K.; Srinivas, K.; Bantilan, M.C. Science of the Total Environment An assessment of yield gains under climate change due to genetic modi fi cation of pearl millet. *Sci. Total Environ.* **2017**, *601–602*, 1226–1237. [CrossRef]

62. Barrett, H.; Browne, A. Export Horticultural Production in Sub-Saharan Africa: The Incorporation of The Gambia. *Geography* **1996**, *81*, 47–56.

63. Takane, T. Smallholders and nontraditional exports under economic liberalization: The case of pineapples in Ghana. *Afr. Study Monogr.* **2004**, *25*, 29–43.

64. IPCC. Summary for Policymakers. In *Climate Change 2007: The Physical Science Basis. Contribution of Working Group I to the Fourth Assessment Report of the Intergovernmental Panel on Climate Change*; Solomon, S., Qin, D., Manning, M., Chen, Z., Marquis, M., Ave, K.B., Eds.; Cambridge University Press: Cambridge, UK; New York, NY, USA, 2007.

65. Wheeler, T.; von Braun, J. Climate change impacts on global food security. *Science* **2013**, *341*, 55–60. [CrossRef]

66. Ramirez-Villegas, J.; Thornton, P.K. *Climate Change Impacts on African Crop Production*; CCAFS Working Paper No. 119; CGIAR Research Program on Climate Change, Agriculture and Food Security (CCAFS): Copenhagen, Denmark, 2015; 27p.

67. Ramirez-Villegas, J.; Lau, C.; Hooker, J.; Jarvis, A.; Ann-Kristin, K.; Arnell, N.; Tom, O. *Climate Analogues: Finding Tomorrow's Agriculture Today CGIAR Research Program on Climate Change*; Agriculture and Food Security (CCAFS): Copenhagen, Denmark, 2011. Available online: https://ccafs.cgiar.org/publications/climate-analogues-finding-tomorrow%E2%80%99s-agriculture-today-0#.XVy3U0G-lPY (accessed on 21 August 2019).

Article

Assessing Future Spatio-Temporal Changes in Crop Suitability and Planting Season over West Africa: Using the Concept of Crop-Climate Departure

Temitope S. Egbebiyi *, Chris Lennard, Olivier Crespo, Phillip Mukwenha, Shakirudeen Lawal and Kwesi Quagraine

Climate System Analysis Group (CSAG), Department of Environmental and Geographical Science, University of Cape Town, Private Bag X3, Rondebosch, 7701 Cape Town, South Africa
* Correspondence: EGBTEM001@myuct.ac.za

Received: 16 July 2019; Accepted: 20 August 2019; Published: 24 August 2019

Abstract: The changing climate is posing significant threats to agriculture, the most vulnerable sector, and the main source of livelihood in West Africa. This study assesses the impact of the climate-departure on the crop suitability and planting month over West Africa. We used 10 CMIP5 Global climate models bias-corrected simulations downscaled by the CORDEX regional climate model, RCA4 to drive the crop suitability model, Ecocrop. We applied the concept of the crop-climate departure (CCD) to evaluate future changes in the crop suitability and planting month for five crop types, cereals, legumes, fruits, root and tuber and horticulture over the historical and future months. Our result shows a reduction (negative linear correlation) and an expansion (positive linear correlation) in the suitable area and crop suitability index value in the Guinea-Savanna and Sahel (southern Sahel) zone, respectively. The horticulture crop was the most negatively affected with a decrease in the suitable area while cereals and legumes benefited from the expansion in suitable areas into the Sahel zone. In general, CCD would likely lead to a delay in the planting season by 2–4 months except for the orange and early planting dates by about 2–3 months for cassava. No projected changes in the planting month are observed for the plantain and pineapple which are annual crops. The study is relevant for a short and long-term adaptation option and planning for future changes in the crop suitability and planting month to improve food security in the region.

Keywords: crop-climate departure; Ecocrop; crop suitability; planting month; CORDEX; West Africa

---

1. Introduction

The West African region has been identified as one of the hotspots with high susceptibility and vulnerability to the impact of climate change and global warming [1]. For example, the global climate is projected to be above 1.5 °C above the pre-industrial level in the next decade [2]. An increase in temperature between 3 °C and 6 °C coupled with a rise in the rainfall variability is projected into the future over West Africa from the AR5 report [3]. Most countries in West Africa heavily rely on agriculture, which is predominantly rainfed, as an important and significant contributor to their economies. It accounts for over 16% of the Gross Domestic Product (GDP) of the region's economy and employs over 60% of the labour force [4–6]. Additionally, West Africa has accounted for about 60% of the total value of the agricultural production in the continent for about 24 years [7]. However, the region has been identified as a hotspot to climate change impacts in the recent time owing to its reducing yields in the total agricultural production since 2007 in comparison to other sub-regions on the continent [7]. Current trends show that there may be further decreases in yields especially in the face of increasing warming and droughts which may lead to food insecurity over the region [8–10].

Findings from the Intergovernmental Panel on Climate Change (IPCC) fifth Assessment Report (AR5) shows widespread impacts from the changing climate to the historical month across all continents [11]. The report reveals a high exposure to climatic events and a low adaptive capacity of the African continent makes it one of the most vulnerable regions of the world. Agriculture is the most and major economic sector of Africa and has been described as the most vulnerable sector to the climate change impact with a great threat to the farming systems, crop production and food security at any level [7,12–14]. For example, past studies e.g., [15–19] have shown the impact of climate change on crop production and yield in Africa and West Africa in particular using different crop models. Sultan et al. [15] showed the decrease in the mean yield of sorghum cultivars due to the impact of climate change resulting from variation in the rainfall pattern and increasing temperature. Jalloh et al. [17] revealed that the impact of climate change will badly affect the production of major staple crops in West Africa particularly sorghum and groundnut in the Sahel. Moreover, Roudier et al. [6] combining the result of 16 published studies, showed that the projected impact of climate change on the crop yield over most African countries is negative (about 11%) with variations among crops, regions and modelling uncertainties posing the challenge for robust assessment of future yields at the regional scale. Further changes in the climate are expected in Africa over the next decades [1], as projections suggest a threat to food security due to the likely increase in climate variability over the next decades in Sub-Saharan Africa (SSA) [7]. As a result, impacts from the changing climate varies from subsectors among regions and different countries in SSA including West Africa but may be more detrimental to the West African region owing to its high susceptibility and low adaptive capacity with further warming [14,20].

The increase in global warming will lead to a new climate regime with a deviation from historical variability with a variation in the timing of emergence for different regions of the world called the climate departure [21,22]. For instance, [21] found that the mean temperature over West Africa will move outside the bounds of historical variability about two decades earlier before the global mean temperature thus making the region a hotspot of climate departure due to the impact of the global warming. On this premise and its direct consequence on rainfed crop production in West Africa, Egbebiyi et al. [23] explored the climate change induced crop realizations of the climate departing from historical variability, developed and proposed the concept called the crop-climate departure (CCD) in the context of recent climate historical variability and future climate projections. The study defines CCD "as a departure from historical crop suitability threshold, whether in terms of variability, mean or both, over a location both in space and in time resulting from climate change (whether radical climatic change or not)" This concept was used to characterize crop suitability across the three agro-ecological zones (AEZs) of West Africa. However, the CCD concept was only tested and applied using three weather stations, within the three AEZs of West Africa. Although these stations are a representation of the three AEZs, nevertheless these cannot be generalized for the entire region, hence there is a need to examine how CCD at different climate windows, near the future (2031–2050) till end of the century (2081–2100) will affect crop suitability over the region using the concept of CCD.

Based on our definition and understanding on CCD, the aim of this present study is to examine the impact of CCD from the historical variability on future changes in crop suitability and month of planting over the entire West African region. Section 2 describes the data and methods used. Results from the study are outlined in Section 3. The discussion of the results and concluding remarks and recommendations for the future are in Sections 4 and 5, respectively.

2. Data and Methodology

2.1. Study Area

The West African (shown in Figure 1) region comprises of 15 countries namely Benin, Burkina Faso, Gambia, Ghana, Guinea Bissau, Guinea, Ivory Coast, Liberia, Mali, Mauritania, Niger, Nigeria, Senegal, Sierra Leone and Togo. It is geographically located at latitude 4–20 °N and 16 °W–20 °E and has rainfed agriculture as its mainstay economy. The region can be divided into three Food and Agriculture Organization (FAO) agro ecological zones (AEZs) namely, Guinea (4–8 °N), Savanna (8–12 °N) and the Sahel (12–20 °N) [24,25]. The region also has some localized highlands (Cameroon Mountains, Jos Plateau, and Guinea Highlands) which influence its climate. The climate of the region is mainly controlled by the West African Monsoon (WAM) which accounts for about 70% of the annual rainfall [24,26]. WAM is an important and dynamic characteristic of the West African climate during the summer month [27].WAM is produced from the reversal of the land and ocean differential heating and dictates the seasonal pattern of rainfall over West Africa between latitudes 9° and 20 °N. It is characterized by winds that blow south-westerly during warmer months (June–September) and north-easterly during cooler months (January–March) of the year [25,27]. It is the major system that influences the onset, variability and pattern of rainfall over West Africa [28], [29]. It alternates between wet (April–October) and dry seasons (November–March) as the rainfall belt follows the migration of Inter-Tropical Discontinuity (ITD) [30] and thus affects the rainfall producing systems with an impact on the rainfed agriculture and influences crops suitability and food production in the region.

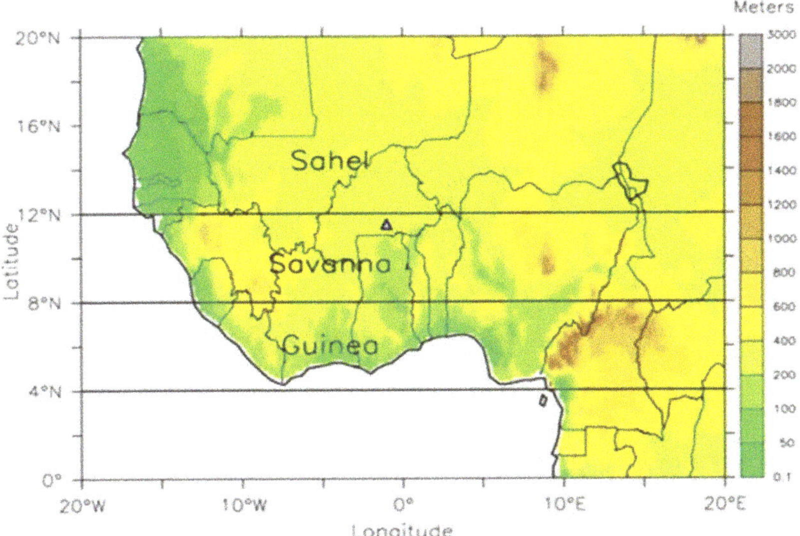

Figure 1. The study area, showing the West African topography and the three Food and Agriculture Organization (FAO) agro-ecological zones, designated as Guinea, Savannah and Sahel, respectively [24,25].

Different crops are grown in various parts of West Africa. Some of the major crops grown in the region are cassava, groundnut, millet, maize, sorghum, yam, plantain, cocoa, rice, wheat [8,26,31,32]. Millet and sorghum account for 64% of the cereal production over the regions in the year 2000 thus making them among the important staple crops in West Africa [26,33]. Cassava is one of the most important staple food crops in terms of production in sub-Saharan West Africa owing to high resilience to drought in the region [26,32,34] This also applies to the Yam production, which account for about

91% of the world's production [26,34,35]. Cereal, maize provides about 20% of the calorie intake in West Africa and is adjudged the most important staple food in the sub-Saharan Africa [26,33]. Other crops such as cocoa and plantain to mention a few contribute significantly to the economy of the region.

2.2. Data

2.2.1. Historical and Future Climate Datasets

For this study, three datasets were used as observations of the present-day climate and the locations where crops are grown as observed from the crop suitability model, Ecocrop output, and modelled simulations of the present and projected crop suitability driven by the observed and projected climate data. The observation dataset was the 0.5° × 0.5° resolution monthly precipitation and minimum and mean temperature gridded dataset for the month of 1901 to 2016 obtained from the Climate Research Unit (CRU TS4.01 version, land only) University of East Anglia [36]. This was used to evaluate the available bias corrected RCMs forced by the 10 CMIP5 global climate models. The bias-corrected climate data were obtained from the Swedish Meteorological and Hydrological Institute, Linköping, Sweden. The modelled climate data were used as inputs into the crop suitability model, Ecocrop [37]. For this study, five different crop types namely; cereals (maize, pearl millet and sorghum), root and tuber (cassava, plantain and yam), legumes (cowpea and groundnut), horticulture (pineapple and tomato) and fruit (mango and orange) were selected based on the FAO 2016 statistics and their economic importance in the region. These different datasets are defined in the sub-sections below.

Temperatures and rainfall are important climate variables used in determining the impacts of climate change at different scales [38,39]. These two climate variables have a significant effect on crop yield [40,41]. While rainfall affects crop production in relation to the photosynthesis and leaf area, the temperature affects the length of the growing season [42,43]. For this study, we used the bias-corrected mean monthly minimum temperature (tmin), mean monthly temperature (tmean) and total monthly precipitation (prec). Data from 10 CMIP5 GCMs downscaled by SMHI-RCA4 are used as input into the crop suitability model (Table 1). We used the RCP8.5 emission scenario for the analysis to investigate the impact of CCD from the historical variability on the crop growth suitability and month of planting over West Africa. We used RCP8.5 because it seems the most realistic emission scenario as seen from the greenhouse gas emission trajectories in comparison to other scenarios and also has the largest simulation ensemble members [44].

Table 1. List of dynamically downscaled Global Climate Models (GCMs) used in the study.

Modelling Institution	Institute ID	Model Name	Resolution
Canadian centre for climate modelling and analysis	CCCMA	CanESM2	2.8° × 2.8°
Centre National de Recherches Meteorolo-Giques/Centre Europeen de Recherche et Formation Avanceesencalcul scientifiqu	CNRMCERFACS	CNRM-CM5	1.4° × 1.4°
Commonwealth Scientific and Industrial Research Organisation in collaboration with the Queensland Climate Change Centre of Excellence	CSIRO-QCCCE	CSIRO-Mk3.6.0	1.875° × 1.875°
NOAA geophysical fluid dynamic laboratory	NOAAGDFL	GFDL_ESM2M	2.5° × 2.0°
UK Met Office Hadley centre	MOHC	HadGEM2-ES	1.9° × 1.3°
EC-EARTH consortium	EC-EARTH	ICHEC	1.25° × 1.25°
Institute Pierre-Simon Laplace	IPSL	IPSL-CM5A-MR	1.25° × 1.25°
Japan agency for Marine-Earth Science and Technology	MIROC	MIROC5	1.4° × 1.4°
Max Planck institute for meteorology	MPI	MPI-ESM-LR	1.9° × 1.9°
Norwegian climate centre	NCC	NorESM1-R	2.5° × 1.9°

2.2.2. Ecocrop—A Crop Suitability Model

The Ecocrop model is a crop suitability model. It uses a crop growth suitability threshold dataset hosted by the FAO [37]. It is a simple mechanistic and empirical model originally developed by Hijmans et al. [37] and based on the FAO-Ecocrop database [45]. It is designed at a monthly scale with the ability to analyse the crop suitability in relation to the climate conditions over a geographical location [37,45]. Ecocrop employs environmental ranges of a crop coupled with numerical assessment of the environmental condition to determine the potential suitable climatic condition for a crop. The suitability rating can be linked to the agricultural yield which is partly dependent on the strength of the climate signal in the agricultural yield [46] The computation of optimal, suboptimal and non-optimal conditions based on these datasets allows for the simulation of the suitability of crops in response to the 12-month climate via t-min, t-mean and prec. [37]. The Ecocrop model evaluates the relative suitability of crops in response to a range of climates including rainfall, temperature and the growing season for optimal crop growth. A suitability index is generated as follows: $0 < 0.20$ (not suitable), $0.20 < 0.4$ (very marginally suitable), $0.4 < 0.6$ (marginally suitable), $0.6 < 0.8$ (suitable), and $0.8 < 1.0$ (highly suitable) [45,47]. The default Ecocrop parameters were assumed. Although those thresholds may vary with different geographical and/or climatic conditions, previous studies have reported a close correlation between the Ecocrop model and the climate change impact projections from other crop models [45,48–50]. A paucity of data over regions of interest like SSA limits the validation of these processes [51]. Nevertheless, the method contributes to the demand for the regional scale assessment of the crop response to future climate projections.

2.3. Methods

We analyzed 10 CMIP 5 GCMs datasets downscaled by CORDEX RCM, RCA4 to assess the impacts of CCD from the historical variability on crop suitability and planting season over West Africa for five different crop types, cereal (maize, pearl millet and sorghum), fruit (mango and orange), horticulture (pineapple and tomato), legume (cowpea and groundnut) and root and tuber (cassava, plantain and yam). We used the RCA4 simulation output for the monthly minimum and mean temperature and total monthly precipitation as input into Ecocrop, a crop suitability model. Using a 20-year moving average at five year time steps, we computed the Suitability Index Value (SIV) for each crop across the 10 downscaled GCMs over West Africa. The Ecocrop suitability output were then used to assess the impact of global warming through CCD from the historical variability on the crop suitability and planting season over a month 1951–2100. Across the agro-ecological zones (AEZs) of West Africa. After the simulation, we computed the mean of the best three consecutive suitability index and best three months of planting window within the growing season across each grid point over the region for the historical and future month. Before examining the RCM-projected changes in the future crop suitability and planting season, we evaluated the capability of the models in simulating the crop suitability spatial distribution and planting date/season during the reference month (1981–2000).

We also used the statistical tool to calculate the trend of change across the three windows compared to the historical month. We assessed the trend of change in the crop suitability and month of planting at each global warming levels for each crop using the Theil-Sen estimator or Sen's slope [52,53]. The Theil-Sen slope estimator is an estimation of the average trend rate only and magnitude of the trend. It is a linear slope that is compatible with the Mann-Kendall test and more robust such that it is less sensitive to outliers in the time series as compared to the standard linear regression trend [54]. The Theil-Sen slope method can detect significant trends with the changing rate than the linear trend [55]. Previous studies [56,57] have used this method in calculating trends.

2.3.1. Simulation Approach and Analysis of suitability

Past studies (e.g., [25,58–60] have evaluated the performance of the RCA4 historical data against the CRU dataset in the past climate. Their results showed that there is a good agreement with a strong correlation (r ≥ 0.6) between the CRU dataset and RCA4 monthly simulated past climate data for both the temperature and precipitation over West Africa. For example, the model replicates the CRU north-south temperature gradient that concurs with previous findings by [58]. Additionally, the RCA4 simulated total monthly rainfall realistically captures the essential features namely, both the zonal pattern and meridional gradient and the rainfall maxima over high topography (i.e., Cameroon Mountains and Guinean Highlands) as observed in CRU which agrees with previous findings by [25,59,60]. The performance of RCA4 in simulating the essential features of West African climate variables, temperature and rainfall, and doubles as the needed input variables for the crop suitability model, Ecocrop makes it suitable and gives confidence in the use of the RCA4 for the crop suitability simulation over the region.

In addition, we compare the Ecocrop simulation over the region with the MIRCA2000 annual harvested area around year 2000 from the global monthly gridded data as described by [61] for six crops, cassava, maize, groundnut, sorghum, millet and plantain available in the MIRCA2000 dataset. The MIRCA2000 dataset provides monthly irrigated and rainfed crops area for 26 crop classes for each month of the year around year 2000 with a spatial resolution about 9.2 km. We compare the spatial agreement between the Ecocrop simulation and MIRCA2000 by using an overlap in the spatial agreement between the two datasets. Although, we admit the short time length of the MIRCA dataset however, it is a useful gridded dataset that has been used to provide information on the crop harvested area across different regions of the world [61] and will be useful to evaluate the simulated Ecocrop spatial suitability distribution at present due to the paucity of the suitability dataset across the globe. To see the overlap and area of agreement in the spatial suitability output of the two datasets, we set the MIRCA2000 annual harvested area dataset as one (1) and the Ecocrop simulated suitable area suitability index value from 0.2 (SIV ≥ 0.2) as two(2). Where the two datasets agree as three(3). The output shows a good agreement between the Ecocrop and MIRCA2000 data for the examined crops with a strong spatial correlation (r > 0.7) (Figure 2). This gives some level of confidence in the use and performance of the Ecocrop simulation over the region.

To assess the impact of CCD from the historical variability on the crop suitability over West Africa, we computed the monthly climatological mean for a 20-year running month, at every five-year timestep for the t-min, t-mean and prec. from 1951–2100. For example, the first 20-year mean computed was 1951–1970, the second 20-year mean was 1956–1975, etc., until the last month 2081–2100. The resulting 12-month values per the 20-year month window was used as an input climatology into the Ecocrop suitability model as developed by the Food and Agriculture Organization, FAO [37] to simulate crop suitability for each downscaled GCM based on the methodologies described in [45]. Ecocrop calculates the crop suitability values in the response climate variables such as a monthly rainfall and temperature datasets and generates an output with a suitability index score from zero (unsuitable) to one (optimal/excellent suitability). It should be noted that this study did not undertake any additional ground-truthing or calibration of the range of climate parameters preferred for either crop and therefore the default EcoCrop parameters were assumed. Suitability index scores were calculated for the range of climate variables reported for the historical baseline (1981–2000) future months, near future (2031–2050), mid-century (2051–2070) and end of century (2081–2100) for the downscaled 10 CMIP5 GCMs that participated in the CORDEX experiment.

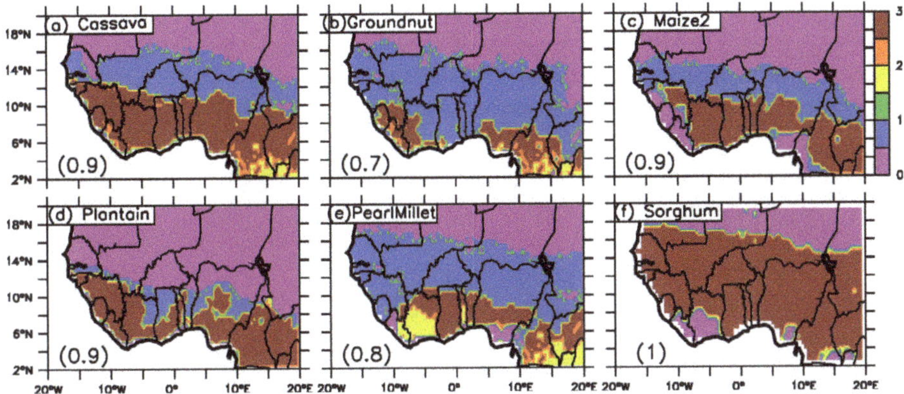

Figure 2. A simulated spatial distribution of the crop harvested area and suitability over West Africa for the year 2000 as simulated by the MIRCA2000 dataset and Ecocrop, respectively. The blue area (represented by 1) are the crop harvested area around the year 2000 as simulated by the MIRCA2000 dataset while the yellow colour represents the suitability index value above0.2 (SIV ≥ 0.2) which is represented by two. The red colour represents the area where the two datasets agree as denoted by three. The number at the left-hand corner represents the spatial correlation (r ≥ 0.7) value between the two datasets. The red colour depicts in Fig. 2a-2f depicts harvested and suitable areas as simulated by MIRCA2000 and Ecocrop from cassava to sorghum respectively. The blue colour depicts MIRCA2000 simulated harvested area only for each crop while yellow means Ecocrop simulated suitable areas for cultivation of each crop in year 2000. The purple colour, 0 depicts non harvested and unsuitable areas as simulated by both MIRCA2000 and Ecocrop for each of the crops around the year 2000.

2.3.2. Assessing the Robustness of Climate Change

We use two conditions (model agreement and statistical significance) to evaluate the robustness of the projected climate change for the three future months. For the model agreement, at least 80% of the simulation must agree on the sign of change. For the statistical significance, at least 80% of the simulations must indicate that the influence of the climate change is statistically significant, at 95% confidence level using a *t* test with regards to the baseline month, 1981–2000. When these two conditions are met then we consider the climate change signal to be significant. [30,44,62,63] have all used the methods to test and indicate the robustness of the climate change signals.

3. Result

3.1. Crop Suitability in the Historical Climate over West Africa

The RCA4 simulated crop suitability from the observed climatology inputs (RCA4-Ecocrop) shows a decreasing mean suitability from south to north over West Africa (north-south suitability gradient). The spatial suitability representation reveals unsuitable or very marginal suitability to the north in the Sahel from lat. 14 °N with a low Suitability Index Value (SIV) value between 0.0–0.4 and a higher suitability to the south in the Guinea-Savanna AEZ with a high SIV (0.6–1.0) sandwiched by an ash/silver suitability line called the Marginal Suitability Line (MSL) with an SIV between 0.41–0.59. In general, the MSL are observed around lat.14 °N in the Sahel AEZ (northern Sahel) for the simulation across the region except for the one observed around lat. 12 °N, the boundary between the Sahel and Savanna AEZ. The RCA4 simulation of all crop types examined, legumes (cowpea and groundnut), root and tuber (cassava, plantain, Yam, white yam), cereals (maize, pearl millet and sorghum) and fruit and horticultural crops (mango, orange, pineapple and tomato) shows that all the crops are very suitable to the south of the MSL but with no or low suitability to the north (Figures 3–6, column 1).

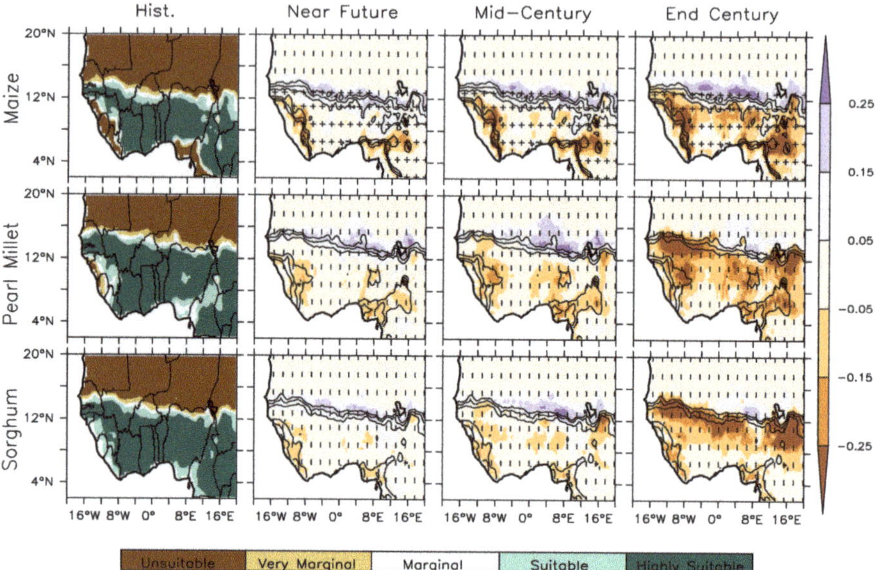

Figure 3. Simulated spatial suitability distribution for the cereal crops, maize pearl millet and sorghum over West Africa for the historical month (1981–2000) (column 1) and the projected change in the crop suitability for the near future month (2031–2050), mid-century (2051–2070) and end of century (2081–2100) (column 2–4, respectively). The vertical strip (|) indicates where at least 80% of the model simulations agrees on the projected sign of change while the horizontal strip (−) indicates where at least 80% of the model simulations agree that the projected change is statistically significant at 99% confidence level. The cross (+) indicates where the two conditions are met, meaning that the change is robust.

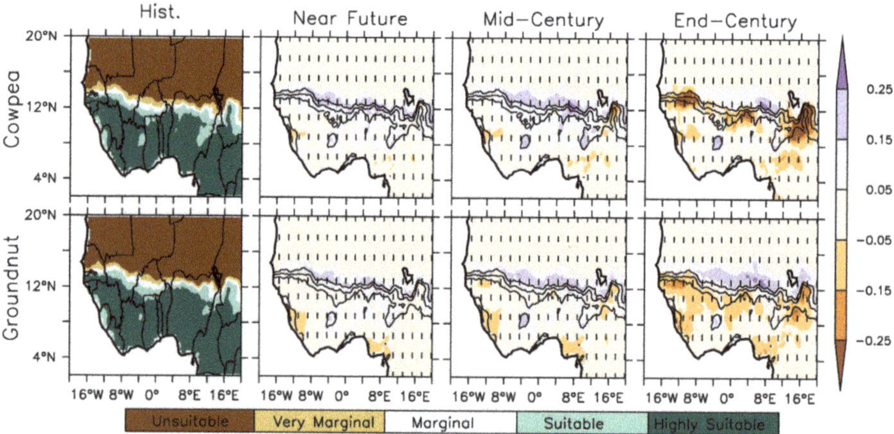

Figure 4. Same as Figure 3 but for the legume crops, cowpea and groundnut.

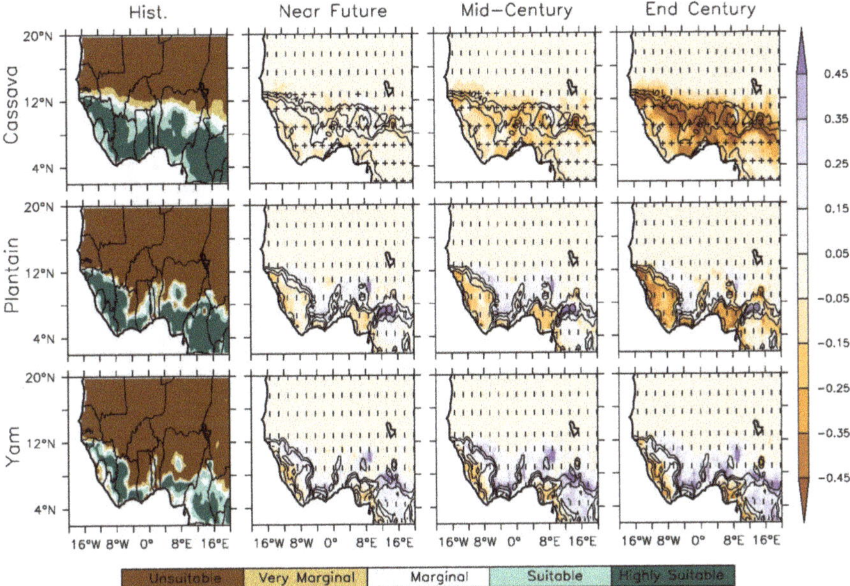

Figure 5. Same as Figure 3 but for the root and tuber crops, cassava, plantain and yam.

Along the coastal areas, legumes and root and tuber crops are suitable along the south-west coast of Senegal to the south-west coast of Cameroon. High SIV are observed for the root and tuber crops, plantain and Yam in the north central part of Nigeria in the Savanna. It is worth mentioning because the surrounding areas are observed to be unsuitable for the cultivation of both crops. For cereals, pearl millet is suitable along the west coast of Senegal and from the south coast Ivory Coast to the south-west coast of Cameroon. Maize is suitable from the south coast of Ivory Coast to the south-west coast of Nigeria. Fruit and horticultural crops are all suitable along the south coast of the Ivory Coast to the south-west coast of Nigeria. Mango and pineapple are suitable along the west coast of Senegal to Gambia while orange and tomato are only suitable along the west coast of Gambia.

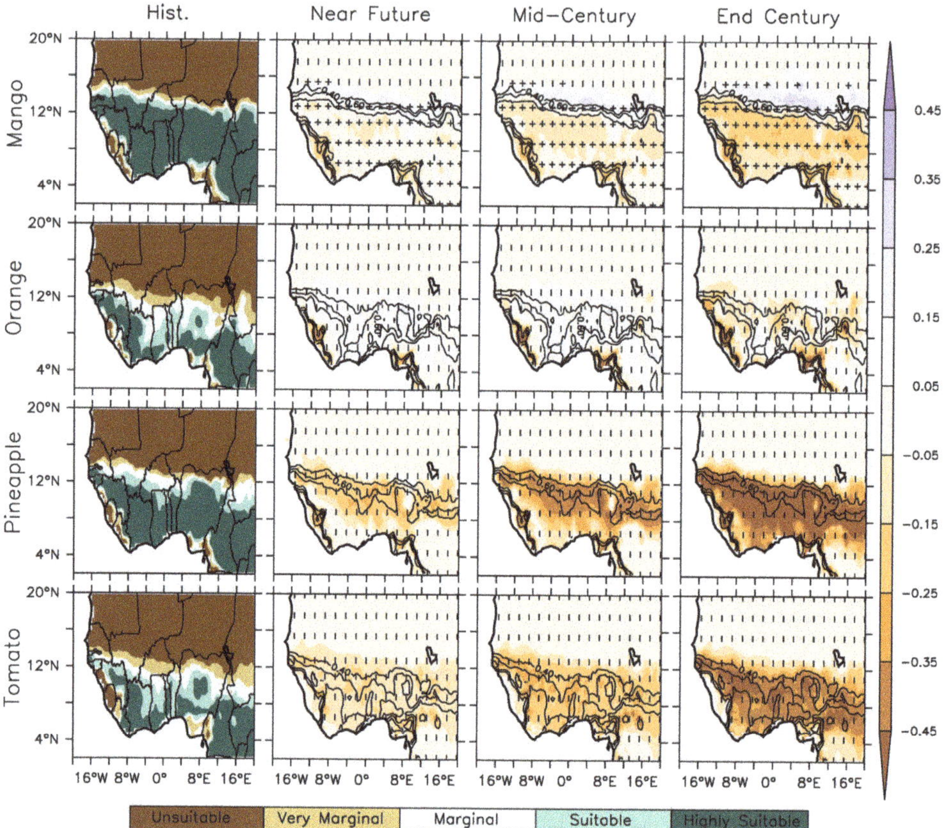

Figure 6. Same as Figure 3 but for the fruit crops, mango and orange and horticultural crops, pineapple and tomato.

RCA4 was also used in simulating the best planting months (PM) from the range of month in a planting window within the Length of Growing Season (LGS) over West Africa for the historical climate (Figures 7–10, column 1). LGS provides information on the start and end of the growing season and can also assist in the simulation process of identifying the best PM within a possible planting window in a growing season over a given location. The simulated planting month represents the first month of the best three months of the planting window. For example, a simulation of April means April–June is the three best PM and varies with crop types across the three AEZs of the region. For the legumes, our simulation shows January–July as the planting windows for cowpea and groundnut over the region (Figure 7, column 1). Jan (January–March) and Feb (February–April) as the best PM for cowpea and groundnut, respectively in the central Guinea and Savanna AEZs except over Sierra Leone, Liberia and the south coast of Nigeria. The month of Feb (Feb-April) was simulated as the best three planting months in the western and eastern Savanna-Sahel AEZs for cowpea, while it was Mar (March–May) over the same area and month for the groundnut. Along the coastal areas, July is simulated as the PM along the southwest coast from southern Sierra Leone to Liberia and the south coast of Nigeria and April along the southwest coast of northern Sierra Leone. For the groundnut, April is the PM along the west coast of Guinea, while May is the PM along the west coast of Sierra Leone and northern Liberia. August and March are the PM at the south coast of Liberia and Nigeria, respectively. The months of December and January are the PMs along the south coast of Ivory Coast to Ghana for the cowpea and groundnut, respectively.

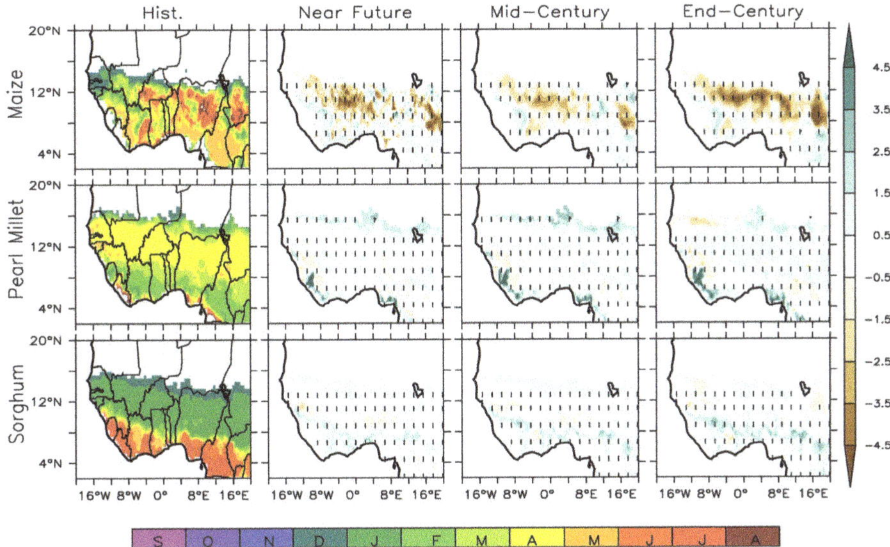

Figure 7. Simulated month of planting for cereals, maize, pearl millet and sorghum over West Africa for the historical month (1981–2000) (column 1) and the projected change in the crop planting month for the near future month (2031–2050), mid-century (2051–2070) and end of century (2081–2100) (column 2–4 respectively). The planting is simulated from September to August. The vertical strip (|) indicates where at least 80% of the model simulations agrees on the projected sign of change while the horizontal strip (−) indicates where at least 80% of the model simulations agree that the projected change is statistically significant at 99% confidence level. The cross (+) indicates where the two conditions are met, meaning that the change is robust.

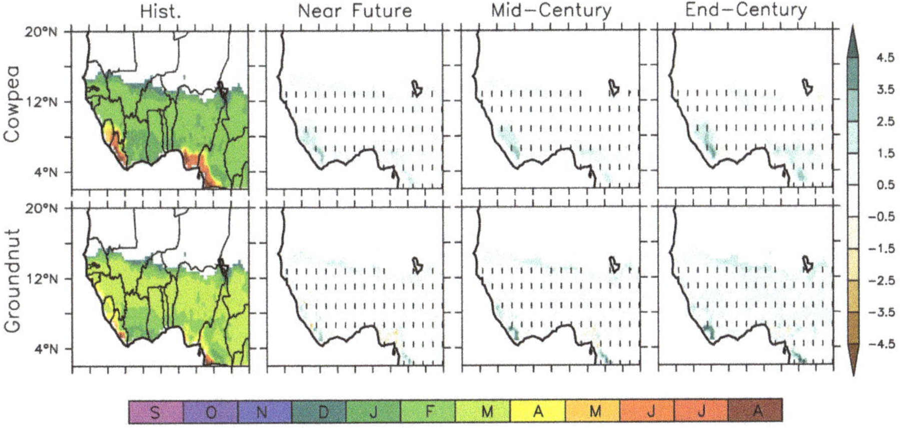

Figure 8. Same as Figure 7 but for the legumes, cowpea and groundnut.

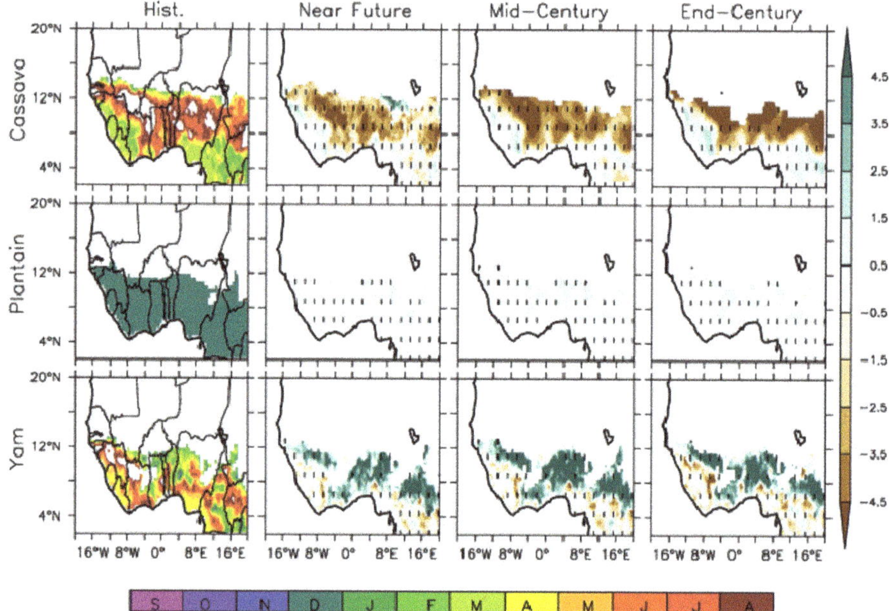

Figure 9. Same as Figure 7 but for the root and tubers, cassava, plantain and yam.

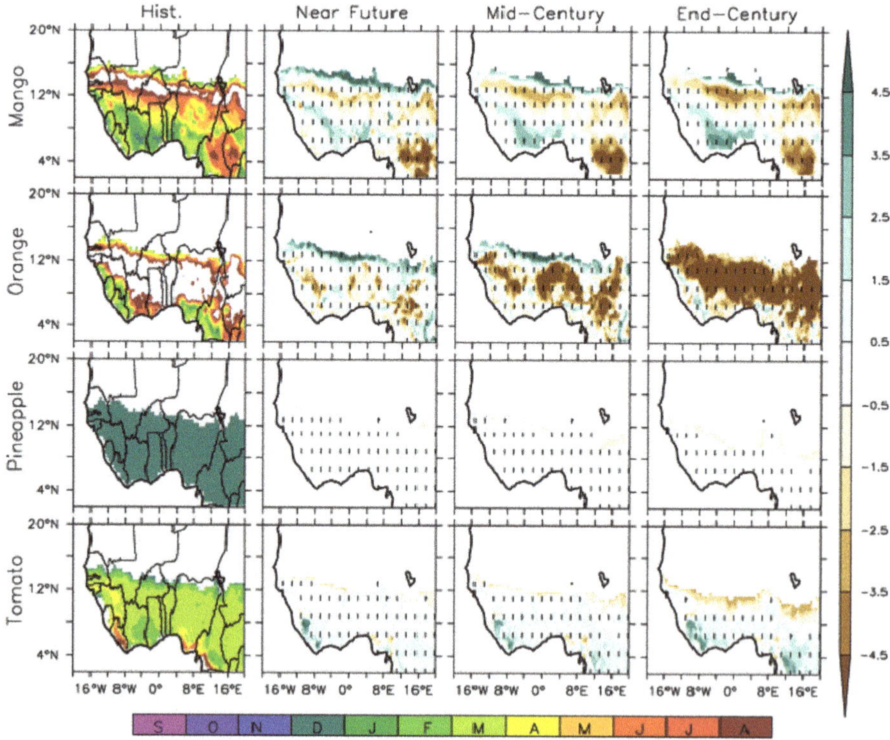

Figure 10. Same as Figure 7 but for the fruit and horticultural crops.

The root and tuber plantain is an annual crop that can be planted at any month of the year (Figure 9). The simulated PM is an overlay of the simulation of other months in the year as the crop may be planted in the suitable zones, Guinea and Savanna at any month/month of the year. For cassava, our simulation shows March (March–May) as the best PM generally over the region (Guinea-Savanna AEZs) except along the south-east coast of Ivory Coast to Ghana with PM in July, northern Guinea to Gambia and south east Senegal and from the boundary of Benin Republic to north west Nigeria in April. The Ecocrop simulation for Yam shows June as the best PM in the central Guinea zone from the south-east Ghana to the south-east of Nigeria and in the north central part of Nigeria as well as Togo in the central Savanna zone. The month of February is observed as the best PM from the south-east Mali to the north-western part of Nigeria in the Savanna. Over the coastal area, April is observed as the PM along the south west coast from Sierra Leone to Liberia and the south coast of Nigeria and June along the west coast of Guinea and from the south coast of Ivory Coast to the south-west coast of Ghana in the past climate.

Our simulation for cereal crops shows February as the PM for pearl millet in Guinea, March and April are the best PMs in the south Savanna and northern savanna and Sahel AEZs, respectively although with exception. For example, in central savanna, from the northern Benin Republic to the north-western Nigeria for millet, the PM is April while in the north-eastern Nigeria in the Sahel it is March compared to April in the Sahel zone. However, for the pearl millet, the PM is April in the western Sahel along the south-west coast of Senegal, June along the west coast of Guinea and January along the south coast of Ivory Coast to the south-west coast of Nigeria. For maize, the PM is simulated to be in May (May–July) in the Guinea and southern Savanna zone of West Africa while it is in December (December–February) in the northern Savanna into the Sahel zone. For sorghum, June is simulated as the PM over Sierra Leone to Liberia and its coastal areas as well as the south coast of Ivory coast and Nigeria while it is May in the central south of Ivory coast and southern Ghana. The crop is simulated to be best cultivated in January in the Savanna-Sahel zones and best in December in the northern Sahel.

The Ecocrop simulation of the best PM for the horticultural crops (Figure 10) in the past climate shows tomato is mainly planted in March over the regions except from the south-east Ivory coast to south-west Ghana and around 14 °N in the Sahel where the best PM in February, along the west and south coast of Liberia and Nigeria, respectively where the best PM is July. Pineapple is an annual crop and it shows similar characteristics as plantain as mentioned above, which can be planted at any month of the year. For the fruit crop, orange shows February as the best PM over Sierra Leone to Liberia and along the west coast from Guinea to Liberia and the south coast of Nigeria. June is observed as the best PM in the south of Ivory coast to Ghana and Nigeria as well as the south coast of the Ivory and Ghana. June is also simulated as the best PM from Guinea Bissau to north-east Nigeria around lat 14 °N in the Sahel. The Ecocrop simulation for mango in the past climate shows February as the PM from the Guinea to southern savanna AEZ, April, May in the northern savanna AEZ and June as the best PM in the southern Sahel AEZ. Along the coastal areas, March is simulated and observed as the best PM from the west coast of Guinea to Liberia and south coast of Nigeria but February over the south coast of Ivory coast and Ghana.

Nevertheless, the evaluation simulations demonstrate that (RCA4-Ecocrop) captures the spatial variation in the suitability with different crops across the three AEZs of West Africa in the present-day climate and can serve as a baseline for evaluating the changes in crop suitability under global warming at different time windows over the region. The model also captures the spatial distribution of the best planting month within a growing season for crops over the region which varies with different months of the year.

3.2. Projected Changes in Tmin, Tmean and Precip over West Africa

An increasing clear trend of warming is projected across West Africa in the future, with predictions of increases of the t-min and t-mean of approximately 1–4.5 °C (Figure 11, Row 2 and 3 respectively). The mean and minimum monthly temperature (t-mean and t-min) is predicted to increase by 1.5–2 °C in the Guinea-Savanna of the regions, about 2–2.5 °C in the Sahel and increases of about 1 °C predicted for the south-west coastal area in the near future month (2031–2050). By mid- century, the t-mean is projected to increase by 2.5 °C and 3.0 °C over the Guinea-Savanna and Sahel, respectively and 3.0 °C increase over the Guinea and 3.5 °C over the Savanna-Sahel for t-min. At the end of the century, a 4.0 °C temperature increase is projected over the Guinea-Savanna zone except the western area and 3.5 increase over the Sahel for t-mean. The projected change in the minimum temperature by the end of the century showed a different pattern over the region as the Guinea zone, southern Guinea-coastal area, is warmer than the Sahel. The projection shows an increase up to 4.5 °C in the southern Guinea (coastal area) and 4 °C inland. A similar characteristic is also observed over the Sahel as the southern Sahel (12–14 °N) is projected as warmer (4.0 °C) than north of 14 °N (3.5 °C) in the Sahel zone. The savanna zone is however different to the Guinea and Sahel as the temperature increases northward over the zone, i.e., southern Savanna (3.5 °C) is lower to the northern Savanna (4.0 °C) except for the western part of the Savanna zone, which is much cooler than the rest with an increase of 2.5 °C. Our findings are consistent with the findings by [30].

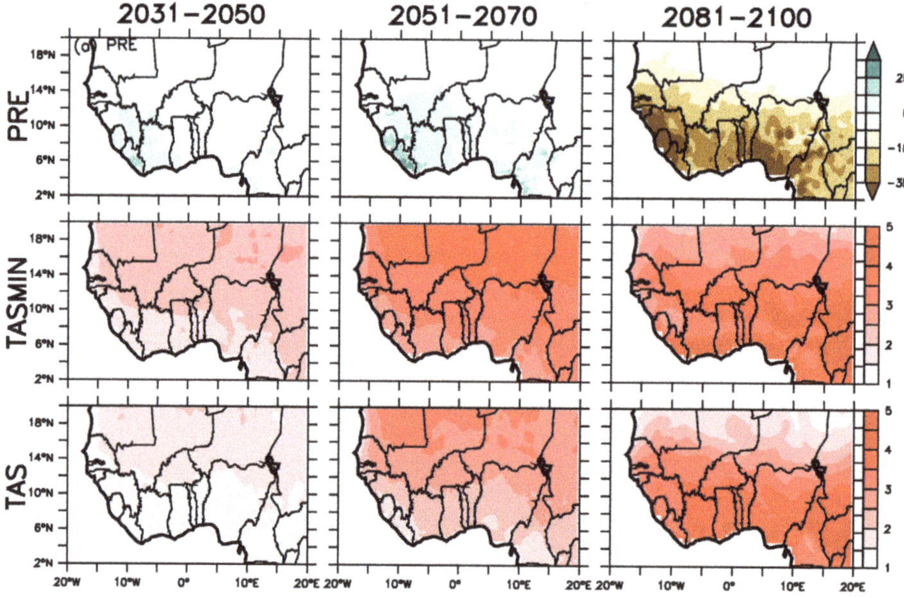

Figure 11. Projected changes in the total monthly rainfall (PRE), minimum (TASMIN) and mean (TAS) monthly temperature over West Africa as simulated by RCA4 for the near future month (2031–2050), mid-century (2051–2070) and the end of the century (2081–2100).

With respect to the predicted effects of climate change on rainfall, it is not a major change in the mean monthly precipitation that is projected over the region except in the south-west Guinea zone extending to the southern Guinea in the Savanna and the south coast of Nigeria (Figure 11, Row 1). The projected increase of about 10 mm extends from the south-west coastal area of Sierra-Leone to Liberia to the south-west coast of Ghana and south coast Nigeria in the near future (2031–2050) compared to the historical climate. By mid-century (2051–2070), the projected change of about 10 mm is expected in the western part of the region from the Guinea zone to the southern Sahel zone and the north-central part of Nigeria. Over the coastal area, an increase up to 25 mm is projected along the south-west coast of Sierra-Leone to Liberia and 10 mm over the south coast of Nigeria. The projection shows that no change is expected in the eastern part of the region by the mid-century. In contrast, by the end of the century, the projected change in the rainfall will be characterized by a decrease in the monthly precipitation across the entire region compared to the baseline. A gradient decrease in the rainfall is projected from south to north with a reduction up to 35 mm over the Guinea-Savanna and about 25 mm over the Sahel. Over the coastal area, a decrease above 35 mm is projected along the west coast of Gambia to northern Liberia and south coast of Nigeria. Our findings are in agreement with [30].

3.3. Impact of CCD on Future Crop Suitability over West Africa

Projected changes in the future crop suitability for all crop types varies across the three future climate windows, from the near future period (2031–2051) to the end of the century (2081–2100) (Figures 3–6, column 2–4). The variation in the impact for the crops can may be linked to the difference in crops response to the different climate window as described in Table 2 below for the three-climate window/period. For the near future, our simulation projects a general no change in the suitability for cereals south of 12 °N except the south coast of Nigeria (Figure 3, column 1). However, a project decreases of about 0.1 SIV is expected in the south coast of Nigeria for all the cereal crop, over Guinea for pearl millet, from Sierra Leone to Liberia for sorghum and from eastern Guinea to Liberia in the western Guinea-Savanna zone. In contrast, an increase in SIV up to 0.2 is expected in the southern Sahel zone for cereals. No suitability change is projected for legumes (Figure 4, column 2) except an increase in SIV of about 0.1 in the southern Sahel (12–14 °N) and up to 0.2 in the central savanna zone, (Figure 6). On the other hand, a projected decrease of 0.1 in SIV is expected along the west coast of Sierra Leone and the south coast of Nigeria for groundnut. The projected increase in SIV provides an increase in the suitable area for the cultivation of both crops. This is so because a 0.2 increase in SIV for the marginally suitable (SIV, 0.4–0.6) areas in the southern Sahel results in the area becoming suitable (SIV, 0.6–0.8) for both crops. The projected decrease in the SIV values along the coastal areas and over Sierra Leone also does not affect the area negatively as the area remains suitable for these crops. For the root and tuber crop (Figure 5, column 2), a projected decrease of about 0.1 SIV is expected for cassava and up to 0.2 in southern Nigeria and along the west coast of Guinea to Liberia for plantain and yam extending to south of Ivory Coast for plantain. A similar magnitude decrease is also expected in the western Guinea-Savanna zone from Guinea to the western Ivory Coast. For the horticulture and fruit crops (Figure 5, column 1), a 0.1 projected decrease in SIV is expected south of 12 °N and the savanna zone for tomato and pineapple, respectively while up to 0.2 SIV decrease is expected in the south coast of Nigeria for mango and orange. However, a projected increase up to 0.2 SIV is expected in the southern Sahel for mango. The projected suitability changes are robust (i.e., at least 80% of the simulation that the climate change is statistically significant at 95% confidence level) for cassava, maize and mango in the near future month (2031–2050) while the changes are consistent for the other nine crops (i.e., at least 80% of the model agree to the sign of change).

Table 2. Projected changes in crop suitability over West African AEZs at different future window periods.

Crops	Near Future (2031–2050)			Mid-Century (2051–2070)			End-Century (2081–2100)		
	Guinea	Savanna	Sahel	Guinea	Savanna	Sahel	Guinea	Savanna	Sahel
Cassava	No change remains suitable	No change remains suitable	No change remains unsuitable	A 0.2 SIV decrease but still suitable	A 0.2 SIV decrease but still suitable	Same as GWL1.5	About 0.4 SIV decrease still suitable	About 0.4 SIV decrease but still suitable	Same as GWL1.5
Plantain	About 0.1 SIV decrease but still suitable	A 0.1 & 0.2 SIV decrease and increase in west and central respectively	No change unsuitable	About 0.2 decrease in SIV but still suitable	A 0.2 SIV decrease and increase in west and central respectively	No change remains unsuitable	About 0.4 SIV decrease may become marginally suitable	About 0.4 and 0.2 SIV decrease to the west and central respectively	No change unsuitable
Yam	Suitable, but not along the coastal area	Only suitable in the west & central Savana	No change, unsuitable	Same as in GWL1.5	Same as GWL1.5	Same as GWL1.5	Same as in GWL1.5	Same as GWL1.5	Same as GWL1.5
Maize	No change but 0.1 decrease SIV northern Cameroon	Suitable, but not along the west coast of Guinea to Sierra Leone	About 0.2 SIV increase, now suitable over the southern Sahel	No change in suitability	About 0.1 SIV decrease but still suitable	Same as GWL1.5	No change in suitability	About 0.2 decrease in SIV but still suitable	Same as GWL1.5 but SIV increase up to 0.3
Pearl millet	No change but very marginal suitability in the south coast Nigeria and north Liberia	No change but about 0.1 SIV decrease in eastern Guinea	About 0.2 SIV increase make northern Sahel suitable	Same as GWL1.5	Same as GWL1.5	Same as GWL1.5	Same as GWL1.5	About 0.3 decrease in SIV but still suitable	A 0.4 SIV decrease in west Sahel but still suitable
Sorghum	No change in suitability	No change in suitability	About 0.2 SIV increase make northern Sahel suitable	No change in suitability	About 0.1 decrease in SIV but still suitable	About 0.1 SIV increase makes Sahel suitable	About 0.1 decrease in SIV but still suitable	About 0.2 SIV decrease west respectively	Above 0.2 SIV decrease but still suitable

Table 2. *Cont.*

Crops	Near Future (2031–2050)			Mid-Century (2051–2070)			End-Century (2081–2100)		
	Guinea	Savanna	Sahel	Guinea	Savanna	Sahel	Guinea	Savanna	Sahel
Mango	No change in suitability	No change in suitability	No change in suitability	About 0.1 decrease in SIV but still suitable	About 0.1 decrease in SIV but still suitable	About 0.1 increase in SIV but still unsuitable	About 0.2 SIV decrease but still suitable	About 0.2 SIV increase but still unsuitable	About 0.2 SIV increase but still unsuitable
Orange	About 0.1 SIV increase	About 0.1 SIV increase	No change in suitability	Same as GWL1.5	Same as GWL1.5	Same as GWL1.5	About 0.2 SIV decrease but still suitable	About 0.2 SIV decrease but still suitable	Same as in GWL1.5
Pineapple	No change in suitability	About 0.2 SIV decrease but still suitable	No change in suitability	About 0.1 decrease but still suitable	About 0.3 decrease but still suitable	Same as GWL1.5	About 0.4 decrease but still suitable	About 0.4 SIV decrease but still suitable	Same as GWL1.5
Tomato	About 0.1 decrease but still suitable	About 0.1 decrease but still suitable	No change in suitability	About 0.3 decrease but still suitable	About 0.3 SIV decrease but still suitable	Same as GWL1.5	About 0.4 decrease but still suitable	About 0.4 SIV decrease but still suitable	Same as GWL1.5
Cowpea	No change in suitability	No change in suitability	About 0.2 SIV increase make southern Sahel suitable	Same as GWL1.5	Same as GWL1.5	Same as GWL1.5	Same as GWL1.5	About 0.1 decrease in SIV but still suitable	Same as GWL1.5
Groundnut	No change in suitability	No change in suitability	About 0.2 SIV increase makes southern Sahel suitable	Same as GWL1.5	Same as GWL1.5	Same as GWL1.5	Same as GWL1.5	About 0.1 decrease in SIV but still suitable	Same as GWL1.5

The Ecocrop suitability simulation by mid-century (2051–2070) shows a projected increase in the magnitude of change of SIV and spatial suitability distribution of suitable areas compared to the past climate for the different crop types. The projected spatial suitability distribution for mid-century shows a similar spatial pattern as the near future period (2031–2050) with an increase in the suitability spatial extent and the magnitude of change in SIV. For cereals (Figure 3, column 3), the projected change is like the spatial suitability pattern as the near future period except for the spatial extension in the suitable area further north in the central Sahel zone for pearl millet. In contrast, a decrease in the suitable area in the western Nigeria for pearl millet and north-west Nigeria for maize and sorghum. The legume (Figure 4, column 3) crops show a similar projected suitability spatial pattern as the near future period except a projected decrease in SIV of about 0.1 and 0.2 of the suitable area is expected in the south-west Chad Republic in the eastern Sahel zone for the groundnut and cowpea, respectively. For the root and tubers (Figure 5 column 3), a decrease of about 0.2 SIV is projected for both the plantain and yam but with a similar spatial suitability pattern as shown for the near future period. However, for cassava about 0.2 decrease SIV is projected over the guinea-Savanna zone but the area remains suitable. For the fruit and horticulture crops (Figure 6, column 3), there are no changes in the projected spatial suitability pattern as observed in the near future period by mid-century. However, there is an increase in the magnitude of change of SIV from 0.1 to 0.2 and 0.2 to 0.3 for the tomato and pineapple, respectively. All the projected suitability changes are statistically significant at 95% confidence level for cassava, maize and mango and are consistent for the other nine crops (i.e., at least 80% of the model agree to the sign of change) by mid-century (2051–2070).

The projected increase in global warming will lead to increasing the magnitude in the projected change for the crop SIV and spatial suitability distribution across different crop types by the end of the century (2081–2100). Cereal (Figure 3, column 4) as projected will be severely affected as more areas becomes less suitable by the end of the century. For legume (Figure 4, column 4), the Savanna zone will be less suitable with a decrease of about 0.1 in SIV while a decrease of about 0.2 SIV is expected along the eastern Sahel zone for groundnut as well as the south coast of Nigeria. Cowpea as projected will be more affected with a decrease of about 0.2 SIV in the northern savanna in the southern Chad Republic and Nigeria with its boundary with south-east Niger Republic in the Sahel and south-west Mali in the western Sahel zones. A decrease up to 0.2 in SIV is expected in the southern Sahel for cereal except maize with an increase of about 0.2 in the central southern Sahel zone. The root and tubers (Figure 5, column 4), show a similar spatial pattern for the decrease in the suitable area as the near future period and mid-century but with an increase in the SIV magnitude of about 0.2, 0.3 and 0.4 for yam, plantain and cassava, respectively. The fruit and horticulture crops (Figure 6, column 4) shows further reduction in the suitable area compared to the near future period with an increase up to 0.4 SIV for the horticulture crop. The Guinea-Savanna will become less suitable with a decrease of 0.1 and 0.2 SIV for orange and mango, respectively. All the projected suitability changes are statistically significant at 95% confidence level for cassava, maize and mango and are consistent for the other nine crops (i.e., at least 80% of the model agree to the sign of change) by the end of the century (2081–2100).

3.4. Impact of CCD on Crop Planting Month over West Africa

At all the three future climate windows, the Ecocrop projected change on the planting month varies for different crop types across the different AEZs of West Africa (Figures 7–10). The impact of CCD resulted in an early or late/delay in the PM for different crops and increases in magnitude across the three zones as described in Table 3 below. It is worth stating that the change in PM describes a change in the best three planting months under the three future windows.

Table 3. Projected changes in time of planting (crop planting months) over West African AEZs at different global warming levels.

Crops	Near Future (2031–2050)			Mid-Century (2051–2070)			End-Century (2081–2100)		
	Guinea	Savanna	Sahel	Guinea	Savanna	Sahel	Guinea	Savanna	Sahel
Cassava	Delayed planting for one month	Early planting by four months	Not applicable	Same as GWL1.5	Same as GWL1.5 but for more area	No planting date	Same as GWL1.5	Same as GWL1.5	No planting date
Plantain	No change in planting date	No change in planting date	No change in planting date	No change in planting date	No change in planting date	No change in planting date	No change in planting date	No change in planting date	No change in planting date
Yam	On month delayed planting	No change in planting date	No change in planting date	Same as GWL1.5	Same as GWL1.5	No change in planting date	Same as GWL1.5	Same as GWL1.5	No change in planting date
Maize	Three months delayed planting	Four months early and delay planting in east and west respectively	No change in planting date	Same as GWL1.5	Same as GWL1.5	No change in planting date	Same as GWL1.5	Same as GWL1.5	No change in planting date
Pearl millet	One-month delayed planting	Two months delayed planting	Two months delayed planting	Same as GWL1.5	Same as GWL1.5	Same as GWL1.5	Same as GWL1.5	Same as GWL1.5	Same as GWL1.5
Sorghum	No change in planting date	No change in planting date	No change in planting date	No change in planting date	No change in planting date	No change in planting date	No change in planting date	No change in planting date	No change in planting date
Mango	Delayed planting for two months	Early planting by four months	One-month delay in southern Sahel zone	Same as GWL1.5	Same as GWL1.5 but for more area	No planting, date	Same as GWL1.5	Same as GWL1.5	No planting date

Table 3. *Cont.*

Crops	Near Future (2031–2050)			Mid-Century (2051–2070)			End-Century (2081–2100)		
	Guinea	Savanna	Sahel	Guinea	Savanna	Sahel	Guinea	Savanna	Sahel
Orange	One-month delayed planting	No change in planting date	No change in planting date	Same as GWL1.5	Same as GWL1.5	No change in planting date	Same as GWL1.5	Same as GWL1.5	No change in planting date
Pineapple	No change in planting date	No change in planting date	No change in planting date	No change in planting date	No change in planting date	No change in planting date	No change in planting date	No change in planting date	No change in planting date
Tomato	One-month delayed planting	No change in planting date	No change in planting date	Same as GWL1.5	Same as GWL1.5	No change in planting date	Two months delayed planting	One-month early planting	No change in planting date
Cowpea	One-month delayed planting	Two months delayed planting	Two months delayed planting	Same as GWL1.5	Same as GWL1.5	Same as GWL1.5	Same as GWL1.5	Same as GWL1.5	Same as GWL1.5
Groundnut	No change in planting date	No change in planting date	No change in planting date	No change in planting date	No change in planting date	No change in planting date	No change in planting date	No change in planting date	No change in planting date

In the near future, cereals crops, pearl millet and sorghum are projected to experience a one-month delay over the region and up to 0.2 along the west coast of Sierra Leone to Liberia and the south coast of Nigeria (Figure 7, column 2). In contrast, the two-month delayed planting is expected over the Savanna-Sahel zone for maize. For the legumes crops, cowpea and groundnut (Figure 8 column 2, see Table 4) no projected change in the PM compared to the past climate is expected over the regions except about one-month delay (i.e., from June to July) in planting over Sierra-Leone and Liberia in the Guinea zone and the southern Sahel zone from Senegal to Chad Republic compared to the planting month (June) over the area. For the root and tuber (Figure 9 column 2), about three to four months early (February/March) the planting is projected for cassava in the near future as compared to June/July, the PM from the historical climate across the region except the north-east Nigeria and the coastal areas (Figure 9, Table 3). No change in the PM is expected in the near future over the coastal areas but about three months delay in planting is projected in the north-eastern part of Nigeria. No change in the PM is projected for plantain, an annual crop which can be planted anytime of the year while a 3–4 months delay is expected for yam except in western Guinea-Savanna and the south coast of Nigeria. For fruits and horticulture (Figure 10, column 2), no projected change in the planting month is expected for tomato and pineapple except a two-month delay over Liberia. Early planting between one to two months is expected in the Guinea-Savanna zone and about three-months delay in the planting of orange in the southern Sahel zone. About two-months and up to four-months delay in planting is projected for mango in the Guinea-Savanna zone and the northern Sahel zone, respectively while a two-month early planting of the crop is expected in the southern Sahel zone. The projected change is consistent for all crops as 80% of the simulation agree to the sign of change.

Table 4. Trends in the projected change in suitability over West Africa for the near future, mid and end of the century periods for different crops.

Crops/Period	2031–2050	2051–2070	2081–2100
Cassava	1.053	1.141	1.497
Cowpea	1.000	1.000	1.002
Groundnut	1.000	1.001	1.030
Maize	1.007	1.021	1.082
Mango	1.013	1.046	1.137
Orange	0.981	0.974	1.089
Pearl millet	1.007	1.022	1.057
Pineapple	1.061	1.216	1.580
Plantain	1.017	1.025	1.215
Sorghum	1.007	1.018	1.032
Tomato	1.219	1.421	1.997
Yam	0.873	0.784	0.779

By mid (2051–2070) and end of the century (2081–2100), most crop types show a similar spatial pattern in the planting month as observed in the near future but with an increase in the magnitude of the delay or early planting period (Figures 7–10, column 3–4). For example, cereal crops show a similar spatial pattern as projected for the near future for sorghum and pearl millet by mid-century (Figure 7, column 3) and the end of the century (Figure 7, column 4) except over Liberia and south coast of Nigeria for pearl millet. These areas are expected to experience a 2–3-month delay in planting. Legume crops, cowpea and groundnut show similar characteristics of no projected change in the PM as the near future period but for an increase in the magnitude a delay period in the south coast of Nigeria and southern Liberia. A delay in planting from one to two months is expected from Sierra-Leone to Liberia and over the south coast of Nigeria for cowpea by mid-century (Figure 8, column 3) and up to three months by the end of century (Figure 8, column 4). A two-month delay in the PM is projected over southern Liberia and a one-month delay in the southern Sahel zone by mid and end of the century (Figure 8, column 3–4). For the root and tuber crops (Figure 9, column 3–4), about a four-month delay in planting is projected over the Savanna zone except the western area of the zone and in the central Guinea zone by mid-century for yam (Figure 9, column 3). A similar pattern is projected by the end of the century for crop (Figure 9, column 4). No change in the planting period is projected for plantain because it is annual crop over these two periods. For cassava a month delay planting by mid-century and up to two-months by the end of the century is projected in the western Guinea-Savanna zone while an early planting is expected in other parts of the Savanna zone and north of the Guinea zone over the two-climate change period. For the fruit and horticulture crops, there is no change in the PM for pineapple being an annual crop. A one-month PM delay is projected for tomato over the region and up to two-months over Liberia by mid and end of the century. However, a projected two-month early planting is expected by the end of the century in the southern Sahel zone. For fruit crops, a four-month early planting compared to the historical climate is projected in the Guinea-Savanna zone with a delay of about three-months in the south Sahel by mid-century. By the end of the century, an early planting of about four-month early compared to the historical climate is projected over the region for orange. Similarly, a two-month early planting is projected for mango in the southern Sahel zone for mid and up to three-months by the end of the century. In contrast, a delay in planting of about two-three months is expected over the Guinea-Savanna zone and up to four-months in the northern Sahel zone. All the projected changes are consistent for all crops as 80% of the simulation agree to the sign of change over the two climate periods.

3.5. Trends in Projected Crop Suitability and Crop Planting over West Africa

We used the Theil-Sen slope to evaluate the trend in the projected suitability and month of planting for the crop types for the near future, mid and end of the century over West Africa (Tables 4 and 5). The trend describes the rate of increase and decrease of the suitable area and SIV with increasing warming over the three-window month. In general, all the crop types show an increasing trend in the projected change in the crop suitability compared to the past climate from the near future to the end of the century when compared to the past climate except for yam (Table 4). The projected change in the suitability index value of suitable areas for tomato showed the highest trend value from 1.219 in the near future month to 1.997 by the end of the century. Compared to other crops, our analysis showed that there was a decreasing trend (from 0.873, the near future to 0.779, end of the century) for yam in the projected suitability change with increasing warming across each time of month from the near future to the end of century over West Africa. Additionally,, there was decrease in the trend between the near future month and mid-century month in the projected change for orange but later increased at end of the century. Moreover, there was no trend in the projected change in the suitability for cowpea over the near future month and mid-century but there was increase in the trend of the projected change for the crop by the end of the century.

Table 5. Trends in the projected change in the month of planting over West Africa for the near future, mid and end of the century periods for the different crops.

Crops/Period	2031–2050	2051–2070	2081–2100
Cassava	1.125	1.171	0.974
Cowpea	0.972	0.957	0.887
Groundnut	0.969	0.952	0.857
Maize	1.000	0.990	0.950
Mango	1.000	0.976	0.909
Orange	1.000	1.111	1.930
Pearl millet	0.980	0.959	0.912
Pineapple	1.000	1.000	1.000
Plantain	1.000	1.000	1.000
Sorghum	1.000	1.000	0.944
Tomato	0.938	0.900	0.851
Yam	1.000	0.924	0.909

Our Theil-Sen slope trend analysis shows a general decreasing trend in the projected change in the planting month compared to the past climate for the different crop types except for orange which gives an increasing trend pattern of the projected planting for all the crop (Table 5). Our trend analysis test show there was no change in the projected change in planting for plantain, pineapple (1.000) and for sorghum for the near future and mid-century month (1.000).

4. Discussion

4.1. Crop Type Sensitivity to CCD and Impact on Food Security

Horticulture, cereals, root and tubers (hereafter HCRT) crops, respectively will be the most impacted by the climate change/departure impact from the historical variability in West Africa. All the five different crop types show a different response to the impact of the global warming induced CCD across the examined three-window month, near the future to the end of the century in West Africa. The variability in the response of the different crop types to CCD is very cardinal to the agricultural production and food security in the region. HCRT are the most negatively affected with decreasing suitability across the three AEZs of West Africa due to the impact of the climate change compared to the legumes and fruit crops. In terms of sensitivity, the HCRT crop suitability show a negative linear relationship with increasing global warming over the region except for cereals with a positive linear relationship in the southern Sahel zone. The negative linear relationship is observed notably over the Guinea-Savanna zone for the HCRT resulting in a decrease in the crop suitable area with increasing warming across the three months examined. The projected negative linear relationship due to an increase in global warming may result in a decrease in the yield of these crop types over West Africa due to a decrease in the crop suitable land [6,64]. For example, previous studies (e.g., Lobell et al. [65], Sultan et al. [15]) have revealed that the impact of climate change will result in a decrease in the yield of cereals by 20% in the near future month over West Africa. Additionally, the result is in line with the findings of [32] that there will be a decline in the suitability and suitable cultivated areas for cassava due to a result of the temperature increases but the crop will remain suitable over the region. In addition, our result also agrees with [66] findings that increasing warming will lead to a decrease in the availability of the suitable land for the cultivation of horticulture with a direct implication on the horticultural production. This agrees with [14] that the variability in the climate will lead to a reduction in the yield quantity of pineapple in Ghana which is one of the key producers of pineapple, which may be linked to the decrease in the suitable areas and SIV as projected in this study.

The projected impacts of CCD on crop suitability will further compound the challenge of food security in West Africa. This is in line with past findings that climate variability and change in the coming decades will further threaten food security in sub-Saharan Africa notably West Africa, a region that plays a major role in the agricultural production [1,7]. West Africa for about 24 years mainly accounts for about 60% of the total value of agricultural outputs within Africa [7]. However, the story has not been the same since 2007 due to instability in the agricultural production over the region and this has been a source of concern [7]. As a result, the projected decrease in crop suitability due to a reduction in the suitable area for crop cultivation coupled with the projected delay in the month of planting will both strongly have a negative impact on the crop yield and agricultural production. This may further plunge the plan for food security in the region into a mirage.

4.2. Impact of CCD on Spatial Suitability Distribution

The impact of CCD will lead to a projected variability in the spatial suitability distribution across the three AEZs for the three future months and different crop types. The magnitude of deviation due to the increase in warming may influence the suitability over the zones as well as crop sensitivity to the projected change in the climate. The crop growth and yield are directly proportional to the climate-crop threshold i.e., climate suitability/threshold [67]. It is important to note that each crop has their climatic or suitability threshold for healthy growth, development and optimal yield and that future changes/departure in the climate generally has a reaching impact on the yield of the crop. This is further buttressed by our finding that CCD may lead to future constraint in the available cultivated area in the Guinea and southern Savanna zones of West Africa. On the other hand, it tends to provide an opportunity in the northern Savanna extending to the southern Sahel. The projected spatial constraint in the suitability and cultivated area will strongly affect the crop production and yield over West Africa. The Guinea-Savanna zone provides and significantly contributes to the agricultural production over the region and a large proportion in the continent [7]. For example, about four of the five different crop types (except the legumes) examined in the study is and will be significantly affected with the projected decrease in SIV and reduction in the cultivated area of the crops. This projected decrease in SIV and the reduction in the spatial distribution of suitable areas for cultivation of major crops such as cassava and the horticulture crops such as pineapple pose a great challenge to the economy of most countries and further raises the challenge of food security in the region. The challenge of food security arising from the projected decrease in the crop suitable area may compound the climatic stress over the region due to the increase in food production to meet the present food demand but with the projected and limited available land for cultivation are not realistic and may become a mirage with the projected increase in the population over the region by mid-century, 2050 [68,69].

On the other hand, crop suitability due to CCD from the historical variability is projected and will lead to an increase in SIV and more suitable area notably in the Southern Sahel. The increase in suitable areas provides an opportunity for more suitable areas in the region for the cultivation of cereals, legumes and mango in the southern Sahel zone (12–14 °N), plantain and yam in the Savanna zone as well as the legume crops in the central savanna zone of West Africa. The projected increase into the Sahel agrees with the previous finding for maize in the Sahel zone with CCD. This shows that the crop spatial suitability distribution and productivity are highly sensitive to variations in the climate such that a departure of the future African climate from the recent range of historical variability will have the most devastating effect on agriculture over the continent [70–72].

4.3. Implication for Socio-Economic Development and Strategy Policy

The above result provides a basis for developing the policy and strategy to reduce future crop loss due to a lack of suitable land and risks of food security over West Africa. At the same time, it advocates for a more proactive response to increase resilience and adaptive options via the urgency and timing of adaptation. For instance, the analysis of crop suitability indicates that a greater proportion of suitable land areas in the West African region may become less suitable or unsuitable in the future from CCD due to global warming, which may enhance a decrease in the crop yield and agricultural production of some crop. On the other hand, the analysis showed an expansion of the suitable area into the Sahel for the cereal and legume crops with CCD, which provide future opportunities for more suitable areas for the cultivation of one of the most staple crops, maize. This will have both positive and negative impacts on regional development and economic activities (e.g., regional trade and international relation in terms of exports and importing goods). The increasing population also implies that the demand for food will be on the increase. However, the projected change in suitability also suggests that a well-planned land use change (through the urgency of adaptation to the CCD) could help reduce the impacts of CCD on the crop yield and food security in the region. Hence, there is a need for the formulation of a strategic policy that can accommodate or encourage such a land-use change. A strategic policy is also required more importantly for the new opportunities such as an expansion into the Sahel for maize and the other crops that may arise out of the impact of CCD over the region. Hence, the results can guide policymakers on how to prioritize their adaptation plan in terms of the urgency of response and redefine mitigation measures to the future impact of CCD on the crop suitability and planting season over West Africa.

5. Conclusions

Summary and Conclusions

In investigating the impact of CCD on the crop suitability and planting month over the entire West African region, we analyzed 10 CMIP 5 GCM datasets downscaled by CORDEX RCM, RCA4 for five different crop types, cereal (maize, pearl millet and sorghum), fruit (mango and orange), horticulture (pineapple and tomato), legume (cowpea and groundnut) and root and tuber (cassava, plantain and yam). The summary from our study are as follows:

We suggest that projected changes in the temperature may lead to an increase between 1–4.5 °C for the minimum and mean temperature over West Africa from the near future to the end of the century. A change of about 10 mm is projected over the western Guinea-Savanna zone and no major changes in other parts of the region and up to 25 mm along the coastal areas (west coast of Sierra-Leone to south-west Ghana and the south coast of Nigeria) for the near future and mid-century. A projected decrease up to 25 mm is expected over the region and up to 35 mm over the coastal area (from the west coast of Gambia to north Liberia) by the end of the century.

Addressing our main objective, the Ecocrop simulated spatial suitability distribution of the crops shows higher suitability are to the south of 14 °N while a lower suitability is to the north. The marginal suitability line (around 12–14 °N) shows the transition between the higher and lower suitability of the crop. Results show that the horticulture crops, pineapple and tomato, respectively are the most negatively affected by the impact of CCD from the historical variability over the region. There is a projected constraint showing a negative linear correlation with increasing warming in the cultivation of most different crop types except for cowpea in the Guinea-Savanna AEZs (south of 14 °N) by the end of the century due to an increasing reduction in the suitable area and crops suitability index value due to the climate departure although most of the crop remains suitable. The impact of CCD will provide opportunities for more suitable areas in the southern Sahel zone for cereals, mango and legumes crops showing a positive linear correlation with increasing warming thus creating more land for cultivation, which can in turn increase the yield and production of the crops. Generally, a projected delay of 1–4 months is expected for most of the crop types with CCD except for orange and cassava as well as maize in the Savanna zone. No projected changes are observed for plantain and pineapple, mainly because they are annual crops.

Statistically, we demonstrated that over 80% of the simulations agree with the sign of the projected change for all the crop types due to the CCD and the changes are statistically significant at 95% confidence interval for maize, cassava and mango. Additionally, we showed there is an increasing trend in the projected crop suitability for all crops except yam with a decreasing trend due to CCD from the historical variability while a decreasing trend is projected for the future change in the month of planting of the crops.

Despite our analysis, the results of this study can be improved and applied to reduce the future impact of crop suitability and risks of food security over West Africa in many ways. For instance, future studies may investigate the impact of CCD on the crop suitability and planting season over the region using more RCMs with different forcing GCMs other than only RCA4. This may help resolve the challenge of uncertainty in the future simulation of the crop suitability and planting season. In addition, the results of the study will be more robust and improve our knowledge on the impact of CCD and its influence on the crop suitability and planting season over West Africa. Further studies on how to reduce the uncertainty will improve the credibility and application of the results. Nevertheless, the present work shows the impact of CCD on the crop suitability and planting season using GCMs downscaled with RCMs. This establishes a premise for future work in advancing our knowledge into how CCD influences the crop suitability and planting season in West Africa.

In conclusion, the application of the concept of CCD in this study has demonstrated future changes in how the crop suitability and planting season can be analyzed. The application of CCD established the impact of climate change on crop suitability over West Africa and further identified spatial variability in the future suitability showing that horticulture, cereal, root and tubers crops will be most negatively affected by the impact of CCD in West Africa. It also identifies the three best planting months in a growing season and the changes in the planting time is about four month delay in the planting season for most crops but early planting for cassava, orange and maize but only in the savanna zone. The application of CCD aims to underpin future works to advance the study of future changes in crop suitability and planting in any region of the world. This type of analysis is important for adaptation options and planning for future changes in the crop suitability and planting period to improve food security.

Author Contributions: T.S.E. was responsible for conceptualization, developing the initial content of the manuscript, including literature search, data analysis and writing of the manuscript. O.C. and C.L. are the supervisors for the research and provided guidance in terms of the article structure, data analysis and finalization of the manuscript. P.M. provided guidance with data computation, S.L. assisted with review and editing of the manuscript, K.Q. assisted with some data processing, analysis and editing of the manuscript.

Funding: This research was supported with funding from the National Research Foundation (NRF, South Africa), Alliance Centre for Climate and Earth Systems Science (ACCESS, South Africa), JW Jagger Centenary Scholarship and Sari Johnson scholarship from the Postgraduate Funding Office, University of Cape Town, South Africa. Interpretation of the findings and conclusion drawn from the study were the responsibilities of the authors and not on any part of NRF, ACCESS, JW Jagger Centenary Scholarship and Sari Johnson scholarship.

Acknowledgments: This study was supported with bursaries from the National Research Foundation (NRF, South Africa), Alliance Centre for Climate and Earth Systems Science (ACCESS, South Africa) and the JW Jagger Centenary Scholarship and Sari Johnson scholarship from the Postgraduate Funding Office, University of Cape Town, South Africa and IMPALA project.

Conflicts of Interest: The authors declare that they have no financial or personal relationships which may have inappropriately influenced them in writing this article.

References

1. IPCC. Summary for Policymakers. In *Climate Change 2013: The Physical Science Basis. Contribution of Working Group I to the Fifth Assessment Report of the Intergovernmental Panel on Climate Change*; Stocker, T.F., Qin, D., Plattner, G.-K., Tignor, M., Allen, S.K., Eds.; Cambridge University Press: Cambridge, UK, 2013.

2. Kirtman, B.; Power, S.B.; Adedoyin, A.J.; Boer, G.J.; Bojariu, R.; Camilloni, I.; Doblas-Reyes, F.; Fiore, A.M.; Kimoto, M.; Meehl, G.; et al. Near-term Climate Change: Projections and Predictability. In *Climate Change 2013: The Physical Science Basis. Contribution of Working Group I to the Fifth Assessment Report of the Intergovernmental Panel on Climate Change*; Stocker, T.F., Qin, D., Eds.; Cambridge University Press: Cambridge, UK; New York, NY, USA, 2013; Chapter 11; pp. 953–1028.

3. Riede, J.O.; Posada, R.; Fink, A.H.; Kaspar, F. What's on the 5th IPCC Report for West Africa? In *Adaptation to Climate Change and Variability in Rural West Africa*; Yaro, J.A., Hessellberg, J., Eds.; Springer: Cham, Switzerland, 2013; Volume 19, pp. 7–24.

4. Benhin, J.K. South African crop farming and climate change: An economic assessment of impacts. *Glob. Environ. Chang.* **2008**, *18*, 666–678. [CrossRef]

5. Schlenker, W.; Lobell, D.B. Robust negative impacts of climate change on African agriculture. *Environ. Res. Lett.* **2010**, *5*, 014010. [CrossRef]

6. Roudier, P.; Sultan, B.; Quirion, P.; Berg, A. The impact of future climate change on West African crop yields: What does the recent literature say? *Glob. Environ. Chang.* **2011**, *21*, 1073–1083. [CrossRef]

7. OECD/FAO. *OECD-FAO Agricultural Outlook 2016–2025: Special Focus on Sub-Sharan Africa*; OECD Publishing: Paris, France, 2016.

8. Nelson, G.C.; Rosegrant, M.W.; Koo, J.; Robertson, R.; Sulser, T.; Zhu, T.; Magalhaes, M. Climate change: Impact on agriculture and costs of adaptation. *Intl. Food Policy Res. Inst.* **2009**, *21*.

9. Nelson, G.C.; Van Der Mensbrugghe, D.; Ahammad, H.; Blanc, E.; Calvin, K.; Hasegawa, T.; Havlik, P.; Heyhoe, E.; Kyle, P.; Lotze-Campen, H.; et al. Agriculture and climate change in global scenarios: Why don't the models agree. *Agric. Econ. (UK)* **2014**, *45*, 85–101. [CrossRef]

10. Ray, D.K.; Foley, J.A. Increasing global crop harvest frequency: Recent trends and future directions. *Environ. Res. Lett.* **2013**, *8*, 044041. [CrossRef]

11. IPCC. Summary for policymakers. In *Climate Change 2014: Impacts, Adaptation, and Vulnerability. Part A: Global and Sectoral Aspects. Contribution of Working Group II to the Fifth Assessment Report of the Intergovernmental Panel on Climate Change*; Field, C.B., Barros, V.R., Dokken, D.J., Mach, K.J., Mastrandrea, M.D., Bilir, T.E., Chatterjee, M., Ebi, K.L., Estrada, Y.O., Genova, R.C., et al., Eds.; Cambridge University Press: Cambridge, UK; New York, NY, USA, 2014; pp. 1–32.

12. Rurinda, J.; Mapfumo, P.; Van Wijk, M.T.; Mtambanengwe, F.; Rufino, M.C.; Chikowo, R.; Giller, K.E. Sources of vulnerability to a variable and changing climate among smallholder households in Zimbabwe: A participatory analysis. *Clim. Risk Manag.* **2014**, *3*, 65–78. [CrossRef]

13. Challinor, A.; Wheeler, T.; Garforth, C.; Craufurd, P.; Kassam, A. Assessing the vulnerability of food crop systems in Africa to climate change. *Clim. Chang.* **2007**, *83*, 381–399. [CrossRef]

14. Williams, P.A.; Crespo, O.; Abu, M. Assessing vulnerability of horticultural smallholders' to climate variability in Ghana: Applying the livelihood vulnerability approach. *Environ. Dev. Sustain.* **2018**, 1–22. [CrossRef]

15. Sultan, B.; Guan, K.; Kouressy, M.; Biasutti, M.; Piani, C.; Hammer, G.L.; McLean, G.; Lobell, D.B. Robust features of future climate change impacts on sorghum yields in West Africa. *Environ. Res. Lett.* **2014**, *9*, 104006. [CrossRef]

16. Parkes, B.; Defrance, D.; Sultan, B.; Ciais, P.; Wang, X. *Projected Changes in Crop Yield Mean and Variability Over West Africa in a World 1.5 K Warmer Than the Pre-Industrial Era*; Copernicus Publications: Gottingen, Germany, 2018; Volume 9, pp. 119–134.

17. Jalloh, A.; Nelson, G.C.; Thomas, T.S.; Roy-Macauley, H. *West African Agriculture and Climate Change: A Comprehensive Analysis*; International Food Policy Research Institute: Washington, DC, USA, 2013; 444p.

18. Ramirez-Villegas, J.; Thornton, P.K. *Climate Change Impacts on African Crop Production*; CCAFS Working Paper no. 119; CGIAR Research Program on Climate Change, Agriculture and Food Security (CCAFS): Copenhagen, Denmark; 127p, Available online: www.ccafs.cgiar.org (accessed on 24 March 2017).

19. Thornton, P.K.; Jones, P.G.; Ericksen, P.J.; Challinor, A.J. Agriculture and food systems in sub-Saharan Africa in a 4°C+ world. *Philos. Trans. R. Soc. A Math. Phys. Eng. Sci.* **2011**, *369*, 117–136. [CrossRef] [PubMed]

20. Adger, W.N. Social Capital, Collective Action, and Adaptation to Climate Change. *Econ. Geogr.* **2003**, *79*, 387–404. [CrossRef]

21. Mora, C.; Frazier, A.G.; Longman, R.J.; Dacks, R.S.; Walton, M.M.; Tong, E.J.; Sanchez, J.J.; Kaiser, L.R.; Stender, Y.O.; Anderson, J.M.; et al. The projected timing of climate departure from recent variability. *Nature* **2013**, *502*, 183–187. [CrossRef] [PubMed]

22. Hawkins, E.; Sutton, R. Time of emergence of climate signals. *Geophys. Res. Lett.* **2012**, *39*, L01702. [CrossRef]

23. Egbebiyi, T.S.; Crespo, O.; Lennard, C. Defining Crop–climate Departure in West Africa: Improved Understanding of the Timing of Future Changes in Crop Suitability. *Climate* **2019**, *7*, 101. [CrossRef]

24. Abiodun, B.J.; Adeyewa, Z.D.; Oguntunde, P.G.; Salami, A.T.; Ajayi, V.O. Modeling the impacts of reforestation on future climate in West Africa. *Theor. Appl. Climatol.* **2012**, *110*, 77–96. [CrossRef]

25. Egbebiyi, T.S. Future Changes in Extreme Rainfall Events and African Easterly Waves Over West Africa. MSc. Thesis, University of Cape Town, Cape Town, South Africa, May 2016.

26. Sultan, B.; Gaetani, M. Agriculture in West Africa in the Twenty-First Century: Climate Change and Impacts Scenarios, and Potential for Adaptation. *Front. Plant Sci.* **2016**, *7*, 1262. [CrossRef] [PubMed]

27. Janicot, S.; Caniaux, G.; Chauvin, F.; De Coëtlogon, G.; Fontaine, B.; Hall, N.; Kiladis, G.; Lafore, J.-P.; Lavaysse, C.; Lavender, S.L.; et al. Intraseasonal variability of the West African monsoon. *Atmos. Sci. Lett.* **2011**, *12*, 58–66. [CrossRef]

28. Omotosho, J.B.; Abiodun, B.J. A numerical study of moisture build-up and rainfall over West Africa. *Meteorol. Appl.* **2007**, *14*, 209–225. [CrossRef]

29. Nicholson, S.E. The West African Sahel: A Review of Recent Studies on the Rainfall Regime and Its Interannual Variability. *ISRN Meteorol.* **2013**, *2013*, 453521. [CrossRef]

30. Klutse, N.A.B.; Ajayi, V.O.; Gbobaniyi, E.O.; Egbebiyi, T.S.; Kouadio, K.; Nkrumah, F.; Quagraine, K.A.; Olusegun, C.; Diasso, U.; Abiodun, B.J.; et al. Potential impact of 1.5 °C and 2 °C global warming on consecutive dry and wet days over West Africa. *Environ. Res. Lett.* **2018**, *13*, 055013. [CrossRef]

31. Paeth, H.; Capo-Chichi, A.; Endlicher, W. Climate Change and Food Security in Tropical West Africa—A Dynamic-Statistical Modelling Approach. *Erdkunde* **2008**, *2*, 101–115. [CrossRef]

32. Jarvis, A.; Ramírez-Villegas, J.; Campo, B.V.H.; Navarro-Racines, C. Is Cassava the Answer to African Climate Change Adaptation? *Trop. Plant Biol.* **2012**, *5*, 9–29. [CrossRef]

33. FAOSTAT. Statistical Yearbook of 2012: Europe and Central Asia; 2012. Available online: http://www.fao.org/3/a-i3621e.pdf (accessed on 1 December 2018).

34. Srivastava, A.K.; Gaiser, T.; Ewert, F. Climate change impact and potential adaptation strategies under alternate climate scenarios for yam production in the sub-humid savannah zone of West Africa. *Mitig. Adapt. Strateg. Glob. Chang.* **2016**, *21*, 955–968. [CrossRef]

35. FAOSTAT. FAO Statistical Yearbook 2014, Africa Food and Agriculture; 2014. Available online: http://www.fao.org/3/a-i3590e.pdf (accessed on 1 December 2018).

36. Harris, I.; Jones, P.D.; Osborn, T.J.; Lister, D.H. Updated high-resolution grids of monthly climatic observations—The CRU TS3.10 Dataset. *Int. J. Climatol.* **2014**, *34*, 623–642. [CrossRef]

37. Hijmans, R.J.; Guarino, L.; Cruz, M.; Rojas, E. Computer tools for spatial analysis of plant genetic resources data: 1. *DIVA-GIS. Plant Genet. Resour. News.* **2001**, *127*, 15–19.

38. Cong, R.-G.; Brady, M. The interdependence between rainfall and temperature: Copula analyses. *Sci. World J.* **2012**, *2012*, 405675. [CrossRef]

39. IPCC. *Meeting Report of the Intergovernmental Panel on Climate Change Expert Meeting on Climate Change, Food, and Agriculture*; Mastrandrea, M.D., Mach, K.J., Barros, V.R., Bilir, T.E., Dokken, D.J., Edenhofer, O., Field, C.B., Hiraishi, T., Kadner, S., Krug, T., et al., Eds.; World Meteorological Organization: Geneva, Switzerland, 2015; 68p.

40. Medori, M.; Michelini, L.; Nogues, I.; Loreto, F.; Calfapietra, C. The impact of root temperature on photosynthesis and isoprene emission in three different plant species. *Sci. World J.* **2012**, *2012*, 525827. [CrossRef]

41. Abbate, P.E.; Dardanelli, J.L.; Cantarero, M.G.; Maturano, M.; Melchiori, R.J.M.; Suero, E.E. Climatic and water availability effects on water-use efficiency in wheat. *Crop Sci.* **2004**, *44*, 474–483. [CrossRef]

42. Olesen, J.E.; Bindi, M. Consequences of climate change for European agricultural productivity, land use and policy. *Eur. J. Agron.* **2002**, *16*, 239–262. [CrossRef]

43. Cantelaube, P.; Terres, J.-M. Seasonal weather forecasts for crop yield modelling in Europe. *Tellus Ser. A Dyn. Meteorol. Oceanogr.* **2005**, *57*, 476–487. [CrossRef]

44. Abiodun, J.B.; Makhanya, N.; Petja, B.; Abatan, A.A.; Oguntunde, G.P. Future projection of droughts over major river basins in Southern Africa at specific global warming levels. *Theor. Appl. Climatol.* **2018**, *137*, 1785–1799. [CrossRef]

45. Ramírez-Villegas, J.; Jarvis, A.; Läderach, P. Empirical approaches for assessing impacts of climate change on agriculture: The EcoCrop model and a case study with grain sorghum. *Agric. For. Meteorol.* **2013**, *170*, 67–78. [CrossRef]

46. Ramírez-Villegas, J.; Lau, C.; Kohler, A.K.; Jarvis, A.; Arnell, N.; Osborne, T.M.; Hooker, J. *Climate Analogues: Finding Tomorrow's Agriculture Today*; CGIAR Research Program on Climate Change, Agriculture and Food Security (CCAFS): Frederiksberg, Denmark, 2011.

47. Hunter, R.; Crespo, O. *Large Scale Crop Suitability Assessment Under Future Climate Using the Ecocrop Model: The Case of Six Provinces in Angola's Planalto Region*; Springer: Cham, Switzerland, 2018.

48. Rippke, U.; Ramirez-Villegas, J.; Jarvis, A.; Vermeulen, S.J.; Parker, L.; Mer, F.; Diekkrüger, B.; Challinor, A.J.; Howden, M.; Howden, S. Timescales of transformational climate change adaptation in sub-Saharan African agriculture. *Nat. Clim. Chang.* **2016**, *6*, 605–609. [CrossRef]

49. Challinor, A.J.; Watson, J.; Lobell, D.B.; Howden, S.M.; Smith, D.R.; Chhetri, N.; Challinor, A.; Howden, S. A meta-analysis of crop yield under climate change and adaptation. *Nat. Clim. Chang.* **2014**, *4*, 287–291. [CrossRef]

50. Vermeulen, S.J.; Challinor, A.J.; Thornton, P.K.; Campbell, B.M.; Eriyagama, N.; Vervoort, J.M.; Kinyangi, J.; Jarvis, A.; Läderach, P.; Ramirez-Villegas, J.; et al. Addressing uncertainty in adaptation planning for agriculture. *Proc. Natl. Acad. Sci. USA* **2013**, *110*, 8357–8362. [CrossRef]

51. White, J.W.; Hoogenboom, G.; Kimball, B.A.; Wall, G.W. Methodologies for simulating impacts of climate change on crop production. *Field Crop Res.* **2011**, *124*, 357–368. [CrossRef]

52. Theil, H. A rank-invariant method of linear and polynomial. *Mathematics* **1950**, *392*, 387.

53. Sen, P.K. Estimates of the Regression Coefficient Based on Kendall's Tau. *J. Am. Stat. Assoc.* **1968**, *63*, 1379–1389. [CrossRef]

54. Wilcox, R.R. Simulations on the Theil-Sen regression estimator with right-censored data. *Stat. Probab. Lett.* **2003**, *39*, 43–47. [CrossRef]

55. Ohlson, J.A.; Kim, S. Linear valuation without OLS: The Theil-Sen estimation approach. *Rev. Acc. Stud.* **2015**, *20*, 395–435. [CrossRef]

56. Wilcox, R.R. A note on the Theil-Sen regression estimator when the regresser is random and the error term is heteroscedastic. *Biom. J.* **1998**, *40*, 261–268. [CrossRef]

57. Peng, H.; Wang, S.; Wang, X. Consistency and asymptotic distribution of the Theil-Sen estimator. *J. Stat. Plan. Inference* **2008**, *138*, 1836–1850. [CrossRef]

58. Gbobaniyi, E.; Sarr, A.; Sylla, M.B.; Diallo, I.; Lennard, C.; Dosio, A.; Dhiédiou, A.; Kamga, A.; Klutse, N.A.B.; Hewitson, B.; et al. Climatology, annual cycle and interannual variability of precipitation and temperature in CORDEX simulations over West Africa. *Int. J. Climatol.* **2014**, *34*, 2241–2257. [CrossRef]

59. Klutse, N.A.B.; Sylla, M.B.; Diallo, I.; Sarr, A.; Dosio, A.; Diedhiou, A.; Kamga, A.; Lamptey, B.; Ali, A.; Gbobaniyi, E.O.; et al. Daily characteristics of West African summer monsoon precipitation in CORDEX simulations. *Theor. Appl. Climatol.* **2016**, *123*, 369–386. [CrossRef]

60. Abiodun, B.J.; Adegoke, J.; Abatan, A.A.; Ibe, C.A.; Egbebiyi, T.; Engelbrecht, F.; Pinto, I. Potential impacts of climate change on extreme precipitation over four African coastal cities. *Clim. Chang.* **2017**, *143*, 399–413. [CrossRef]

61. Portmann, F.T.; Siebert, S.; Döll, P. MIRCA2000-Global monthly irrigated and rainfed crop areas around the year 2000: A new high-resolution data set for agricultural and hydrological modeling. *Glob. Biogeochem. Cycles* **2010**, 1–24. [CrossRef]

62. Nikulin, G.; Lennard, C.; Dosio, A.; Kjellström, E.; Chen, Y.; Hänsler, A.; Kupiainen, M.; Laprise, R.; Mariotti, L. Cathrine Fox Maule The effects of 1.5 and 2 degrees of global warming on Africa in the CORDEX The effects of 1.5 and 2 degrees of global warming on Africa in the CORDEX ensemble Manuscript version: Accepted Manuscript. *Environ. Res. Lett.* **2018**, *13*, 065003.

63. Maure, G.A.; Pinto, I.; Ndebele-Murisa, M.R.; Muthige, M.; Lennard, C.; Nikulin, G.; Dosio, A.; Meque, A.O. The southern African climate under 1.5 °C and 2 °C of global warming as simulated by CORDEX regional climate models. *Environ. Res. Lett.* **2018**, *13*, 065002. [CrossRef]

64. Ahmed, K.F.; Wang, G.; Yu, M.; Koo, J.; You, L. Potential impact of climate change on cereal crop yield in West Africa. *Clim. Chang.* **2015**, *133*, 321–334. [CrossRef]

65. Lobell, D.B.; Burke, M.B.; Tebaldi, C.; Mastrandrea, M.D.; Falcon, W.P.; Naylor, R.L. Prioritizing Climate Change Adaptation Needs for Food Security in 2030 Region. *Science* **2008**, *319*, 607–610. [CrossRef]

66. Malhotra, S.K. Horticultural crops and climate change: A review. *Indian J. Agric. Sci.* **2017**, *87*, 12–22.

67. Luo, Q. Temperature thresholds and crop production: A review. *Clim. Chang.* **2011**, *109*, 583–598. [CrossRef]

68. UNDP. *The 2030 Agenda for Sustainable Development*; A/RES/70/1; UNDP: New York, NY, USA, 2015; Volume 16301, pp. 13–14.

69. FAO. *The State of Food Security and Nutrition in the World 2018. Building Climate Resilience for Food Security and Nutrition*; Licence: CC BY-NC-SA 3.0 IGO; FAO: Rome, Italy, 2018.

70. Lobell, D.B.; Gourdji, S.M. The Influence of Climate Change on Global Crop Productivity. *Plant Physiol.* **2012**, *160*, 1686–1697. [CrossRef]
71. Taylor, K.E.; Stouffer, R.J.; Meehl, G.A. An overview of CMIP5 and the experiment design. *Bull. Am. Meteorol. Soc.* **2012**, *93*, 485–498. [CrossRef]
72. Zhang, X.; Cai, X. Climate change impacts on global agricultural water deficit. *Geophys. Res. Lett.* **2013**, *40*, 1111–1117. [CrossRef]

Article

Managing New Risks of and Opportunities for the Agricultural Development of West-African Floodplains: Hydroclimatic Conditions and Implications for Rice Production

Aymar Yaovi Bossa [1,2,*], Jean Hounkpè [1,2], Yacouba Yira [1,3], Georges Serpantié [4], Bruno Lidon [5], Jean Louis Fusillier [5], Luc Olivier Sintondji [2], Jérôme Ebagnerin Tondoh [6] and Bernd Diekkrüger [7]

[1] West African Science Service Centre on Climate Change and Adapted Land Use (WASCAL), Ouagadougou, Burkina Faso; hounkpe.j@wascal.org (J.H.); yira.y@wascal.org (Y.Y.)
[2] National Water Institute, University of Abomey Calavi, Cotonou, Benin; o_sintondji@yahoo.fr
[3] Applied Science and Technology Research Institute–IRSAT/CNRST, P.O. Box 7047, Ouagadougou, Burkina Faso
[4] Institute for Research and Development—IRD-UMR GRED-UPV, 34090 Montpellier, France; georges.serpantie@ird.fr
[5] Centre for International Cooperation in Agronomic Research for Development—CIRAD-UMR G-eau, 34090 Montpellier, France; bruno.lidon@cirad.fr (B.L.); jean-louis.fusillier@cirad.fr (J.L.F.)
[6] UFR des Sciences de la Nature, Université Nangui Abrogoua, 02 BP 801 Abidjan 02, Cote D'Ivoire; jetondoh@gmail.com
[7] Department of Geography, University of Bonn, Meckenheimer Allee 166, 53115 Bonn, Germany; b.diekkrueger@uni-bonn.de
[*] Correspondence: bossa.a@wascal.org

Received: 29 November 2019; Accepted: 6 January 2020; Published: 10 January 2020

Abstract: High rainfall events and flash flooding are becoming more frequent, leading to severe damage to crop production and water infrastructure in Burkina Faso, Western Africa. Special attention must therefore be given to the design of water control structures to ensure their flexibility and sustainability in discharging floods, while avoiding overdrainage during dry spells. This study assesses the hydroclimatic risks and implications of floodplain climate-smart rice production in southwestern Burkina Faso in order to make informed decisions regarding floodplain development. Statistical methods (Mann-Kendall test, Sen's slope estimator, and frequency analysis) combined with rainfall—runoff modeling (HBV model) were used to analyze the hydroclimatic conditions of the study area. Moreover, the spatial and temporal water availability for crop growth was assessed for an innovative and participatory water management technique. From 1970 to 2013, an increasing delay in the onset of the rainy season (with a decreasing pre-humid season duration) occurred, causing difficulties in predicting the onset due to the high temporal variability of rainfall in the studied region. As a result, a warming trend was observed for the past 40 years, raising questions about its negative impact on very intensive rice cultivation packages. Farmers have both positive and negative consensual perceptions of climatic hazards. The analysis of the hydrological condition of the basin through the successfully calibrated and validated hydrological HBV model indicated no significant increase in water discharge. The sowing of rice from the 10th to 30th June has been identified as optimal in order to benefit from higher surface water flows, which can be used to irrigate and meet crop water requirements during the critical flowering and grain filling phases of rice growth. Furthermore, the installation of cofferdams to increase water levels would be potentially beneficial, subject to them not hindering channel drainage during peak flow.

Keywords: inland valley development; hydroclimatic hazard; water control structure; sustainable rice production

1. Introduction

The future of West Africa, and its economic, political, and social balance, depend on the ability of the agricultural sector to adapt and ensure food security under multiple pressures, such as climate change and demographic growth. In Africa, only 12.5 million hectares are irrigated out of a total of 202 million hectares of cultivated land, or 6.2%. The proportion of irrigated land in the south of Saharan Africa is even smaller, with only 5.2 million hectares, or 3.3%, of cultivated land being irrigated [1]. The increase in population will have serious implications in terms of agricultural production and the availability of natural resources. Adaptive strategies to help cope with the potential decrease in crop yields include promoting the extensive development of inland valleys in West Africa. This is because of their great potential as rice-based production systems due to the high and secure water availability and soil fertility [2]. As such, the West African floodplains are privileged places for agricultural intensification, but play a diminishing role in the face of droughts that affect rainfed crops. Key factors of concern for the agricultural development of floodplains include flood hazards, surface flow deficits due to dry spells, and early flood recession. The valorization of floodplains faces numerous technical, social, and economic constraints that involve an intensification of crops and hence new risks linked to climate change. These are characterized by increased irregularities in rainfall, onsets of extreme floods, and long-lasting droughts.

As a landlocked country, Burkina Faso is vulnerable to climate variability [3]. This variability not only occurs at a daily, seasonal, and interannual scale, but can also be multidecadal. A break in annual rainfall was observed during the 1970s, irrespective of latitude or longitude, in West Africa [4]. It was especially prominent in the savanna, which includes the study area of Dano. The causes of this prolonged drought, subcontinental in scale, remain controversial and undoubtedly multifactorial and multiscalar. Many authors have shown, in addition to the global natural variations (i.e., astronomical, oceanic, and volcanic), the anthropogenic effects at different scales. These are observed at a regional (increase in the albedo effect because of the rapid urbanization of the Sahel, and deforestation of the lower coast reducing real evapotranspiration (ETR) and increasing flow), intercontinental (European air pollution of the 1970s, favoring regional cooling, i.e., anticyclonic conditions in regulatory pathways), and global (greenhouse gas-related climate change and its effects on heat and excessive events) level [5]. A change in hydroclimatic conditions can substantially modify the hydrological regime of an inland valley and its drainage area [6]. This modification can result in flooding or drying conditions in the lowlands, implying a possible reduction in its productivity. Furthermore, land cover and land use change can alter the floodplain and impact its ecosystem [7], but investigating this is beyond the scope of this study. Given the uncertainties in climate model predictions (especially for precipitation), analyzing the current climate hazards using observed data and their possible implications for the future is required.

Burkina Faso's agricultural sector continues to generate approximately one-third of the country's GDP and employs 80% of the population [8], despite the harsh climatic condition. Notwithstanding the importance of agriculture in the economy of Burkina Faso, the sector is facing many challenges, including threats from many natural disasters, such as floods, droughts, and violent winds, which lead to low crop and livestock productivity [9]. Since 1970, investment has been made by the government of Burkina Faso to address the issue posed by hydroclimatic risks. This includes developing rice production intensification policies in inland valleys that encompass physical development, the social organization of production, material support, organization of the rice sector, subsidies, and legal connotations. Subsequently, 10% of the inland valleys suitable for agriculture have been developed in southwestern Burkina Faso. However, as reported by the regional agriculture extension service, up to 30% of the developed inland valleys have been abandoned due to increasing hydroclimatic hazards.

There is therefore a need to describe the seasonal, average, and frequency characteristics of the climate that can impact rice production in the region.

The objective of this work is to analyze the hydroclimatic hazards by considering the period of 1922–2017 and their implications for rice production in southwest Burkina Faso to support agricultural policies for adequate water infrastructural development. Two research questions are considered: (i) What are the current trends in climatic and hydrological hazards and what are their implications for food production in inland valleys? (ii) What are farmers' perceptions of the hydroclimatic risks in the region, and what strategies have consequently been developed to face the challenges encountered?

2. Materials and Methods

This section is divided into five sub-sections, which are the study area and data used, the various modalities of lowland development (traditional vs. modern development), climate-related local knowledge and hydroclimatic variables' analysis, rainfall–runoff modeling and frequency analysis, and water availability evaluation during the critical phase of rice development.

2.1. Study Area and Data Used

The case study areas are the Lofin catchment and Lofin inland valley, located in the municipality of Dano in the southwest region (région du Sud-Ouest) of Burkina Faso, West Africa (Figure 1). Dano is situated in a tropical climate region with a unimodal rainfall regime (Figure 2). The mean annual rainfall is approximately 921 mm, with a standard deviation of 106 mm for the period 1980–2018. The annual rainfall regime is characterized by the alternation of two contrasting seasons: a dry season from November to March, in which rainfall is almost absent (58 mm in October, the driest month), and a rainy season from April to October (average 238 mm in August, the wettest month) [10]. The climatic water demand (ET0) is more stable during the year, but it varies, on average, from 123 mm in August to 175 mm in April [10]. Long-cycle crops (120 days), such as rice sown after the 10th of July, are at risk of water stress in the middle and end of the growing cycle. The average annual temperatures between 1970 and 2013 ranged from 25 to 33 °C and from 25 to 31 °C at the Boromo and Gaoua stations, respectively (Figure 2). The annual insolation varies between 6 and 8 h/day, and the air humidity ranges from 35% to 80%. The dominant vegetation comprises shrubs and/or the tree savanna type, and resulting successional vegetation from the degradation of cleared forests [11]. This is due to both human activities and the dry period since 1970.

In this region, wetlands have historically been used as pasture in the dry season. Rice was one of the first crops cultivated in these areas (Figure 3). From the 20th century onwards, the agricultural use of the inland valleys, referred to locally as 'bas-fonds', was fostered because of population growth and migratory flows. Currently, in addition to rice, wetland use has been diversified with other crop types, such as vegetables, fruits, and cereals [12]. Rice products are mainly intended for consumption, with an increasing share of rice in the local food.

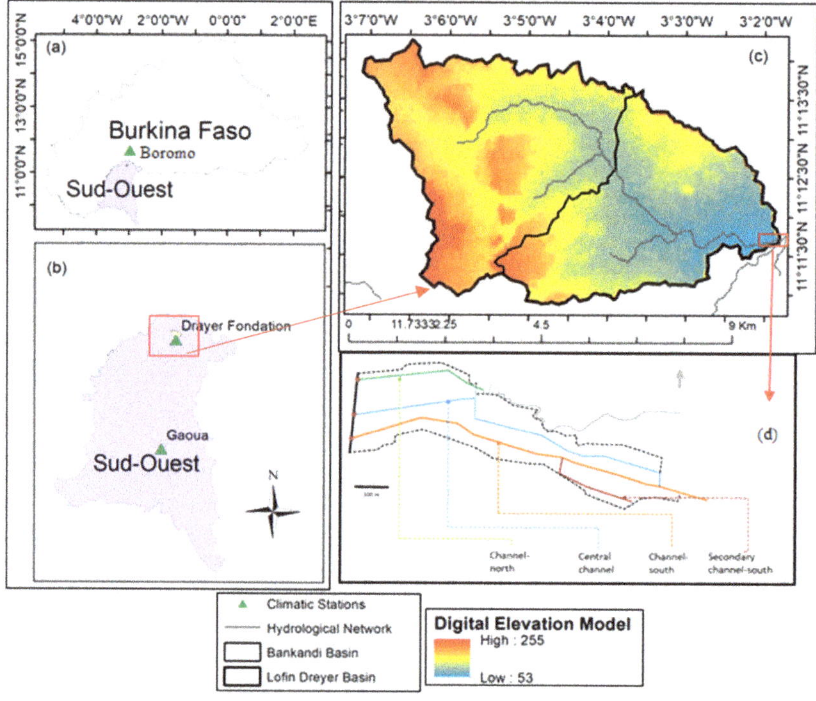

Figure 1. (**a**) Location of the study area in Burkina Faso. (**b**) The southwest region. (**c**) The Lofin catchment. (**d**) The Lofin inland-valley with irrigation/drainage channels.

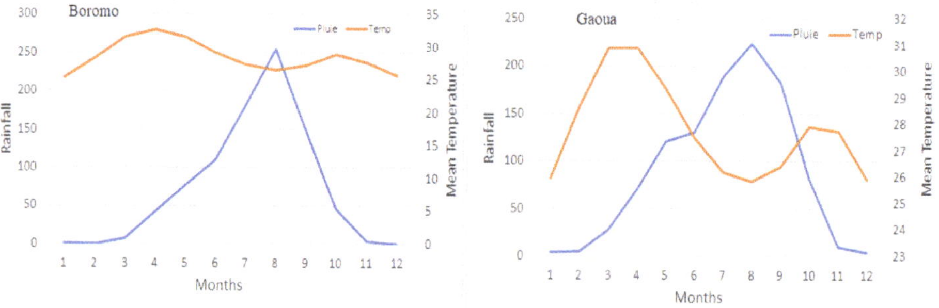

Figure 2. Rainfall and temperature at Boromo and Gaoua climatic stations.

The data used in this study are rainfall data of the Boromo and Gaoua stations from 1922 to 2016, and rainfall data of the Dreyer Foundation from 2017 to 2018. Discharge data of the Lofing-Radier station from 2017 to 2018 were measured by WASCAL (www.wascal.org) during project implementation. Other climate data, such as the minimum and maximum temperature, sunshine duration, wind speed, and minimum and maximum relative humidity from 1922 to 2016 of the Boromo and Gaoua stations were used and obtained from the Burkina Faso national meteorological directorate (including long-term rainfall data).

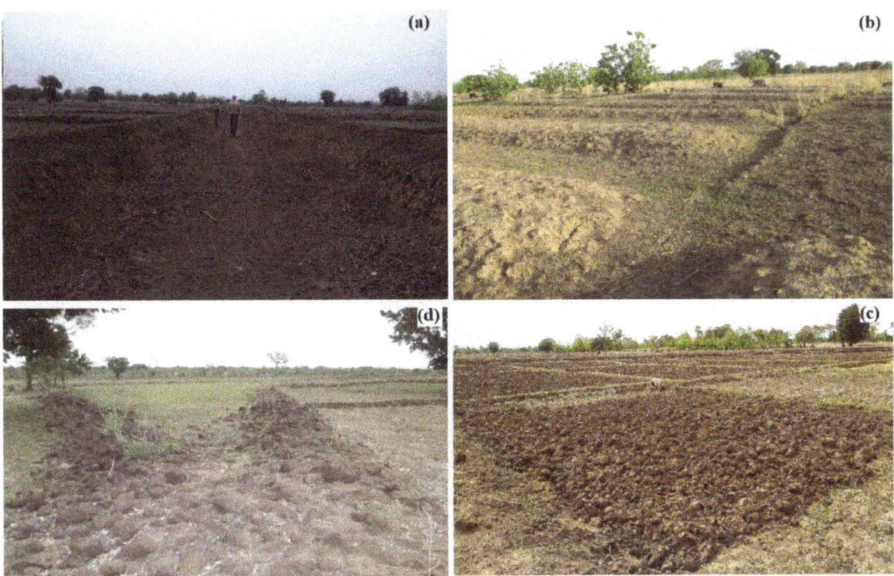

Figure 3. Different types of lowland development, including a rice field model with sprinkler drains (**a**,**b**), cyclopean concrete pouring dikes with a central cofferdam (**c**), and compacted clay bunds protected by geotextiles and rocks following the level curves with openings (**d**).

2.2. Various Modalities of Lowland Development: Traditional vs. Modern Development

Four types of lowland development have been observed and can be classified into two main groups: traditional and modern lowland designs. Traditional lowland development is a combination of the techniques developed by farmers for managing the agrarian space, the water, and the various types of production. The objective pursued by the farmers is to grow crops while minimizing the risks associated with drought and flooding. Schemes designed by farmers for the control and management of water in the lowlands include, firstly, large ridges arranged perpendicularly to the water flow direction, and secondly, large ridges in the shape of a contour dike with gaps.

Baffles are formed that not only slow the flow of water favoring infiltration, but that also enable the management of a water level in the grooves between the ridges. Upland crops (maize or tobacco) are placed on the ridges. Sorghum and taro are placed on the flank of the ridges. Rice, a water-demanding crop, is sown and transplanted into the furrows. This polyculture system is adapted to local conditions, has a low associated cost, and is more resilient to the risk of climatic disasters.

In contrast, modern lowland developments are those designed and implanted by external organisations, such as funded projects and programs, and NGOs. The principle objectives of such developments are multi-functional, aiming to

- Partially control water through the installation of hydraulic structures;
- Distribute the water at the landscaped site;
- Optimize the drainage of flood waters;
- Avoid and minimize the adverse effects of water shortages due to dry spells during the crop season; and
- Support non-seasonal crops, if possible.

To achieve these objectives, several types of development were designed, of which three (3) types of models are described. First is the rice field model with sprinkler drains (Figure 3). This model consists of channeling runoff by following preferential paths marked by the differentiation of surface elevation.

The canals are used for irrigation and drainage. The individual plots are partitioned by small bunds which the producers can open to irrigate their plants. This model is promoted in the area by the Dreyer Foundation. Second are compacted clay bunds protected by geotextiles and rocks that follow the level curves with openings. This model is a flood spreading arrangement, with the possibility of drainage being provided by the openings. The third model is the cyclopean concrete pouring dikes with a central cofferdam. This model is based on the threshold for slowing the flow of water on the course bed, which leads to a substantial change in the height of the water level, and is then managed using the cofferdam.

With these three models of landscape control, adding garden plants arranged with wells is necessary. The modern development models of lowlands strongly alter the hydrology of the valley bottom. Although this may be advantageous, it also adds new constraints that can become risks, depending on the physical and social environment.

2.3. Climate-Related Local Knowledge and Hydroclimatic Variables' Analysis

A survey of farmers' perceptions on climate and climatic changes was conducted in the Lofing lowland in 2017 using a questionnaire. A total of 17 farmers were randomly selected, and a questionnaire was administered individually. The hydroclimatic variables were statistically analyzed using the quantile method, Mann-Kendall test, and Sen's slope estimator. Different time steps were considered to aggregate the time series over 10-day, monthly, and annual time scales. Reference evapotranspiration was computed using the Food and Agriculture Organisation (FAO) ETo calculator software based on the Penman–Monteith formula [13]. The rainfall onset and cessation dates were defined using the Franquin (1969) [14] method. At the 10-day scale, the rainfall onset corresponds to the date from which rainfall is greater than half of the potential evapotranspiration (R > ET0/2). The methodological framework of the study is presented in Figure 4. It shows the different steps fulfilled to perform this study. The Mann-Kendall test [15,16] is a nonparametric trend detection method widely applied to hydroclimatic variables [17–19]. The null hypothesis H_0 for the test is that there is no trend in the time series, while for the alternative hypothesis H_1, there is a significant trend in the time series at the 0.05 significance level. This is a robust test in the sense that it does not make any assumptions about the distribution of variables. In addition, the Sen slope method [20] is considered for estimating the magnitude of the slope if a trend is detected in the time series.

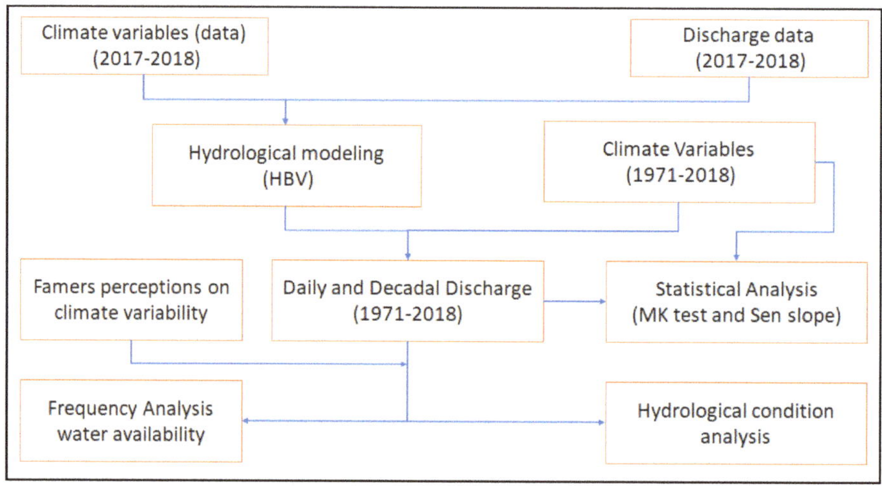

Figure 4. Methodological framework of the study.

2.4. Rainfall–Runoff Modeling and Frequency Analysis

To further access the hydrological aspect of the study area, an HBV model [21] was calibrated and validated for the Lofing basin at the outlet of Lofing Radier. HBV is a conceptual, lumped, and time-continuous hydrological model that simulates discharge using rainfall and potential evaporation as inputs. The HBV model has four main component routines: (1) snow (not used); (2) soil moisture (computes actual evapotranspiration, soil moisture, and groundwater recharge); (3) response function (calculates runoff and groundwater levels); and (4) routing (calculates the distribution of runoff for a given time series). Model calibration used 2018 data, and validation was performed with data from 2017. Performance criteria included the Nash–Sutcliffe efficiency (NSE) normalized statistic [22] combined with the coefficient of determination (R^2). The past hydrological condition of the basin was then simulated using climatic data from the Boromo and Dano stations (Figure 1) for the period 1971–2018 (Figure 4).

2.5. Water Availability Evaluation during the Critical Phase of Rice Development

The flowering and grain filling growth stages of rice are the most critical phases of its development. Water stress during these phases can be seriously detrimental to the rice yield [10]. In the Dano region, rice varieties with a growing period of 120 days are the most cultivated. By considering the following 10-day periods of rice sowing seed (21–30 June, 1–10 July, and 11–20 July), the critical periods for rice development ranged from 11–20 September to the 11–20 October. The total discharge of water during these 10-day periods was obtained. Trends in these data collection periods were evaluated using the Mann-Kendall and Sen slope tests. The availability of water in terms of discharge, level, and volume needed for effective rice development in the Lofing inland valley was then assessed.

3. Results and Discussion

3.1. Farmers' Perceptions of Changes in Climatic Hazards in the Lofing Inland Valley

Farmers have both positive and negative consensual perceptions of climatic hazards. According to them, no severe drought has been observed since 1974, except in 1984, when a famine was experienced. Since 1996, extreme rains able to destroy houses have not been observed, except in 2015, when a long-lasting and heavy rainfall event was recorded. Events perceived negatively were deemed to be of the greatest importance. Farmers consider that the heat waves initially experienced mainly in April have shifted to March and that they last almost the entire year. The weather is therefore perceived as becoming warmer. Furthermore, the farmers perceive there to be changes in annual rainfall patterns because, according to them, rainfall was previously well-distributed throughout the rainy season. Now, they believe that the dry season is longer, and the rainy season is shorter and irregular. Dry spells have become more frequent during crop growth and mainly occur during the rice grain formation stage. According to the interviewed farmers, there is currently a decrease in rainfall amount per event, with more thunderstorm winds, but limited rainfall, in comparison to past observations.

The events best-described by the farmers are the catastrophic years, the increasing heat, and the effect of wetland development over the last two years. The consensual perceptions of the degrading rainy season should be considered with caution, given the vagueness of the compared periods and the intensity of the variations. For example, it is virtually impossible to attribute a precise date to references such as "previously", "formerly", or "around 2000", when farmers stated that there have been more severe droughts in the past than at present. Was there a period of a small series of very regular years that was idealized and would now be "referenced"? Did the farmers identify recent problematic years to more comprehensively judge the past, thereby forgetting about the reality of the variability and the change in the process? An in-depth analysis of several timescales of the long climatological series, including a frequency analysis, is required to answer these questions.

Some perceptions are not consensual, namely, the rainfall onset date in 2017 and the effect of the dike and channel rehabilitation in 2017. In fact, 10 of the 17 people interviewed reported an early rainfall onset, 3/17 interviewees reported a normal onset, and 4/17 interviewees indicated a late rainfall onset in 2017. The least consensual perceptions are paradoxically related to the climate or the water regime of the year. These perceptions address, on the one hand, the location of the respondent's plot in relation to the newly constructed dike, and on the other hand, the expectations that are relative to a farmer's specific needs and workplan. The perceptions of climate risk also do not have the same levels of concern among individual respondents. Few have seen "no change" to the climate. In terms of the motivation for sowing rice rather than transplanting in 2017, respondents cited the climatic risk. However, the perception of climate risk varies, according to the respondents' gender and the level of development of the lowlands (Table 1). Indeed, excluding the developed lowlands, the climate risk was mentioned by 100% of women as the reason for rice transplanting. This result might be due a lack of knowledge about agricultural rainfall onset identification. Early sowing is adopted by farmers to free themselves from rainfall and hydrological hazards (i.e., uncertainty about the moisture conditions of the lowland region). Despite existing water control structures in the developed lowlands (which are designed to enable transplanting), up to 20% of farmers do not wait for good sowing conditions. Men are more restricted by their other farming operations. Rice is of a lower priority, and where there is a competition of labor against cotton, the lowlands are often abandoned. Indeed, the climate risk is less important to them, as it is shifted toward alternative activities. This local information is extremely valuable as it both identifies the concerns of local farmers and provides new information regarding their perceptions and likely responses as a result. It is nevertheless necessary to compare the local farmers' perceptions with measured data, which is independent of the farmers' gender, knowledge, and situation.

Table 1. Reasons for the choice of sowing (rather than transplanting) in 2017, according to gender and situation.

Reasons	Operational Constraints	Social Organization	Climate Risk	Lack of Know-How
Undeveloped lowland by men (%)	43	0	43	14
Undeveloped lowland by women (%)	0	0	100	0
Developed lowland by men (%)	63	13	25	0
Developed lowland by women (%)	80	0	20	0

3.2. Analysis of Rainfall over the Last 40 Years at the Regional Scale

The mean rainfall recorded at Dano during 2013–2017 by the Dreyer Foundation (951 mm) is similar to that of the period 1970–2013 at Boromo-Gaoua (962 mm). For that reason, the Boromo-Gaoua rainfall stations, which are the closest to Dano, have been used to provide a detailed understanding of the local climatic pattern. Increased water exceedance events (water available to recharge the reserve and water flows) have been observed in Boromo since 1984 (Figure 5a). There has been no increase in rainfall; however, an increase in water excess implies a change in rainfall regime and/or land use. More rainfall in a shorter time period results in an increase in flood risk and less actual evapotranspiration. In Gaoua, the change was small, but similar, to the change depicted in Boromo, except that the water excess did not increase. Figure 5b shows the climatic balance of the last five years in Dano. There was a high seasonal and interannual variability of rainfall during this period. In 2013, there was no pre-humid period, the water excess was limited, and an early cessation of the season occurred, while in 2014, the pre-humid period occurred in mid-July. In 2015 and 2016, there was no pre-humid period, and the risk of inundation was high. A pre-humid period was detected in early September 2017, with limited water excess.

The period of rainfall uncertainty is longest during the pre-humid season. The hazard zone corresponds to rainfall being less than half of the potential evapotranspiration between the 25 and 75 quantiles of ten days of rainfall. The season profile is asymmetric, meaning that the cessation of the season is more predictable than its onset and that the early rainy season provides more information on the rainy season duration. An ideal opportunity for an informed choice of season length, mainly if there is the potential to irrigate, is provided based on this data.

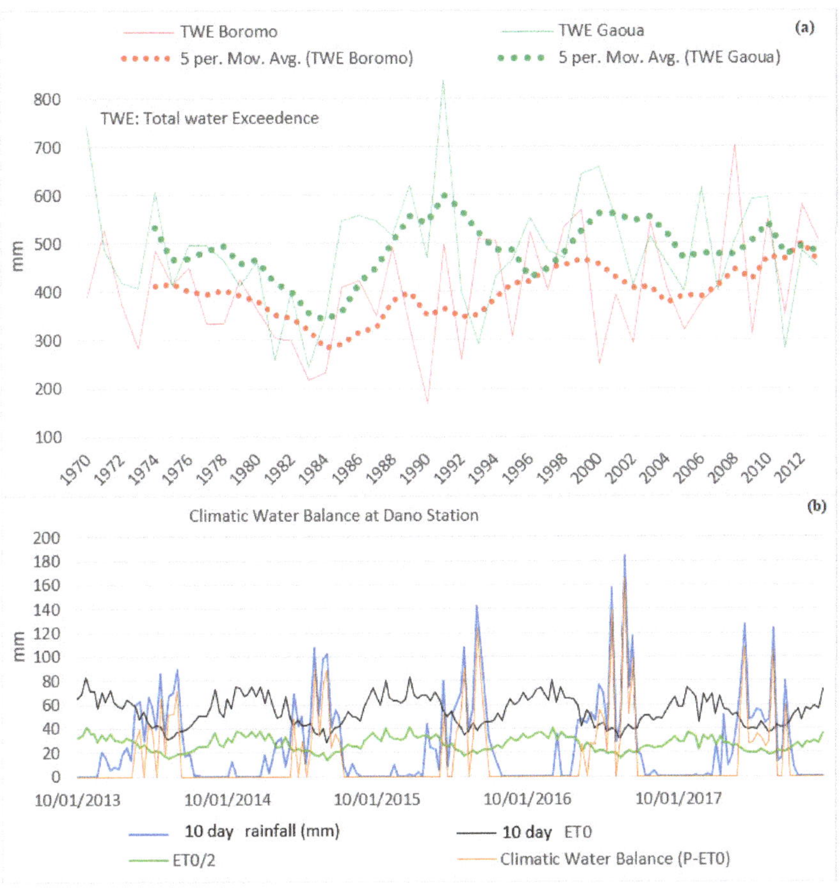

Figure 5. Water excess at Boromo and Gaoua rainfall stations from 1970 to 2012 (**a**) and the climatic water balance at Dano station (2013:2017) (**b**).

3.3. Temperature Analysis at the Regional Scale

The Sen slopes estimated for each month by considering the minimum and the maximum temperatures at Boromo and Gaoua from 1970 to 2013 indicate a tendency toward a warmer atmosphere. During the rainy season (August to October), the average minimum temperature and the average maximum temperature increased at a rate of 0.031–0.035 and 0.005–0.016 °C yr^{-1}, respectively, which is in line with the global observation [23,24]. The minimum temperature increased faster (around two times) than the maximum temperature during the rainy season. Climate change causes increasing air temperatures and evapotranspiration, increases the risk of intense rainstorms, and increases the risk of heat waves associated with drought [25]. An increase in temperature was also observed in the Beninese part of the Niger basin by Badou et al. (2017). An increase in temperature implies a higher level of evapotranspiration, a higher water demand for crops, and a lower level of water exceedance. An increasing temperature will exhibit a larger impact on the grain yield than on vegetative growth and will reduce the ability of the crop to efficiently fill the grain or fruit [26]. This output has the potential to inform famers and stakeholders in framing appropriate policies for rice intensification in the region. The minimum relative humidity (RH) increases from July to October at Boromo and from August to December at Gaoua. The maximum RH decreases from May to September at Boromo and from July to September at Gaoua.

3.4. Potential of Watering/Drainage of the Channels' System in the Lofing Inland Valley

High seasonal and interannual variabilities of discharge were observed at the Lofing-Radier outlet (Figure 6). Throughout the growing season, the river provides enough water to satisfy the requirements of rice (100 L s^{-1} is needed to irrigate 30 ha) (Figure A1). Irrigation should be possible whenever necessary, mainly during the dry spell period. Irrespective of the rice sowing date, the critical period (end of the rice cycle) requiring irrigation varies between the 10th of September and 20th of October. Channel dimensioning is challenging in rice cultivated in inland valleys. There must be a trade-off between the necessities of the discharge peak flow, while maintaining a water level in the channels required for direct irrigation (through, for instance, the use of cofferdams), but also maintaining the wetness of cultivated parcels. The drainage capacity of the channels is large enough for discharging the peak flows arriving from the Dano basin at the outlet of Lofing, while its irrigation potential is problematic. Although the inflow into the channel system was adequate, the water level in the channels is problematic. The minimum water level required in one of the main channels for irrigation is 35 cm. In 2017, the water level in the channels rarely exceeded 35 cm (Figure A2), suggesting that the installation of cofferdams to increase water levels would be beneficial. The implementation of cofferdams, however, may result in additional problems, such as hindering the channel drainage function during peak flow. Frequency analysis has shown that the likelihood of obtaining a high flow during the critical period is low, and mainly occurs after the 21st of September. To ensure that irrigation at the end of the rice cycle is sufficient, different mobile cofferdam types require testing to ensure their efficiency and acceptability in terms of cost and the capacity of farmers to implement the technology.

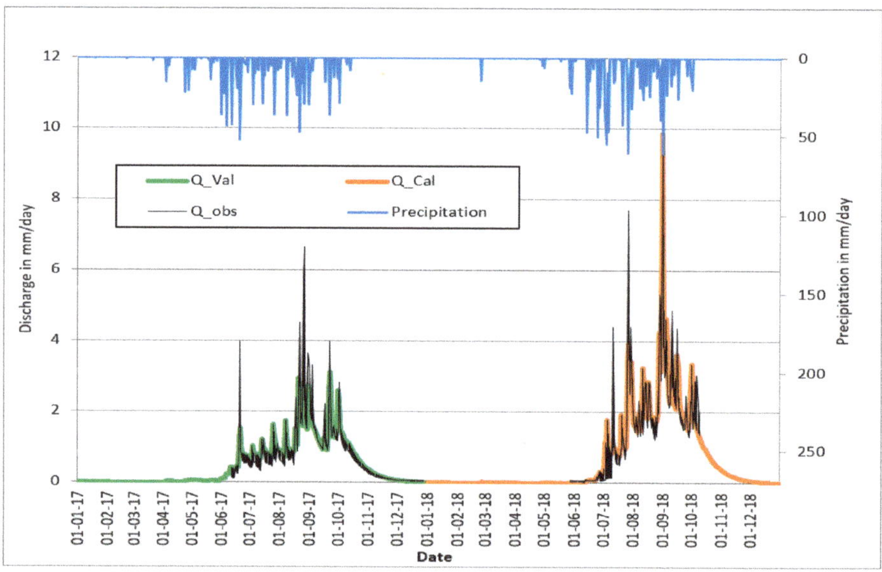

Figure 6. Observed and simulated discharges produced by the HBV model (Q_Val, Q_Cal, and Q_obs corresponding to the validation discharge, the calibration discharge, and the observed discharge, respectively).

3.5. Model Calibration and Validation

The HBV model was calibrated for the year 2018, and the simulated discharge was compared to the observed discharge using the numerical and visual criteria. The observed and simulated discharges are similar (Figure 6). NSE values for calibrated/validated data were 0.75/0.70, with a coefficient of determination of 0.75/0.73 and a logarithm of NSE of 0.74/0.85. The high values imply a strong performance of the HBV model for the two years of observed discharge. Both high discharge and recession discharge (lower discharge) series were accurately simulated, although it is acknowledged that a greater number of discharge observations are required to increase the accuracy in the future.

3.6. Hydrological Condition of the Lofing Upstream River

The calibrated and validated HBV model simulated discharge of the Lofing-Radier River for the period 1971–2018. The water balance components, precipitation, discharge, and actual evapotranspiration are shown in Figure 7. The results of the Mann-Kendall trend test applied to the discharge statistic are shown in Table 2. No statistically significant trend at the 5% level was found in the total annual discharge for the period 1971–2018, indicating that the hydrological regime of the catchment did not vary at the annual scale. Nevertheless, a small annual increasing rate of the discharge of 0.0074 mm per day (2.7 mm yr^{-1}) was evident, implying a constant availability of water at the annual scale. The flowering and grain filling of rice seeded between the 20th and 30th of June occurred between the 11th and 20th of September. This stage of rice growing is critical since water stress experienced in this period may drastically reduce the rice yield. No significant increase in the water stress level is found during this period. Between the 21st and 30th of September, the third quartile displays an increase in the risk of excess water levels. In October, there is an increase in water resource availability, which is mainly beneficial for rice production during this critical stage.

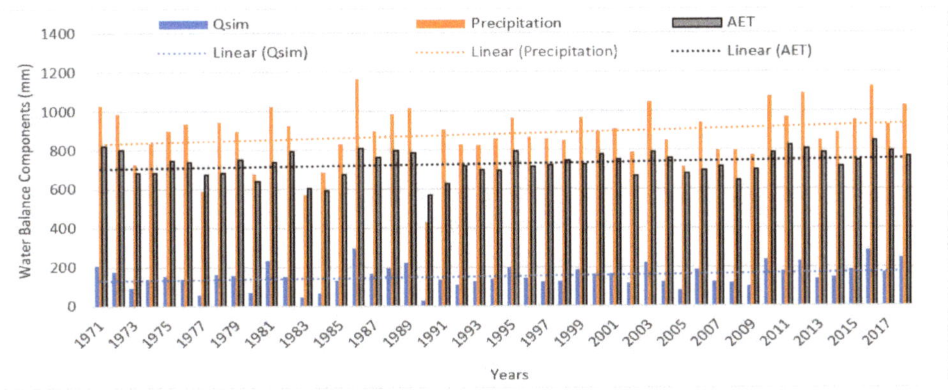

Figure 7. Water balance components: precipitation, simulated discharge (Qsim), and actual evapotranspiration (AET) for the period 1971–2018.

Table 2. Sen slope (SS) and total 10-day water discharges (see Section 3.3) in different conditions for 1971–2018. * indicates a significant trend obtained through the Mann-Kendall test.

	11–20 September		21–30 September		1–10 October		11–20 October	
Test implemented (1971–2018)	SS/MK *	10 day water (mm)	SS/MK *	10 day water (mm)	SS/MK *	10 day water (mm)	SS/MK *	10 day water (mm)
1st Quartile (Dry Condition)	0.006	<12.7	0.009	<11.7	0.009 *	<9.2	0.006 *	<6.4
Median (Normal Condition	0.009	19.5 [12.7, 24.1]	0.010	16.8 [11.7, 20.5]	0.010 *	12.6 [9.2, 5.7]	0.007 *	9.2 [6.4, 10.4]
3rd Quartile (Wet Condition)	0.014	>24.1	0.016 *	>20.5	0.013 *	>15.7	0.008 *	>10.4
Sum	0.120	-	0.125	-	0.127 *	-	0.070 *	-

The water level in the river at the gauging station decreased from the 10th of September to 10th of October (Table 2). Therefore, if there is enough rainfall, sowing the rice during 10–30 June will be optimal to take advantage of the higher surface water flows that can be mobilized to irrigate and meet the crops' water requirements during the critical phases of flowering and formation-filling of the grains. Lower flow rates can be utilized to irrigate the crop during the critical phases if sown between the1st and 20th of July.

4. Conclusions

Rainfall events exceeding 100 mm and flash flooding are becoming more frequent, leading to severe damage to crop production and water infrastructure. Special attention must therefore be given to the design of water control structures to ensure their flexibility and sustainability in discharging floods while avoiding overdrainage during dry spells. In this study, we analyzed the hydroclimatic conditions of the study area Dano, Burkina Faso, and the implication for rice production in the region. There was no significant increase in annual rainfall for the period of 1970–2013; however, an increasing delay in the onset of the rainy season (with a decreasing pre-humid season duration) was observed. This causes difficulties in predicting the onset due to the high temporal variability of rainfall in the studied region. As a result, a warming trend was observed for the past 40 years, raising questions about its negative impact on very intensive rice cultivation packages. During the rainy season (August to October), the average minimum and maximum temperatures increased by 0.031 and 0.016 °C yr^{-1}, respectively, comparable to global observations. The maximum relative humidity decreased due to this increase in temperature, while the sunshine duration also decreased. Farmers have both positive and negative consensual perceptions of climatic hazards. The HBV hydrological model indicated no significant increase in water discharge; however, the total 10-day water level observed between the 11th of September and 20th of October, corresponding to the critical flowering and grain filling phases of rice growth, showed an increasing trend for the period 1971–2018.

The sowing of rice during the 10–30 June has been identified as optimal in order to benefit from the higher surface water flows, which can be used to irrigate and meet the crop water requirements during the critical phase outlined. The installation of cofferdams to increase water levels would be beneficial, subject to them not hindering channel drainage during peak flow, although water flow after the 21st of September was generally insufficient to be deemed an issue. To ensure that irrigation at the end of the rice cycle is sufficient, different mobile cofferdam types require testing to ensure their efficiency and acceptability in terms of cost and the capacity of farmers to implement the technology. The results of this study will be useful to rural communities, as well as decision makers, in framing agricultural risk management in the study region and devising policy for rice intensification in lowland areas. Further data collection is required to improve the HBV model output and to account for climate and land change effects on rice production in Dano, Burkina Faso.

Author Contributions: Conceptualization, A.Y.B. and G.S.; methodology, A.Y.B., G.S., J.H., and B.D.; formal analysis, A.Y.B., J.H., Y.Y., G.S., and B.D.; writing—original draft preparation, review, and editing, A.Y.B., J.H., G.S., Y.Y., B.L., J.L.F., L.O.S., J.E.T., and B.D. All authors have read and agreed to the published version of the manuscript.

Funding: This research received no external funding.

Acknowledgments: The authors are grateful for the financial support provided by the French Agency for Development (AFD) under the auspices of the AGRICORA initiative and GENERIA project. They thank the German Federal Ministry of Education and Research (BMBF) for supporting the WASCAL program.

Conflicts of Interest: The authors declare no conflicts of interest.

Appendix A

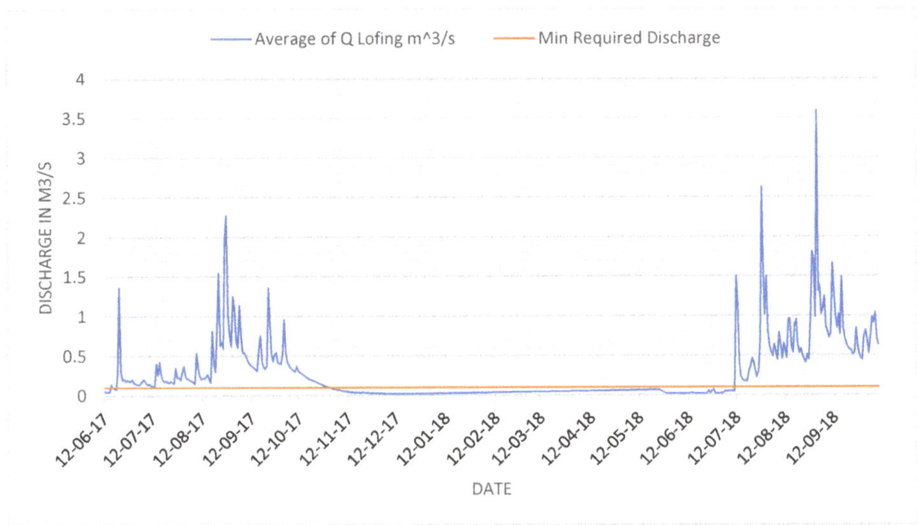

Figure A1. Comparison of the available discharge and irrigation water requirement for the 30 ha Lofing inland-valley.

Appendix B

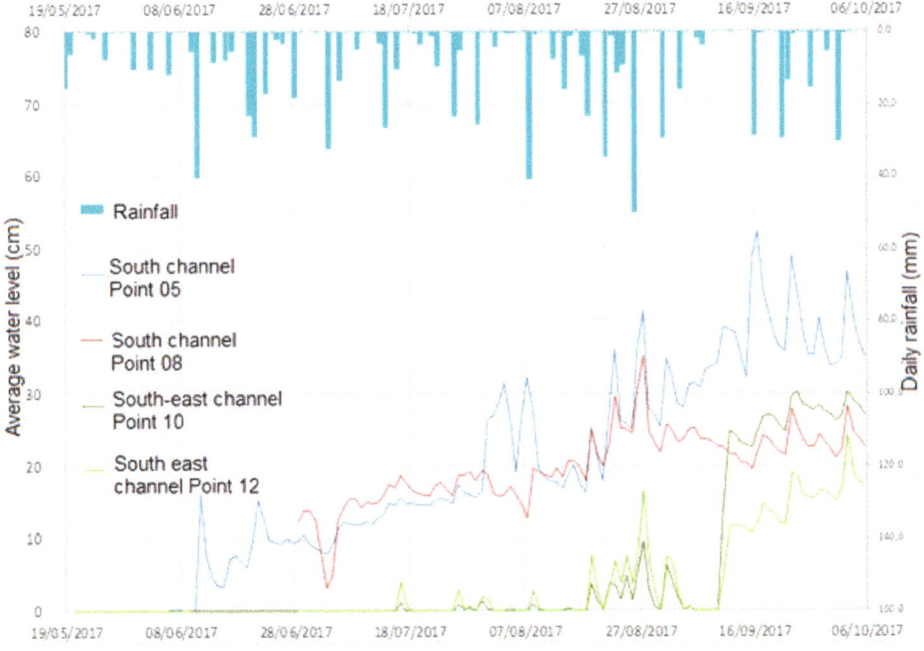

Figure A2. Comparison of the observed and required water height in the irrigation/drainage channels of Lofing inland-valley. Points represent the water level measurement locations.

References

1. Faurès, J.; Sonou, M. *Les aménagements hydro-agricoles en Afrique Situation actuelle et perspectives*; FAO: Rome, Italy, 2005.
2. Danvi, A.; Giertz, S.; Zwart, S.J. Rice Intensification in a Changing Environment: Impact on Water Availability in Inland Valley Landscapes in Benin. *Water* **2018**, *10*, 74. [CrossRef]
3. Ouédraogo, M.; Dembele, Y.; Somé, L. Perceptions et stratégies d'adaptation aux changements des précipitations: Cas des paysans du Burkina Faso. *Cah. Agric.* **2010**, *21*, 87–96. [CrossRef]
4. Badou, D.F.; Kapangaziwiri, E.; Diekkrüger, B.; Hounkpè, J.; Afouda, A. Evaluation of recent hydro-climatic changes in four tributaries of the Niger River Basin (West Africa). *Hydrol. Sci. J.* **2017**, *62*, 715–728. [CrossRef]
5. Jacob, D.; Kotova, L.; Teichmann, C.; Sobolowski, S.P.; Vautard, R.; Donnelly, C.; Koutroulis, A.G.; Grillakis, M.G.; Tsanis, I.K.; Damm, A.; et al. Van Earth's Future Climate Impacts in Europe Under + 1.5 °C Global Warming. *Earth's Future* **2018**, *6*, 264–285. [CrossRef]
6. Yacouba, Y.; Aymar, B.Y.; Fusillier, J.; Thomas, Y.B. Failure of inland valleys development: A hydrological diagnosis of the Bankandi valley in Burkina Faso. *Model. Earth Syst. Environ.* **2019**, *5*, 1733–1741. [CrossRef]
7. Leemhuis, C.; Thonfeld, F.; Näschen, K.; Steinbach, S.; Muro, J.; Strauch, A.; Ander, L.; Daconto, G.; Games, I. Sustainability in the Food-Water-Ecosystem Nexus: The Role of Land Use and Land Cover Change for Water Resources and Ecosystems in the Kilombero Wetland, Tanzania. *Sustainability* **2017**, *9*, 1513. [CrossRef]
8. USAID Agriculture and Food Security. Available online: https://www.usaid.gov/burkina-faso/agriculture-and-food-security (accessed on 29 April 2019).
9. Yameogo, T.B.; Bossa, A.Y.; Torou, B.M.; Fusillier, J.; Da, D.E.C.; Yira, Y.; Serpanti, G.; Some, F.; Dama-balima, M.M. Socio-Economic Factors Influencing Small-Scale Farmers' Market Participation: Case of Rice Producers in Dano. *Sustainability* **2018**, *10*, 4354. [CrossRef]
10. Raherizatovo, T. Conception d'aménagements à maîtrise partielle de l'eau pour l'irrigation du riz dans les bas-fond du Sud-Ouest du Burkina Faso. Master's Thesis, CIRAD Montpellier—UMR G-EAU, Montpellier, France, 2018.
11. Bellefontaine, R.; Gaston, A.; Petrucci, Y. *Aménagement des forêts naturelles des zones tropicales sèches*; FAO: Rome, Italy, 1997.
12. Da, S.J. *Etude des usages et de la regénération d'une plante alimentaire au Sud-Ouest du Burkina Faso*; Polytechnic University of Bobo-Dioulasso: Bobo-Dioulasso, Burkina Faso, 2009.
13. Allen, R.; Pereira, L.; Raes, D.; Smith, M. *Crop Evapotranspiration—Guidelines for Computing Crop Water Requirements—FAO Irrigation and Drainage Paper 56*; FAO: Rome, Italy, 1998.
14. Franquin, P. Analyse agroclimatique en régions tropicales, saison pluvieuse et saison humide, applications. *Cah. ORSTOM* **1969**, *9*, 65–95.
15. Mann, H. Nonparametric tests against trend. *Econometrica* **1945**, *13*, 245–259. [CrossRef]
16. Kendall, M. *Rank Correlation Methods*; Griffin: London, UK, 1975.
17. Ahmad, I.; Tang, D.; Wang, T.; Wang, M.; Wagan, B. Precipitation trends over time using Mann-Kendall and spearman's Rho tests in swat river basin, Pakistan. *Adv. Meteorol.* **2015**, *2015*, 431860. [CrossRef]
18. Shadmani, M.; Marofi, S.; Roknian, M. Trend Analysis in Reference Evapotranspiration Using Mann-Kendall and Spearman's Rho Tests in Arid Regions of Iran. *Water Resour. Manag.* **2011**, *26*, 211–224. [CrossRef]
19. Zhang, W.; Yan, Y.; Zheng, J.; Li, L.; Dong, X.; Cai, H. Temporal and spatial variability of annual extreme water level in the Pearl River Delta region, China. *Glob. Planet. Chang.* **2009**, *69*, 35–47. [CrossRef]
20. Nash, J.E.; Sutcliffe, J.V. River flow forecasting through conceptual models: Part I—A discussion of principles. *J. Hydrol.* **1970**, *10*, 282–290. [CrossRef]
21. Sen, P.K. Estimates of the Regression Coefficient Based on Kendall's Tau. *J. Am. Stat. Assoc.* **1968**, *63*, 1379–1389. [CrossRef]
22. Bergström, S. *The HBV Model—Its Structure and Applications*; RH No 4; SMHI: Norrköping, Sweden, 1992.
23. IPCC Climate Change. 2001: The scientific basis. In *Contribution of Working Group I to the Third Assessment Report of the Intergovernmental Panel on Climate Change*; Houghton, J.T., Ding, Y., Eds.; Cambridge University Press: Cambridge, UK, 2001; p. 881.
24. Arnell, N.W. Relative effects of multi-decadal climatic variability and changes in the mean and variability of climate due to global warming: Future streamflows in Britain. *J. Hydrol.* **2003**, *270*, 195–213. [CrossRef]

25. Touré, H.A.; Kalifa, T.; Kyei-Baffour, N. Assessment of changing trends of daily precipitation and temperature extremes in Bamako and Ségou in Mali from 1961–2014. *Weather Clim. Extrem.* **2017**, *18*, 8–16. [CrossRef]
26. Hat, J.L.; Prueger, J.H. Temperature extremes: Effect on plant growth and development. *Weather Clim. Extrem.* **2015**, *10*, 4–10.

MDPI

St. Alban-Anlage 66

4052 Basel

Switzerland

Tel. +41 61 683 77 34

Fax +41 61 302 89 18

www.mdpi.com

Climate Editorial Office

E-mail: climate@mdpi.com

www.mdpi.com/journal/climate